APPLIED GENETICS OF LEGUMINOSAE BIOTECHNOLOGY

ISBN 1-4020-1485-6

Published by Kluwer Academic Publishers,
P.O. Box 17, 3300 AA Dordrecht, The Netherlands.

Sold and distributed in North, Central and South America
by Kluwer Academic Publishers,
101 Philip Drive, Norwell, MA 02061, U.S.A.

In all other countries, sold and distributed
by Kluwer Academic Publishers,
P.O. Box 322, 3300 AH Dordrecht, The Netherlands.

Printed on acid-free paper

Printed and bound in Great Britain by MPG Books Limited, Bodmin, Cornwall.

Dedicated to our beloved teacher
Professor H. S. Srivastava
(1946–2001)

CONTENTS

PART II: FORAGE LEGUMES

PART III: TREE LEGUMES

FOREWORD

Legumes certainly include many very important crop plants that contribute highly critical protein to the diets of many around the world. Many important forages and green manure crops are legumes. Legumes are also very large contributors to the vegetable oil and animal feed protein sectors. One characteristic of legumes that could become even more important as world energy sources decline and nitrogen fertilizer prices increase is nitrogen fixation, something few other plants can do. Thus legumes have a unique and very important niche in agriculture.

While some of the small seeded forage legumes have been relatively easy to work with in tissue culture as far as culture initiation, plant regeneration and transformation are concerned, most large seeded legumes, like soybean, have been recalcitrant. Today, however, many laboratories are inserting genes into soybean and producing unique plants for both commercial and scientific uses. These advancements have taken a large amount of research effort and still require time and labour.

There are possible alternatives to the usual transformation methods using *Agrobacterium tumefaciens* co-cultivation or particle bombardment of tissue cultures or direct DNA uptake by protoplasts that could circumvent the need for *in vitro* culture techniques. These include the pollen tube method which has been reported to be successful with several species but has not generally been thoroughly documented by molecular techniques or by reproducibility in other labs. Another method is the floral infiltration with *A. tumefaciens* method pioneered with *Arabidopsis thaliana*. This has been successful with some other species including *Medicago truncatula* so might be an approach amenable to other legumes.

Most of us are interested in our own favorite legume species but there could be great importance to developing a 'model' legume especially in the genomics era. Most plant phenomena can be studied with the most important plant model, *Arabidopsis*, but this species does not fix nitrogen so this critical event is ignored. The legumes *Lotus japonicus* and *M. truncatula* are being groomed as possible model systems. Large-scale efforts are under way to obtain expressed sequence tags (ESTs) from many legumes and in time some genomes will be sequenced. Transformation methods are crucial for functional genomic studies to determine what the genes do in the plant, so we need to continue to improve the methods so that any lab can transform their plant of interest.

Thus as the chapters of this monograph demonstrate, progress is being made in legume biotechnology and this is being put into practice for crop improvement both by improving breeding techniques and by increasing the number and types of genes that can be used by the breeders and other researchers. The initial commercial success of these steps is demonstrated by the fact that over half of the US soybean crop grown in 2001 will be Roundup herbicide resistant. This has allowed farmers to have very effective weed control for systems such as no-till that is environmentally friendly by minimizing both soil erosion and the energy needs for tillage.

There are many other crop improvements coming that will be beneficial not only for the environment and farmers of the world, but also to consumers of both developed and developing countries. Examples include the high vitamin A and high iron rice that clearly should benefit many. I also look forward to plants that can produce increasingly needed energy and chemical feedstocks to help decrease in CO_2 in the atmosphere.

The chapters of this monograph were written by many experts from around the world and should be of great utility for not only researchers working in the area like plant physiologists, molecular biologists, geneticists and genomicists, but also for students, technocrats and planners wishing to know the state of the science.

JACK M. WIDHOLM
Professor of Plant Physiology
University of Illinois
Department of Crop Sciences
ERML, 1201 W. Gregory
Urbana, IL 61801, USA
e-mail: widholm@uiuc.edu

PREFACE

Leguminosae is the second most important family of flowering plants next to Graminae as a source of human and animal food. These plants are a rich source of inexpensive protein which forms an important part of vegetarian diets of increasing populations of developing countries. In recent years, there has been an upsurge in their use in other parts of the world due to their extra-nutritional qualities and industrial uses. Leguminous plants are a crucial part of agriculture ecosystem in all parts of the world; they restore atmospheric nitrogen and soil fertility and thus reduce the demands of nitrogen fertilizers. Production of leguminous crops has been stagnant during the last four decades due to their prominent susceptibility to pests and pathogens, sensitivity to abiotic stresses and poor nutrient use efficiency, etc. The nutritional imbalance and presence of antimetabolites in their seeds are other gray areas which require urgent attention. Recently, legumes have been examined for production of secondary metabolites (nutraceutical).

Improvement of leguminous crops through conventional breeding has met with limited success due to the narrow genetic variability in their germplasm and sexual incompatibility with related and wild species. Recent advances in molecular biology and biotechnology have made it possible to transfer desirable genes across the sexual boundaries. The transgenic plants so produced can be used either in their own right or as a gene pool for inclusion in traditional breeding programmes. Genetic transformation of these crops has been difficult and challenging to date due to their poor regeneration ability and low competency of regenerable cells to genetic transformation. In recent years gene transfer in a wide range of leguminous species has been attempted. The transformation and *in vitro* regeneration are highly dependent upon genotype, physiological status of the target explant, hormonal and nutritional signals, host recognition and use of suitable promoters, etc. The three major gene transfer techniques, i.e. *Agrobacterium*-mediated, protoplast-based methods and biolistic have been used to transform legumes. *Agrobacterium*-mediated transformation, however, has been the method of choice as it offers a simple, rapid and precise mode of DNA transfer, low copy integration and is less expensive. On the other hand, the direct gene transfer techniques often result in multiple copy integration, transgene silencing and reduced fertility of trangenics. In addition, *in planta* techniques either using *Agrobacterium* or electroporation of intact meristems hold promise of much higher frequencies and more simple means of tissue culture independent production of transgenics.

β-glucuronidase (GUS) gene (*uidA*) has been the most commonly used reporter of transient and stable transformations in legumes; however, its assay is destructive in nature. Some of the non-destructive alternatives to GUS marker gene, the fire fly luciferase (*luc*) and jellyfish green fluorescent protein (*gfp*) genes are being used. In most of the leguminous crops kanamycin resistance gene (*nptII*) has been the selectable marker of choice, though hygromycin resistance (*hpt*) and phosphinothricin resistance (*bar*) genes have also been successfully used in some of them. Besides testing other herbicides like glyphosate resistance and imazapyr resistance, positive selection strategies implying cytokinin gene (*ipt*), phosphomannose isomerase gene (*pmi*) and xylose isomerase gene (*xyl*) also need to be tested for leguminous crops.

The choice of effective promoter for high and tissue specific expression of the introduced genes is critical for devising high frequency transformation systems in leguminous crops. The 35S promoter of cauliflower mosaic virus (CaMV) has provided constitutive expression of selectable marker and *uidA* reporter genes in a majority of the legumes. The other promoters used are tissue specific promoters, e.g. legumine B-4 (leB4), concanavaline A (CoA), sunflower seed albumin (*sfa8*) gene, etc. The native promoters from leguminous plants can be more ideal and they should be isolated and cloned.

Most of the legumes have been transformed with reporter and selectable marker genes only. The generation of fertile transgenics with normal phenotype and transfer of transgenes to the next generation have been achieved only in a few cases. Transgenic plants with agronomically important traits, e.g. herbicide, insect, virus and other biotic and abiotic stresses and improved nutrient use efficiency, nutritional quality and digestibility of seeds and forage, have been developed only in a limited number of species. Attempts have also been made to understand gene expression in legumes using transgenic approach. Soybean is one legumious crop being cultivated in nearly 63% of the total area under transgenics. The concerted efforts are needed to improve *in vitro* regeneration, development of highly infective disarmed *Agrobacterium* strains, optimal selection system, use of more potent promoters, isolation and identification of genes for quantitative and qualitative desirable traits, stability and inheritance of transgenes to subsequent progenies, public acceptance of GMO legumes and minimizing health and environmental risks. Recently there has been an initiative in identification, isolation and cloning of genes through genome sequencing projects. Among legumes, the genome sequencing has been taken for two model legumes, *Medicago truncatula* (Samuel Roberts Noble Foundation Inc., Ardmore, USA) (http://www.noble.org) and *Lotus japonicus* (at the Kasuza Centre, Japan). The complete sequence of the whole transcriptionally active part of the genome is projected to be completed in the near future.

This volume includes nineteen chapters presenting the current status, state of the art, applications and limitations of *in vitro* regeneration and genetic transformation in major grain, forage and tree leguminous species. The grain legumes include soybean, pea, common bean, pigeonpea, chickpea, cowpea, mungbean, urdbean, lentil, *Vicia* sp., guar, winged bean and peanut. Major forage legumes are lupins, *Medicago*, white clover and *Lotus* and tree species included in this volume are *Albizzia*, *Leucaena*, *Acacia* and *Pseudoacacia*. Each chapter of this volume was prepared by the distinguished scientists who have made pioneering and significant contributions to their respective field of study. We are indebted to them for their diligence in preparing their comprehensive and insightful accounts. We are grateful to Professor J. M. Widholm (USA) for sparing his valuable time for writing the Foreword.

We are also grateful to Professors J. Finer, W. A. Parrott, I. K. Vasil, D. P. S. Verma, M. Chrispeels, Peggy Ozias Akins, P. Christou, H. J. Jacobson, M. Ishimoto, S. Sopory and D. Pental, and to Professor Marcel Hofman, the series editor, 'Focus on Biotechnology' for their kind suggestions and encouragement. We extend our warmest appreciation and thanks to Kluwer Academic Publishers for their keen interest in bringing out this title with quality work. We are thankful to our research scholars and family members for their understanding and patience during preparation of this title. We appreciate the efforts made by Mr. Ashok Datta and Ms. Reema of *LaPrints* New Delhi, India for preparing the camera ready version.

Rohtak, India
June, 2002

Pawan K. Jaiwal
Rana P. Singh

PART I
GRAIN LEGUMES

RECENT ADVANCES IN SOYBEAN TRANSFORMATION

RANDY D. DINKINS, M. S. SRINIVASA REDDY, CURTIS A. MEURER, CARL T. REDMOND AND GLENN B. COLLINS*
Department of Agronomy, University of Kentucky, Lexington, KY, USA 40546-0091
**e-mail: gcollins@pop.uky.edu*

Abstract

Soybean (*Glycine max* (L.) Merrill) transformation continues to be problematic even with the efforts of numerous researchers in the field. In spite of this low efficiency, soybean transformation is now routine utilizing the somatic embryogenic and meristematic tissue culture methods. A comparison, including the advantages and disadvantages, of each method are presented in this review. Soybean transformation is currently done using both *Agrobacterium tumefaciens* and particle bombardment. We review some of the current selectable marker genes in use, and we also describe additional non-standard transformation techniques that have been, or are being attempted, to produce transgenic soybeans.

1. Introduction

Genetic transformation of most crops previously classified as recalcitrant has been overcome and is now feasible. Soybean (*Glycine max* (L.) Merrill) tissue culture and transformation, in contrast, continues to be laborious and the frequency of recovery of transgenic plants is still very low. In this review, we highlight some of the recent advances in soybean tissue culture and transformation systems and identify some of the major obstacles that must be resolved before a more reliable and efficient soybean transformation system will be available. We also describe some of the alternatives that have been tried with little or no success, yet with additional research have the potential for leading to a breakthrough in soybean transformation. A recent review of soybean transformation protocols, co-authored by our group, contains many of the procedures that are still in use, with some modifications (Trick *et al.*, 1997). In the present review, we emphasize those protocol changes that have resulted in higher transformation frequencies as well as some of the pitfalls that have been identified. Another older review (Christou, 1992) is also available, and contains important background information in soybean tissue culture and transformation.

P. K. Jaiwal and R. P. Singh (eds.), Applied Genetics of Leguminosae Biotechnology, 3–21.

Plant transformation involves the delivery of target DNA sequences to cells or tissues that are both transformation competent and capable of regeneration into a full grown fertile plant. The transformation process will not be successful where either of these criteria are not met. Soybean transformation has been difficult to achieve due to difficulties in both of the above areas. There are no reports that describe an organogenic regeneration procedure with soybean tissues similar to that of the tobacco leaf disk regeneration and transformation system. All the present successes with soybean have relied on proliferation of pre-existing meristematic cells, either in the apex or in the cotyledonary node region, or via somatic embryogenesis. Both organogenic and embryogenic tissue culture methods have been used successfully to regenerate fertile soybean plants (Hinchee *et al.*, 1988; McCabe *et al.*, 1988; Parrott *et al.*, 1994; Di *et al.*, 1996; Maughan *et al.*, 1999; Zhang *et al.*, 1999; Simmonds and Donaldson, 2000; Yan *et al.*, 2000). The methodology of each tissue culture system is described in more detail below. Both are equally cumbersome and yield low frequencies of transgenic soybean plants.

Difficulties have also been encountered in the delivery and stable integration of the foreign DNA into soybean. *Agrobacterium*-mediated transformation and particle bombardment have both been used to produce fertile transgenic soybean plants although the efficiency is low for both methods. The low efficiency of *Agrobacterium*-mediated transformation has earned soybean the title as one of the most recalcitrant of the major agronomic crops. Thus, *Agrobacterium*-mediated transformation of soybean has lagged behind that of other dicotyledonous crop species. We have included in this review two protocols that are widely used for soybean and these methods have been used successfully in our laboratory. The first protocol is the transformation of embryogenic tissue in conjunction with particle bombardment (Protocol 1), and the second utilizes the cotyledonary node explant and *Agrobacterium tumefaciens* to deliver the foreign gene (Protocol 2).

2. Soybean tissue culture and regeneration

2.1. MERISTEMATIC CELL PROLIFERATION AND PLANT REGENERATION

Apical or cotyledonary node meristematic tissue explants can be induced to regenerate whole plants very efficiently in soybean. These explants will produce a large number of plants when explants are placed on a high cytokinin medium (Cheng *et al.*, 1980; Lazzeri *et al.*, 1985; Barwale *et al.*, 1986; Wright *et al.*, 1986, 1987). Plants have been obtained that were initiated in the meristem region of the apical dome (McCabe *et al.*, 1988; Christou *et al.*, 1989; Aragao *et al.*, 2000), as well as the small meristematic region of the cotyledonary node (Barwale *et al.*, 1986; Wright *et al.*, 1986) and epicotyl (Wright *et al.*, 1987; Dan and Reichert, 1998). After initiation and elongation, shoots are easily excised and can be rooted on an IBA/auxin medium. Transgenic plants have been obtained via particle bombardment of apical cells (McCabe *et al.*, 1988), as well as by *Agrobacterium*-mediated transformation of cotyledonary node tissues (Hinchee *et al.*, 1988; Di *et al.*, 1996; Zhang *et al.*, 1999) and germinating seeds (Chee *et al.*, 1989). However, for the routine production of transgenic plants, the use of meristematic cells presents a problem

Protocol 1. Summary of solid medium cycling protocol

Stage	Description
Initiation	Establish somatic embryogenic cultures by excising immature (4–6 mm length) zygotic cotyledons (cultivar: Jack) and plating them on media (MS basal media with 40 mg/L of 2,4-dichlorophenoxyacetic acid (2,4-D); 2% sucrose, 2 g Phytagel, pH 7.0 flat side up!) (Fig. 1A). Incubate the plates at 23 °C in a culture room (23:1 light to dark photoperiod). Embryo development will begin in about 30 days. Transfer young embryos to proliferation media (MS-based media with 20 mg of 2,4-D; 2% sucrose, pH 5.8) for a continual re-initiation of somatic embryos and subculture to fresh media every 21 days (Fig. 1B). For best results, select rapidly proliferating, tightly packed embryo cultures using a standard stereoscope.
Transformation	Place the selected somatic embryo cultures on moistened filter paper in sterile Petri plates and dry in a laminar flow hood 15 minutes prior to shooting. Transform somatic embryos using a DuPont PDS1000 gene gun with gold/DNA microprojectile preparation. We have modified the standard gold/DNA microprojectile preparation. To prepare the gold/DNA microprojectiles for nine shots (0.83 mg of gold and 2.77 μg of DNA per shot), weigh 7.5 mg of 1.0 micron gold (BioRad), rinse in 100% ethanol and resuspend in 111 μl of 100% ethanol. Divide this into three aliquots of 35 μl each and process each aliquot separately. Centrifuge each aliquot at 5000 rpm for 10 sec, add 1 ml water and without resuspending the gold centrifuge at 2000 rpm for 5 min. Decant ethanol/water mixture and resuspend the gold in 220 μl of water and add 25 μl of DNA (8.33 μg), 250 μl of 2M $CaCl_2$ (Sigma) and 100 μl of 0.1 M spermidine (Sigma) and incubate in ice for 2 min. Gently vortex for 10 min and centrifuge at 1000 rpm for 5 min. Remove liquid and quickly add 600 μl of 100% ethanol and gently vortex for 1 min. Pellet gold/DNA at 1000 rpm for 5 min, gently resuspend in 36 μl of ethanol and incubate on ice for one hour. Flick the tube and resuspend with pipet and spread 10 μl on each carrier. Shoot cultures using 1550-psi rupture disk (BioRad) and a focusing device described by Torisky *et al.* (1996) at 11.0 cm distance from DNA to target tissue.
Selection	Immediately transfer shot tissue to a non-selection media (MS-based media with 20 mg of 2,4-D; 2% sucrose, pH 5.8) for 7 days. The selection of transformed material is obtained by transferring embryo cultures to low pressure selection media (MS basal media with 20 mg of 2,4-D; 2% sucrose, 15 mg hygromycin (Sigma), pH 5.8) for 14 days and follow transferring to a high pressure selection media (MS-based media with 20 mg of 2,4-D; 2% sucrose, 30 mg hygromycin pH 5.8) for 135 days (subculturing every 24–28 days). Although cultures will appear to turn brown after 45 days (Fig. 1C), continue to transfer entire tissue groups. After 100 days, green embryos (putative transformed) will begin to appear (Fig. 1D). We have observed that growth at this stage is faster in low light (5–10 μE than in high light > 50 μE).
Conversion	Transfer surviving embryogenic cultures to maturation media (MS basal with 6% maltose, and 0.5 gm of activated charcoal, pH 5.8) for 21 days. Final maturation is obtained by transferring individual embryos onto maturation media (MS-basal media with 6% maltose, pH 5.8) until embryos have grown to 1.0 cm in length and cotyledons are visible, and have lost their green color. Desiccate groups of 6–8 embryos in plastic Petri plates for three to seven days. Remove embryos as they become creamy-white in color and transfer to germination media (MS- based media, 3% sucrose, pH 5.8). Plantlets should begin for form shoots and roots in 14–21 days with some embryos requiring up to 60 days (Fig. 1E). Transfer plantlets to Phytatrays (Sigma) containing B_5 media (1% sucrose, 3 g Phytagel, pH 5.8). After 14 days in Phytatrays or as plantlets grow to the top of the tray, transfer the putatively transformed plants to humidity domes with moistened sterilized artificial soil (ProMix) (Fig. 1F). After 10 days, plants should be acclimated to soil and ready to grow in the greenhouse (Figs. 1G and H).

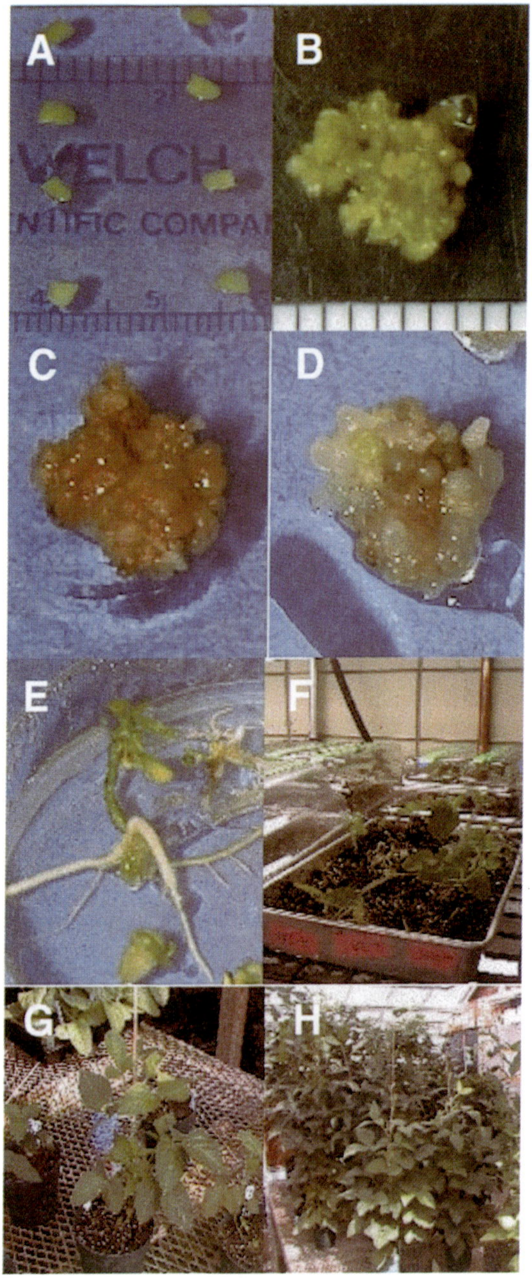

Figure 1. Solid media cycling. A. Initiation of zygotic embryogenic cotyledons on MS-D40 media. B. Proliferating embryos on MS-D20 media. C. Embryos after 50 days in selection (MS-D20, 30 mg hygromycin). D. Putative transgenic embryo in selection (MS-D20, 30 mg hygromycin). E. Conversion of embryo to plantlets in MSO3 media. F. Plants in humidity dome. G. Acclimation of small plants to greenhouse conditions. H. Mature plants in greenhouse.

Protocol 2. Basic cotyledonary node protocol

Stage	Description
Seed sterilization	Soybean seed are pre-washed by brief agitation in sterile water with one drop of detergent added for wetting. Seed are briefly air dried and then sterilized with chlorine gas generated between 100 ml of chlorine bleach and 5 ml 12N HCl, within a sealed desiccator for 12 hours. Seeds may be stored in sealed, dry petri plates.
Seed germination	Seed are germinated for 1 to 4 days on germination media (media consisting of B_5 salts and vitamins, pH 5.8, solidified with 5 g/L agarose). Transfer germinated seedlings to a sterile petri plate and remove the seed coat and the radicle to 5 mm below the junction with the seed cotyledon. The seed are bisected to yield two identical explants (Fig. 2A).
Explant preparation	Bisected explants are placed adaxial side up and macerated with a scalpel through the primary meristem and cotyledonary node region to disrupt primary shoot morphogenesis and provide a wound site for the *Agrobacterium* infiltration.
	NOTE: Most laboratories using this protocol readily admit that the wounding step is the most critical component of the protocol. Insufficient wounding will result in a failure to provide adequate access for the *Agrobacterium* infiltration to the plant tissue and will also fail to disrupt primary shoot development. Excessive wounding will result in nearly complete disruption of all shoot morphogenesis.
Co-cultivation	Transfer the explants to a prepared *Agrobacterium* suspension in liquid plant media (consisting of B_5 salts and vitamins, 5 μM N^6-benzylaminopurine (BAP), 30 mg/L sucrose, pH 5.5) augmented with 100 μM acetosyringone and then co-cultivated on the above media solidified with 5 g/L agarose, for 36–72 hours at 20 °C.
Selection	Following co-culture, transfer explants to counter selection medium (media consisting of B_5 salts and vitamins, 5 μM BAP, 30 mg/L sucrose, pH 5.7, solidified with 5 g/L agarose, augmented with 100 mg/L vancomycin, 100 mg/L timentin and 500 mg/L mefoxin) and with the appropriate selective agents (5 mg/L glufosinate when using the *bar* gene, 100 Following co-culture, transfer explants to counter selection mg/L kanamycin monosulfate when using *nptII* gene).
	Subculture explants every 10 days to fresh counter selection media for a total of 3 subcultures. Without selection agents added to the media, the cotyledonary node will become a very prolific explant.
	After 30 days of selection, transfer explants to the counter selection media containing lower levels of antibiotics and selective agents (1 mg/L glufosinate or 25 mg/L kanamycin monosulfate) to permit shoot elongation.
Elongation and conversion	Elongated shoots (Fig. 2B) are excised and placed in maturation solid media (media consisting of B_5 salts and vitamins without BAP, pH 5.7, solidified with 5 g/L agarose). This media is augmented with 10 μM indolebutyric acid (IBA) to promote rooting. Rooted shoots are transferred to sterile soil and acclimated to non-sterile conditions.

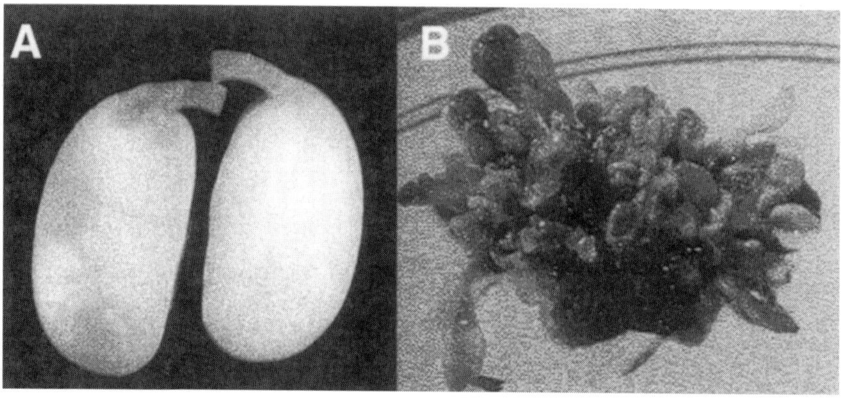

Figure 2. Cotyledonary node protocol. A. Bisected prepared soybean seed prior to infection with Agrobacterium culture. B. Putative transformed shoots. Numerous shoots regenerate from the cotyledonary node region.

with targeting the transgenes to the cells and selection of cell lines that will eventually differentiate into a new plant. Problems associated with transgene chimeras from meristematic protocols have been described by Christou *et al.* (1989), Meurer *et al.* (1998), and Zhang *et al.* (1999). Thus, while the tissue culture procedure associated with regeneration of plants from the meristematic cells and tissues is probably the easiest and most robust as far as soybean tissue culture is concerned, there are continued problems associated with regeneration of transgenic plants from these tissues as will be discussed below in further detail.

2.2. SOMATIC EMBRYOGENESIS PROTOCOLS

The somatic embryogenic protocols involve inducing an explant to produce somatic embryos in tissue culture. Somatic embryos have both a shoot and root meristem and will germinate to form whole plants. Somatic embryogenic procedures involve several changes in media during the stages of (a) somatic embryo induction, (b) somatic embryo growth, (c) maturation, (d) germination and (e) conversion to whole plants.

There have been numerous reports of soybean somatic embryo production in the literature (Christianson *et al.*, 1983; Lippmann and Lippmann, 1984; Lazzeri *et al.*, 1985; Ranch *et al.*, 1985; Finer, 1988; Hartweck *et al.*, 1988; Hepher *et al.*, 1988; Parrott *et al.*, 1988); however, somatic embryos produced using earlier protocols, were not successful in producing transgenic soybean plants with stable inheritance of the transgene by the progeny. Finer (1988) described a somatic embryogenic system that used immature cotyledons as explants on high levels of 2,4-D (40 mg/L) which eventually led to production of transgenic soybean lines, and is the basis for most embryogenic systems in use today. Somatic embryos are produced from single cells or very small numbers of cells on or near the surface of newly initiated or pre-existing embryos (Finer and McMullen, 1991; Liu *et al.*, 1992, 1996; Hazel *et al.*, 1998). This combination of single (or few) cells and surface origin of proliferating embryos provides an attractive target for transgene introduction and selection.

Somatic embryo induction using immature embryos is a direct process with success being strongly affected by cultivar selection, immature zygotic cotyledon size, and media composition (Parrott *et al.*, 1989; Bailey *et al.*, 1993a; Santarem *et al.*, 1997). Recent studies evaluating different cultivars have indicated that the cultivar 'Jack' remains the most reliable public genotype for production of high numbers of quality somatic embryos (Meurer *et al.*, 2000). Other cultivars with some common parentage with 'Jack', such as 'Fayette' and 'Dwight', also appear to be good candidates using the somatic embryogenic protocols; however, the latter cultivars continue to be less responsive than the cultivar 'Jack'. These cultivars all share a level of relatedness, yet, a genetic basis for the somatic embryogenic response has not been clearly demonstrated. Attempts to elucidate a genetic basis have been unsuccessful as somatic embryogenesis is a complex trait (Bailey and Parrott, 1993b; Tian *et al.*, 1994; Simmonds and Donaldson, 2000). Overall, late maturing cultivars, particularly those adapted to the southern United States and semi-tropical regions such as the cultivars 'Benning' and 'Stonewall', have performed poorly using existing somatic embryo induction protocols. This has been a major area of disappointment, especially in the regions where these adapted cultivars are grown, as any transgenic trait introduced in an early maturing cultivar, will require additional generations of backcrossing into a late maturing adapted cultivar. Thus, a major area of needed research is to identify somatic embryogenic protocol parameters that will result in an increased response from a broader range of soybean genotypes and cultivars.

Once somatic embryos have been induced on solid medium containing a high level of auxin, such as 2,4-D, continued proliferation and selection of transgenic somatic embryos following transformation is done on either solid, liquid or alternating sequence of solid and liquid media. Each method has advantages and disadvantages. Liquid medium protocols are initiated by placing clumps of the somatic embryos in liquid medium such as FN (Finer and Nagasawa, 1988; with 2,4-D at 5–10 mg/L) and more recently FNL (Samoylov *et al.*, 1998a), and require weekly transfers. While this system often results in a rapid selection (7–10 weeks) and identification of transgenic lines its prolonged use will induce undesirable mutations. Singh *et al.* (1998) demonstrated that the liquid tissue culture process was capable of inducing permanent cytogenetic and genetic aberrations in as little as 8 months with some cultivars. This heritable change was directly responsible for reduced fertility of tissue-cultured lines. Recent research has concentrated on how to maintain cultures in liquid proliferation long enough to identify transformed lines without exceeding the culture time likely to cause abnormalities. The solid media protocol in place in the authors' laboratory (see Protocol 1) limits the culture time to one year.

Several problems have been associated with liquid medium procedures. These include repetitive medium changes, and potential problems associated with potential culture contamination, as well as problems associated with conversion and maturation of the somatic embryos. Further treatment of the modifications in protocols to circumvent problems associated with embryo conversion and maturation are discussed below. A modification of the liquid suspension culture protocol that circumvents the necessity for continual manipulation of the cultures is a solid medium proliferation culture system. This procedure is similar to the liquid medium protocol but results in slower growth rates. Transfer of the embryogenic cultures on the solid medium is monthly compared to weekly for the liquid. Additionally, all embryogenic cultures initiated in the authors' laboratory, and other laboratories that collaborate with the authors (J. J. Finer, W. A. Parrott, L. O. Vodkin and J. Widholm, personal communications) are initiated on solid media with slightly divergent protocols for transformation, selection, and maturation. Thus, one of the major advantages of the solid medium protocol is the simplicity of the tissue culture manipulations as compared to the liquid suspension cultures. Nevertheless, one of the major disadvantages appears to be that the length of time required for selection of transgenic embryo lines (6–8 months) on solid medium can result in a high frequency of infertile and/or abnormal transgenic plants as discussed above (unpublished observations). One strategy to circumvent this problem has been to shorten the selection time. However, this has led to an increase in the incidence of false positives or escapes (unpublished observations). Thus there appears to be a fine line (roughly six months on selection) between obtaining true transgenic embryogenic lines that will result in whole transformed fertile plants and transformed plants that have reduced fertility and are abnormal. One strategy to circumvent the infertility problem has been to discard cultures that are greater than one year of age.

Embryogenic cultures are maintained on selection until lines are shown to be transgenic by either GUS staining or PCR, and these are then placed onto maturation medium for embryo development. Somatic embryos that proliferate in either liquid or solid selection media must eventually undergo histodifferentiation and physiological maturation of the somatic embryo, followed by proper activation of the root and apical meristem during germination to permit proper conversion (Slawinska and Obendorf, 1991; Bailey *et al.*, 1993a, 1993b; Samoylov *et al.*, 1998a). Media preparations for somatic embryo maturation, germination, and conversion usually do not contain exogenous plant growth regulators, but do require an added osmoticum, such as sorbitol (Walker and Parrott, 2001). The improvements in osmoticum as well as adding exogenous amino acids have been shown to greatly enhance and shorten the time for the recovery of soybean plants via somatic embryogenic methods (D. R. Walker and W. A. Parrott, personal communication). These recent modifications in media recipes circumvent some of the problems associated with conversion and maturation that were observed in the previous liquid suspension protocols (Santarem *et al.*, 1997; Samoylov *et al.*, 1998a, 1998b; Walker and Parrott, 2001). Taken together, the recovery of fertile transgenic soybean plants using the somatic embryogenic procedures is now common, although it still requires 6–9 months and the frequency of recovery of transgenic plants continues to be low. Thus, liquid suspension cultures and selection of transgenic somatic embryos is now a viable alternative to the solid medium selection and promises to shorten the time in selection.

In summary, the somatic embryogenic procedure has been shown to produce fertile soybean plants at a high frequency. The major problem has been in regenerating fertile transgenic soybean plants. This is primarily due to the long period associated with selection of the transgenic embryogenic lines. The selection time required, minimally four months, and more often six to eight months, is one of the contributing factors to a high frequency of abnormal and infertile regenerated transgenic soybean plants. Further discussion on selection schemes and agents will be done below, but the protocols presently in use by us, and a number of different laboratories, have shown that it is feasible to recover normal, fertile transgenic soybean plants using this procedure (Parrott *et al.*, 1994; Maughan *et al.*, 1999; Simmonds and Donaldson, 2000; Yan *et al.*, 2001).

2.3. MERISTEMATIC CELL PROLIFERATION VS. SOMATIC EMBRYOGENESIS PROTOCOLS

The regeneration of fertile transgenic soybean plants has been accomplished using both the meristematic cell proliferation techniques and somatic embryogenesis protocols. The preference to a given system is dependent on each individual investigator. Meristematic cell proliferation protocols are much simpler and the plants initially regenerated are recovered much quicker than with the somatic embryogenic protocols. Cotyledonary node meristematic cells have been shown to be competent for transformation and this transformation protocol has received considerable attention for that reason. Meristematic protocols, including cotyledonary nodes, do not require maintenance of donor plants at the correct growth stage, continued selection of proliferating embryogenic tissues, culture times in excess of 6 months, and dependence on a very limited subset of cultivars. These limitations are all concerned with the somatic embryogenesis protocols. Many researchers have evaluated a large number of soybean genotypes using the meristematic cell protocols with the majority of genotypes providing at least some capability to regenerate multiple shoots (Byrne *et al.*, 1987; Delzer *et al.*, 1990; Meurer *et al.*, 1998). Proliferative shoot clumps can be initiated directly from germinating seed with the caveat that only a select group of cells are capable of regeneration and targeting of transformation must therefore be precise. Additionally, the targeted meristematic cells are often in different stages and transformation often results in the recovery of chimeras (Meurer *et al.*, 1998). Somatic embryogenic procedures, in contrast, rarely yield chimeric lines and can often yield multiple clones of the same transformation event, thus it is necessary to maintain the integrity of the lines to ensure that independent transgenic lines are produced (Yan *et al.*, 2000). In addition, once a cell line is transformed, continued culture of some embryos arising from a transformation is possible while other clones can be undergoing conversion or growth in the greenhouse.

3. Soybean transformation

3.1. PARTICLE BOMBARDMENT

Particle bombardment technology is based on the acceleration of DNA-coated particles into cells such that the particles (usually referred to as microprojectiles) penetrate the cells, followed by the disassociation of the DNA from the particle and integration of the DNA into the host chromosome. Since the initial demonstration of the technology by Klein *et al.* (1987) in onion cells, particle bombardment has been utilized to introduce DNA into many plant species, including soybean. In fact, soybean was one of the initial crops where whole fertile transgenic plants were recovered (McCabe *et al.*, 1988). Since 1989 numerous reports of transgenic soybean plants recovered using particle bombardment have appeared (Christou *et al.*, 1989; Falco *et al.*, 1995; Stewart *et al.*, 1996; Maughan *et al.*, 1999). The equipment, procedures and tissue preparation for particle bombardment of soybean tissues are relatively simple, and have been described by others in detail (Vain *et al.*, 1993; Trick *et al.*, 1997; Finer *et al.*, 1999). As there are minor variations in particle bombardment

6. Conclusions and future prospects

Soybean transformation is now possible, with public sector and private company laboratories reporting successes. Soybean transformation is now entering the era in which the emphasis is no longer on whether transformation can or will occur, but on the analysis of the transgenic soybean plants for commercial and agronomic traits (Parrott *et al.*, 1994; Padgette *et al.*, 1995; Falco *et al.*, 1995; Di *et al.*, 1996; Denbow *et al.*, 1998; Zeitlin *et al.*, 1998; Maughan *et al.*, 1999). Using established procedures it is possible to establish a soybean transformation laboratory with the expectation of recovery of fertile transgenic soybean plants. Yet, in spite of the advances in transformation protocols, soybean transformation continues to be problematic and the success rate and efficiency are very low.

References

Aragão F J L, Sarokin L, Vianna G R and Rech E L (2000) Selection of transgenic meristematic cells utilizing a herbicidal molecule results in the recovery of fertile transgenic soybean (*Glycine max* (L.) Merril) plants at high frequency. *Theor. Appl. Genet.*, **101**: 1–6.

Bailey M A, Boerma H R and Parrott W A (1993a) Genotype effects on proliferative embryogenesis and plant regeneration of soybean. *In Vitro Cell Dev. Biol. Plant*, **29**: 102–108.

Bailey M A, Boerma H R and Parrott W A (1993b) Genotype-specific optimization of plant regeneration from somatic embryos of soybean. *Plant Sci.*, **93**: 117–120.

Bailey M A, Boerma H R and Parrott W A (1994) Inheritance of tumor formation in response to *Agrobacterium tumefaciens* in soybean. *Crop Sci.*, **34**: 514–519.

Barwale U B, Kerns H K and Widholm J M (1986) Plant regeneration from callus cultures of several soybean genotypes via embryogenesis and organogenesis. *Planta*, **167**: 473–481.

Bechtold N, Ellis J and Pelletier G (1993) *In planta Agrobacterium*-mediated gene transfer by infiltration of adult *Arabidopsis* plants. *C.R. Acad. Sci. Paris Life Sci.*, **316**: 1194–1199.

Byrne M C, McDonnell R E, Wright M S and Carnes M G (1987) Strain and cultivar specificity in the *Agrobacterium*-soybean interaction. *Plant Cell Tiss. Org. Cult.*, **8**: 3–15.

Chang S S, Park S K, Kim B C, Kang B J, Kim D U and Nam H G (1994) Stable genetic transformation of *Arabidopsis thaliana* by *Agrobacterium* inoculation *in planta*. *Plant J.*, **5**: 551–558.

Chee P P, Forber K A and Slightom J L (1989) Transformation of soybean (*Glycine max*) by infecting germinating seeds with *Agrobacterium tumefaciens*. *Plant Physiol.*, **91**: 1212–1218.

Cheng, T Y, Saka H and Voqui-Dinh T H (1980) Plant regeneration from soybean cotyledonary node segments in culture. *Plant Sci. Lett.*, **19**: 91–99.

Chowrira G M, Akella V, Fuerst P E and Lurquin P F (1996) Transgenic grain legumes obtained by *in planta* electroporation-mediated gene transfer. *Mol. Biotech.*, **5**: 85–96.

Christianson M L, Warnick D A and Carlson P S (1983) A morphologically competent soybean suspension culture. *Science*, **222**: 632–634.

Christou P (1992) Soybean and other *Glycine* species. In: *Genetic Engineering and In Vitro Culture of Crop Legumes*. Technomic Publishing Co., Inc. Lancaster and Basel, 37–127.

Christou P, Swain W F, Yang N S and McCabe D E (1989) Inheritance and expression of foreign genes in transgenic soybean plants. *Proc. Natl. Acad. Sci. USA*, **86**: 7500–7504.

Clough S J and Bent A F (1998) Floral dip: a simplified method for *Agrobacterium*-mediated transformation of *Arabidopsis thaliana*. *Plant J.*, **16**: 735–743.

Dan Y and Reichert N A (1998) Organogenic regeneration of the soybean from hypocotyl explants. *In Vitro Cell Dev. Biol. Plant.*, **34**: 14–21.

Delzer B W, Somers D A and Orf J H (1990) *Agrobacterium tumefaciens* susceptibility and plant regeneration of 10 soybean genotypes in maturity groups 00 to II. *Crop Sci.*, **30**: 320–322.

Denbow D M, Grabau E A, Lacy G H, Kornegay E T, Russell D R and Umbeck P F (1998) Soybeans transformed with a fungal phytase gene improve phosphorus availability for broilers. *Poultry Sci.*, **77**: 878–881.

Desfeux C, Clough S J and Bent A F (2000) Female reproductive tissues are the primary target of *Agrobacterium*-mediated transformation by the *Arabidopsis* floral-dip method. *Plant Physiol.*, **123**: 895–904.

Dhir S K, Dhir S and Widholm J M (1991a) Plantlet regeneration from immature cotyledon protoplasts of soybean (*Glycine max* L.). *Plant Cell Rep.*, **10**: 39–43.

Dhir S K, Dhir S, Hepburn A and Widholm J M (1991b) Factors affecting transient gene expression in electroporated *Glycine max* protoplasts. *Plant Cell Rep.*, **10**: 106–110.

Di R, Purcell V, Collins G B and Ghabrial S A (1996) Production of transgenic soybean lines expressing the bean pod mottle virus coat protein precursor gene. *Plant Cell Rep.*, **15**: 746–750.

Donaldson P A and Simmonds D H (2000) Susceptibility to *Agrobacterium tumefaciens* and cotyledonary node transformation in short-season soybean. *Plant Cell Rep.*, **19**: 478–484.

Falco S C, Guida T, Locke M, Mauvais J, Sanders C, Ward R T and Webber P (1995) Transgenic canola and soybean seeds with increased lysine. *Bio/Technology*, **13**: 577–582.

Finer J J (1988) Apical proliferation of embryogenic tissue of soybean (*Glycine max* (L.) Merrill). *Plant Cell Rep.*, **7**: 238–241.

Finer J J, Finer K R and Ponappa T (1999) Particle bombardment mediated transformation. In: *Current Topics in Microbiology and Immunology; Plant Biotechnology: New Products and Applications* (Eds J Hammond, P McGarvey and V. Yusibov). Springer-Verlag, Berlin, Germany, 60–80.

Finer J J and McMullen M D (1991) Transformation of soybean via particle bombardment of embryogenic suspension culture tissue. *In Vitro Cell Dev. Biol. Plant*, **27**: 175–182.

Finer J J and Nagasawa A (1988) Development of an embryogenic suspension culture of soybean (*Glycine max* (L.) Merrill). *Plant Cell Tiss. Org. Cult.*, **15**: 125–136.

Gritz L and Davies J (1983) Plasmid-encoded hygromycin B resistance: the sequence of hygromycin B phosphotransferase gene and its expression in *Escherichia coli* and *Saccharomyces cerevisiae. Gene*, **25**: 179–188.

Hadi M Z, McMullen M D and Finer J J (1996) Transformation of 12 different plasmids into soybean via particle bombardment. *Plant Cell Rep.*, **15**: 500–505.

Hansen G, Das A and Chilton M D (1994) Constitutive expression of the virulence genes improves the efficiency of plant transformation by *Agrobacterium. Proc. Natl. Acad. Sci. USA*, **91**: 7603–8607.

Hansen G, Shilito R D and Chilton M D (1997) T-strand integration in maize protoplasts after codelivery of a T-DNA substrate and virulence genes. *Proc. Natl. Acad. Sci. USA*, **94**: 11726–11730.

Hartweck L M, Lazzeri P A, Cui D and Collins G B (1988) Auxin-orientation effects on somatic embryogenesis from immature soybean cotyledons. *In Vitro Cell Dev. Biol. Plant*, **24**: 821–828.

Hazel C B, Klein T M, Anis M, Wilde H D and Parrott W A (1998) Growth characteristics and transformability of soybean embryogenic cultures. *Plant Cell Rep.*, **17**: 765–772.

Hepher A, Boulter M E, Harris N and Nelson R S (1988) Development of a superficial meristem during somatic embryogenesis from immature cotyledons of soybean (*Glycine max* (L.) Merrill). *Ann. Bot.*, **62**: 513–519.

Hiei Y, Komari T and Kubo T (1997) Transformation of rice mediated by *Agrobacterium tumefaciens. Plant Mol. Biol.*, **35**: 205–218.

Hinchee M A W, Connor-Ward D V, Newell C A, McDonnell R E, Sato S J, Gasser C S, Fischhoff D A, Re D B, Fraley R T and Horsch R B (1988). Production of transgenic soybean plants using *Agrobacterium*-mediated gene transfer. *Bio/Technology*, **6**: 915–922.

Hood E E, Jen G, Kayer L, Kramer J, Fraley R T and Chilton M D (1984) Restriction endonuclease map of pTiBo542, a potential Ti plasmid vector for genetic engineering of plants. *Bio/Technology*, **2**: 702–709.

Hood E E, Gelvin S B, Melchers L S and Hoekema A (1993) New *Agrobacterium* helper plasmids for gene transfer to plants. *Transgenic Res.*, **2**: 208–218.

Horsch R B, Fry J E, Hoffman N L, Eicholtz D, Rogers S G and Fraley R T (1985) A simple and general method for transferring genes into plants. *Science*, **227**: 1229–1231.

Hu W Y and Wang L (1999) *In planta* soybean transformation technologies developed in China: procedure, confirmation and field performance. *In Vitro Cell Dev. Biol. Plant.*, **35**: 417–420.

Ishida Y, Saito H, Ohta S, Hiei Y, Komari T and Kumashiro T (1996) High efficiency transformation of maize (*Zea mays* L.) mediated by *Agrobacterium tumefaciens. Nature Biotechnol.*, **14**: 745–750.

Joersbo M, Donaldson I, Kreiberg J, Petersen S G, Brunstedt J and Okkels F T (1998) Analysis of mannose selection used for transformation of sugar beet. *Mol. Breed.*, **4**: 111–117.

Kaeppler H F, Gu W, Somers D A, Rines H W and Cockburn A F (1990) Silicon carbide fiber-mediated DNA delivery into plant cells. *Plant Cell Rep.*, **9**: 415–418.

Klein T M, Wolf E D, Wu R and Sanford J C (1987) High-velocity microprojectiles for delivering nucleic acids into living cells. *Nature*, **327**: 70–73.

Kovacs L and Pueppke S G (1993) The chromosomal background of *Agrobacterium tumefaciens* Chry5 conditions high virulence on soybean. *Mol. Plant-Microbe Interact*, **6**: 601–608.

Lazzeri P A, Hildebrand D F and Collins G B (1985) A procedure for plant regeneration from immature cotyledon tissue of soybean. *Plant Mol. Biol.*, **3**: 160–167.

Lin W, Odell J T and Screiner R M (1987) Soybean protoplast culture and direct gene uptake and expression by cultured soybean protoplasts. *Plant Physiol.*, **85**: 856–861.

Lippman B and Lippman G (1984) Induction of somatic embryos in cotyledonary node tissue of soybean (*Glycine max* (L.) Merrill). *Plant Cell Rep.*, **3**: 215–218.

Liu W, Moore P J and Collins G B (1992) Somatic embryogenesis in soybean via somatic embryo cycling. *In Vitro Cell Dev. Biol. Plant.*, **28**: 153–160.

Liu W, Torisky R S, McAllister K P, Avdiushko S, Hikdebrand D F and Collins G B (1996) Somatic embryo cycling: evaluation of a novel transformation and assay system for seed-specific gene expression in soybean. *Plant Cell Tiss. Org. Cult.*, **47**: 33–42.

Livak K J, Flood J A, Marmaro J, Giusti W and Deetz K (1995) Oligonucleotides with fluorescent dyes at opposite ends provide a quenched probe system useful for detecting PCR product and nucleic acid hybridization. *PCR Methods and Applications*, **4**: 357–62.

Luo G, Hepburn A and Widholm J (1994) A simple procedure for the expression of genes in transgenic soybean callus tissue. *Plant Cell Rep.*, **13**: 632–636.

Maughan P J, Philip R, Cho M J, Widholm J M and Vodkin L O (1999) Biolistic transformation, expression and inheritance of bovine ß-casein in soybean (*Glycine max* L.). *In Vitro Cell Dev. Biol. Plant*, **35**: 334–349.

Mauro A O, Pfeiffer T W and Collins G B (1995) Inheritance of soybean susceptibility to *Agrobacterium tumefaciens* and its relationship to transformation. *Crop Sci.*, **35**: 1152–1156.

McCabe D E, Swain W F, Martinell B J and Christou P (1988) Stable transformation of soybean (*Glycine max* L.) by particle acceleration. *Bio/Technology*, **6**: 923–926.

Meurer C A, Dinkins R D and Collins G B (1998) Factors affecting soybean cotyledonary node transformation. *Plant Cell Rep.*, **18**: 180–186.

Meurer C A, Dinkins R D, Redmond C T, McAllister K P, Tucker D M, Walker D R, Parrott W A, Trick H N, Essig J S, Franz H M, Finer J J and Collins G B (2000) Embryogenic response of multiple soybean (*Glycine max* (L.) Merrill) cultivars across three locations. *In Vitro Cell Dev. Biol. Plant.*, **37**: 62–67.

Negrotto D, Jolley M, Beer S, Wenck A R and Hansen G (2000) The use of phosphomannose-isomerase as a selectable marker to recover transgenic maize plants (*Zea mays* L.) via *Agrobacterium* transformation. *Plant Cell Rep.*, **19**: 798–803.

Owens L D and Cress D E (1985) Genotypic variability of soybean response to *Agrobacterium tumefaciens* strains harboring Ti or Ri plasmids. *Plant Physiol.*, **77**: 87–94.

Padgette S R, Kolacz K H, Delannay X, Re D B, LaVallee B J, Tinius C N, Rhodes W K, Otero Y I, Barry G F, Eichholz D A, Peschke V M, Nida D L, Taylor N B and Kishore G M (1995) Development, identification, and characterization of a glyphosate-tolerant soybean line. *Crop Sci.*, **35**: 1451–1461.

Palanichelvam K, Oger P, Clough S J, Cha C, Bent A F and Farrand S K (2000) A second T-region on the soybean-supervirulent Chrysopine-type Ti plasmid pTiChry5, and construction of a fully disarmed vir helper plasmid. *Mol. Plant Microbe Interact.*, **13**: 1081–1091.

Parrott W A, All J N, Adang M J, Bailey M A, Boerma H R and Stewart Jr C N (1994) Recovery and evaluation of soybean (*Glycine max* (L.) Merrill) plants transgenic for a *Bacillus thuringiensis* var. kurstaki insecticidal gene. *In Vitro Cell Dev. Biol. Plant*, **30**: 144–149.

Parrott W A, Dryden G, Vogt S, Hildebrand D F, Collins G B and Williams E G (1988) Optimization of somatic embryogenesis and embryo germination in soybean. *In Vitro Cell Dev. Biol. Plant*, **8**: 817–820.

Parrott W A, Hoffman L M, Hildebrand D F, Williams E G and Collins G B (1989) Recovery of primary transformants of soybean. *Plant Cell Rep.*, **7**: 615–617.

Ponappa T, Brzozowski A E and Finer J J (1999). Transient expression and stable transformation of soybean using jellyfish green fluorescent protein. *Plant Cell Rep.*, **19**: 6–12.

Ranch J P, Oglesby L and Zielinski A C (1985) Plant regeneration from embryo-derived tissue cultures of soybeans. *In Vitro Cell Dev. Biol. Plant*, **21**: 653–657.

Rashid H, Yokoi S Toriyama K and Hinata K (1996) Transgenic plant production mediated by *Agrobacterium* in *Indica* rice. *Plant Cell Rep.*, **15**: 727–730.

Samoylov V M, Tucker D M and Parrott W A (1998a) A liquid medium-based protocol for rapid regeneration from embryogenic soybean cultures. *Plant Cell Rep.*, **18**: 49–54.

Samoylov V M, Tucker D M and Parrott W A (1998b) Soybean (*Glycine max* (L.) Merrill) embryogenic cultures: the role of sucrose and total nitrogen content on proliferation. *In Vitro Cell Dev. Biol. Plant.*, **34**: 8–13.

Santarem E R, Pelissier B and Finer J J (1997) Effect of explant orientation, pH, solidifying agent and wounding on initiation of soybean somatic embryos. *In Vitro Cell Dev. Biol. Plant.*, **33**: 13–19.

Santarem E R, Trick H N, Essig J S and Finer J J (1998) Sonication-assisted *Agrobacterium*-mediated transformation of soybean immature cotyledons: optimization of transient expression. *Plant Cell Rep.*, **17**: 752–759.

Sato S, Newell C, Kalacz K, Tredo L, Finer J and Hinchee M (1993) Stable transformation via particle bombardment in two different soybean regeneration systems. *Plant Cell Rep.*, **12**: 408–413.

Simmonds D H and Donaldson P A (2000) Genotype screening for proliferative embryogenesis and biolistic transformation of short-season soybean genotypes. *Plant Cell Rep.*, **19**: 485–490.

Singh R J, Klein T M, Mauvais C J, Knowlton S, Hymowitz T and Kostow C M (1998) Cytological characterization of transgenic soybean. *Theor. Appl. Genet.*, **96**: 319–324.

Slawinska J and Obendorf R L (1991) Soybean somatic embryo maturation: composition, respiration and water relations. *Seed Sci. Res.*, **1**: 251–262.

Stachel S E, Messens, E, Van Montagu M and Zambryski P (1985) Identification of the signal molecules produced by wounded plant cells which activate the T-DNA transfer process in *Agrobacterium tumefaciens*. *Nature*, **318**: 624–629.

Stewart Jr C N, Adang M J, All J N, Barmy H R, Cardineau G, Tucker D M and Parrot W A (1995) Genetic transformation, recovery and characterization of fertile soybean (*Glycine max* (L.) Merrill) transgenic for a synthetic *Bacillus thuringiensis cryIAc* gene. *Plant Physiol.*, **112**: 121–129.

Tian L N, Brown D C W, Voldeng H and Webb J (1994) *In vitro* response and pedigree analysis for somatic embryogenesis of long-day photoperiod adapted soybean. *Plant Cell Tiss. Org. Cult.*, **36**: 269–273.

Tinland B (1996) The integration of T-DNA into plant genomes. *Trends Genet.*, **6**: 178–184.

Thompson C J, Movva N R, Tizard R, Crameri R, Davies J E, Louwereys M and Botterman J (1987) Characterization of the herbicide-resistance gene *bar* from *Streptomyces hygroscopicas*. *EMBO J.*, **6**: 2519–2523.

Tjokrojusumo D, Heinrich T, Wylie S, Potter R and McComb J (2000) Vacuum infiltration of *Petunia hybrida* pollen with *Agrobacterium tumefaciens* to achieve plant transformation. *Plant Cell Rep.*, **9**: 792–797.

Torisky R, Fellers J P and Collins G B (1996) A focusing device for tissue transformation with the Du Pont/Bio Rad PDS1000 helium microprojectile system. *Plant Mol. Biol. Rep.*, **14**: 124–133.

Torisky R S, Kovacs L, Avdiushko S, Newman J D, Hunt A G and Collins G B (1997) Development of a binary vector system for plant transformation based on the supervirulent *Agrobacterium tumefaciens* strain Chry5. *Plant Cell Rep.*, **17**: 102–108.

Townsend J A and Thomas L A (1993) An improved method of *Agrobacterium*-transformation of cultured soybean cells. *Patent WO94/02620.*

Trick H N, Dinkins R D, Santarem E R, Di R, Samoylov V, Meurer C A, Walker D R, Parrott W A, Finer J J and Collins G B (1997) Recent advances in soybean transformation. *Plant Tiss. Cult. Biotech.*, **3**: 9–26.

Trick H N and Finer J J (1998) Sonication-assisted *Agrobacterium*-mediated transformation of soybean (*Glycine max* (L.) Merrill) embryogenic suspension culture tissue. *Plant Cell Rep.*, **17**: 482–488.

Vain, P, Keen N, Murillo J, Rathus C, Nemes C and Finer J J (1993) Development of the particle inflow gun. *Plant Cell Tiss. Org. Cult.*, **33**: 237–246.

Walker D R and Parrott W A (2001) Normalizing soybean somatic embryo development in liquid medium. *Plant Cell Tiss. Org. Cult* (in press).

Wang Y C, Klein T M, Fromm M, Cao J, Sanford J C and Wu R (1988) Transient expression of foreign genes in rice, wheat and soybean cells following particle bombardment. *Plant Mol. Biol.*, **11**: 433–439.

Wei Z M and Xu Z H (1990) Plant regeneration from protoplasts of immature cotyledons of *Glycine soja* Seib. et Zucc. *Plant Sci.*, **70**: 101–104.

Wordragen M F and Dons H J M (1992) *Agrobacterium tumefaciens*-mediated transformation of recalcitrant crops. *Plant Mol. Biol. Rep.*, **10**: 12–36.

Wright M S, Koehler S M, Hinchee M A and Carnes M G (1986) Plant regeneration by organogenesis in *Glycine max*. *Plant Cell Rep.*, **5**: 150–154.

Wright M S, Williams M W, Pierson P E and Carnes M G (1987) Initiation and propagation of (*Glycine max* (L.) Merrill) plant from tissue-cultured epicotyls. *Plant Cell Tiss. Org. Cult.*, **8**: 83–90.

Yan B, Reddy, M S S, Collins G B and Dinkins R D (2001) *Agrobacterium tumefaciens*- mediated transformation of soybean (*Glycine max* (L.) Merrill) immature zygotic cotyledon explants. *Plant Cell Rep* (in press).

Zeitlin L, Olmsted S S, Moench T R, Co M S, Martnell B J, Paradkar V M, Russell C, Queen C, Cone R A and Whaley K J (1998) A humanized monoclonal antibody produced in transgenic plants for immunoprotection of the vagina against genital herpes. *Nat. Biotechnol.*, **16**: 1361–1364.

Zhang Z, Xing A, Staswick P and Clemente T E (1999) The use of glufosinate as a selective agent in *Agrobacterium*-mediated transformation of soybean. *Plant Cell Tiss. Org. Cult.*, **56**: 37–46.

GENETIC TRANSFORMATION IN PEA

JAN E. GRANT* AND PAULINE A. COOPER
New Zealand Institute for Crop and Food Research Ltd., Private Bag 4704, Christchurch, New Zealand
**e-mail: grantj@crop.cri.nz*

Abstract

Genetic engineering in peas (*Pisum sativum* L.) is discussed in relation to *Agrobacterium tumefaciens*-mediated transformation and direct gene transfer via electroporation. There are several methods of *A. tumefaciens*-mediated transformation applied routinely in a number of laboratories around the world. All have been used to transfer potentially useful genes into peas. Each of these methods are compared with regard to ease of transformation, efficiency and genotype independence. Some transgenic peas have been field tested and could be suitable for commercial release depending on the regulatory environment and patent claims on the genes and technologies. A number of useful genes that have been transferred into peas are discussed. In addition transgenic peas have been studied as 'model systems' for nodule development, and as bioreactors.

1. Introduction

Pisum sativum L. is cultivated as food for human consumption and as a feed for animals and is an important component of arable crop rotations. There are pea cultivar types – garden or vining peas (including edible pod types) which are used for canning, freezing, dehydrating or as a fresh product, usually for human consumption. Field and forage pea types include maple, blue, white, dun and marrowfat. Cultivated peas typically have white or purple flowers, are self-pollinating and have a chromosome number 2n=14. Peas are grown worldwide largely in temperate areas and at higher altitudes or during cooler seasons of warmer regions (Gane, 1985).

Peas, as with many legumes, were regarded as a crop 'recalcitrant' to transformation. As peas are a minor crop compared with other grain legumes, e.g. soybean, much less effort has been expended to establish transformation systems. As with many crops, initial reports for pea transformation concentrated on suitable regeneration systems, transient and stable expression in callus or various explants, and suitability for *Agrobacterium*

P. K. Jaiwal and R. P. Singh (eds.), Applied Genetics of Leguminosae Biotechnology, 23–34.

infection (Filipone and Lurquin, 1989; de Kathen and Jacobsen, 1990; Shaerer and Pilet, 1991; Lulsdorf *et al.*, 1991; Nauerby *et al.*, 1991). Incremental progress was made.

The breakthrough in pea transformation came in 1990 when Puonti-Kaerlas and co-workers reported the achievement of transformed plants and stable inheritance of the transgene (Puonti-Kaerlas *et al.*, 1992). Since then a number of reports have been published for different and improved methods of pea transformation.

Many factors are important in successful transformation and many facets of the process are not fully understood, so plant transformation, especially in legumes, remains more an art than an exact science (Birch, 1997; Gelvin, 2000). Transformation requires that target regenerable cells from the host plant are available, that there are suitable methods for introducing the DNA into such cells, and that transformed cells containing the introduced DNA can be selected and regenerated in a reliable manner. Birch (1997) has a list of criteria for practical plant transformation that includes ready availability of target tissue, genotype independence, high efficiency, technical simplicity, efficient selection/screening of putative transformants, avoidance of chimeras, capacity to remove reporter/selectable marker genes that are not the target trait being introduced, simple integration patterns and copy number for ease of subsequent genetics, stable and predictable expression, and absence of valid patent claims on products.

2. Methods and efficiency for gene transfer

In peas, most of the effort towards transformation has used *Agrobacterium tumefaciens*. This was considered the optimal technique as peas were shown to be susceptible to *Agrobacterium* infection, and regeneration systems were available (Filipone and Lurquin, 1989; de Kathen and Jacobsen, 1990; Shaerer and Pilet, 1991; Lulsdorf *et al.*, 1991; Nauerby *et al.*, 1991). Only one method using direct gene transfer has been reported (Chowrira *et al.*, 1996). This was employed in response to the difficulties of *in vitro* regeneration methods and used electroporation of intact meristematic tissue of intact axillary buds.

2.1. *AGROBACTERIUM* TRANSFORMATION METHODS

Puonti-Kaerlas *et al.* (1990) used cultivars Stivo and Puget and recovered transgenic plants after co-cultivation of explants from shoot cultures and epicotyls of 10 day old seedlings. This procedure used *Agrobacterium* strain GV3101. Hygromycin at 15 mg/L as the selection agent allowed the recovery of transgenic plants whereas kanamycin at 75 mg/L did not. The efficiency for producing shoots for each of the cultivars was 1–15% for Stivo and 0.2% for Puget. The shoots were rooted prior to transfer to the glasshouse. The process, which included a lengthy callus phase, took 15 months from explant to seed-bearing primary regenerant. All the primary regenerants that produced viable seed were tetraploid (Puonti-Kaerlas *et al.*, 1992), probably due to the long callus phase required in this procedure. This method is not used by others.

Schroeder *et al.* (1993) co-cultivated longitudinal slices of immature embryos of cultivars Greenfeast and Rondo and used phosphinothricin as the selection agent.

This method took approximately 9 months from explant to seed-bearing primary regenerant. Between 1.5 and 2.5% of starting explants gave rise to transformed plants. These authors considered explant source, bacterial strain, choice of selectable marker and the presence of hormones during co-cultivation to be important factors for consistent production of transgenic pea plants. The explant source was obtained from immature pea seeds 2–5 days after maximum seed fresh weight was reached and when the pod colour had begun to change. The embryo axis was dissected and cut under a stereo microscope into 3–5 transverse segments. The *Agrobacterium* strain used was AGL1 and the selectable marker was phosphinothricin at 10 mg/L. Shoots surviving the selection regime were rooted and transferred to the glasshouse.

Davies *et al.* (1993) obtained transgenic plants by injecting the C58/3 strain of *Agrobacterium* into the cotyledonary node of cultivar Puget. After co-cultivation, explants were grown on a medium containing kanamycin as the selection agent. Approximately 1.4% of injected seed gave rise to transgenic plants. The time taken from inoculation to transfer to soil was about 4 months. Another 2–3 months would be required to obtain mature seed from such plants. This method has been unreliable for consistent production of transformed plants (D. R. Davies, personal communication).

The method of Grant *et al.* (1995) for transformation of peas used immature cotyledons from both glasshouse and field grown peas, the AGL1 strain of *Agrobacterium* and the *bar* gene as the selectable marker. The cultivars Bolero, Trounce, Huka and Bohatyr were transformed with efficiencies of 0.7–2%. The cotyledons were taken from 'eating stage' peas where the seed had just reached maximum fresh weight and the pods were still green and smooth. After co-cultivation for 6 days the explants were selected on 10 mg/L phosphinothricin. Subsequently, the co-cultivation time was reduced to 4 days and the level of phosphinothricin to 8 mg/L to allow recovery of transgenic plants from more cultivars.

These four reports all showed the introduced DNA to be stably inherited in subsequent generations with no evidence of chimerism reported. Since these reports, improvements on the initial successful methods with *Agrobacterium*-mediated transformation have continued, resulting in increases in the number of cultivars successfully transformed, the number of selection agents that work efficiently and finally reports of useful genes being successfully field trialled.

Bean *et al.* (1997) improved the system that Davies *et al.* (1993) developed for the cultivar Puget by using the EHA105 strain of *Agrobacterium*, and the *bar* gene with low levels of phosphinothricin selection (2.5 mg/L for 3 weeks followed by 5 mg/L). Bean *et al.* (1997) regarded the less severe selection as critical but they did find chimeras amongst their transgenic plants. They found that rooting of putative transgenic shoots was erratic and therefore used the method of grafting these shoots using 2 week old non-transgenic peas. This method produced 1.1% transgenics per seed explant inoculated.

The initial pea transformation successes were obtained using hygromycin and phosphinothricin selection and it was thought that kanamycin, irrespective of the promoter sequence, was ineffective for selection. However, in 1998b, Grant *et al.* demonstrated that kanamycin at 75 mg/L was effective for selecting transformed peas at efficiencies of 0.8–3.4%. This time a much broader range of cultivars was used including Bolero, Hadlee, Courier and Crown as well as a semi-leafless maple pea and a vegetable pea breeding line. In addition, hygromycin (2 mg/L) was also shown to be effective in this system with the cultivars Crown and Lincoln, giving efficiencies of 0.6% and 2.8% respectively. This report and the previous reports indicate the genotype independence of this method. In our laboratory we have now transformed over 28 breeding lines or cultivars of pea. There is variation in the efficiency of transformation for each genotype ranging from 0.2% to 7% (J. E. Grant, unpublished data). This method is not optimized for individual genotypes so that current advanced breeding lines developed by breeders can be easily incorporated into the transformation programme.

Morton *et al.* (2000a,b) have detailed improvements on the original method developed by Schroeder *et al.* (1993) and have also shown that kanamycin at 75–100 mg/L is effective in their transformation system. The method is also effective for hand harvested mature pea seeds that are germinated for 48 hours. To overcome cultivar variation in rooting efficiency of transgenic shoots, Morton *et al.* (2000b) grafted the shoots onto Greenfeast rootstock resulting in a 4–6 week reduction in time for production of mature transgenic plants. The efficiency of their transformations using kanamycin selection was not reported.

Nadolska-Orczyk and Orczyk (2000) investigated some of the factors that influence the efficiency of pea transformation. They tested three *Agrobacterium* strains and the pGPTV series of vectors (Becker *et al.*, 1992) which use the *nos* promoter on the selectable marker gene. Using the cultivar Laser and 50 mg/L kanamycin selection, bacterial strain EHA105 was found to be best. Transgenic plants were recovered from kanamycin and phosphinothricin selection but not from hygromycin or methotrexate selection. It is difficult to get a real picture of efficiency from Nadolska-Orczyk and Orczyk's (2000) work because they appeared to get a large number of escapes on selection with both kanamycin and phosphinothricin, 65% and 72% escapes respectively.

Polowick *et al.* (2000) also investigated some factors that may influence transformation of pea. They used the method of Schroeder *et al.* (1993) to test different vectors on a range of Canadian breeding lines. For selection of transgenic shoots they used phosphinothricin (10 mg/L), kanamycin (40 mg/L) or chlorsulfuron (10 mg/L). Efficiency of transformation was 0–6.4% with an average of 0.6%. However, because of the small sample size they calculated that there was no significant difference between plant genotype or vector selection. In addition, like Schroeder *et al.* (1993) and Grant *et al.* (1995, 1998b), these authors found that by rooting the transgenic shoots on the appropriate selection medium effectively eliminated non-transgenic shoots that may have 'escaped' on the regeneration media.

Miller *et al.* (1999) also used a modified Schroeder *et al.* (1993) method which uses embryonic axes from mature seed. They used the *Agrobacterium* strain AGL1 and reported that the critical factors for them were the plant genotype, the explant, drying the explant after co-cultivation and using thiadiazuron (TDZ) in the regeneration medium. They reported one experiment where they verified eight transformed shoots from fifty seeds. However, these were not confirmed by Southern analysis and experience in our laboratory has shown that there can be wide experiment to experiment variation, so these results would need to be repeated.

Recent developments in our laboratory have been to improve the efficiency and to develop co-transformation methods to ensure that the gene of interest is transferred and not the marker or reporter genes. We have carried out co-transformation experiments with pea using *Agrobacterium* strain AGL1, with one plasmid per bacterium. The bacteria were mixed together and transformed into the cultivars Bolero and Lincoln. When the explants were selected on 75 mg/L of kanamycin, 33% of the independently transformed lines were transformed with both plasmids. *Agrobacterium* with both plasmids in the one bacterium is currently being tested as Komari *et al.* (1996) found this to be more efficient for delivery and integration of both T-DNAs into tobacco cells.

The methods described above are operating at a similar level of efficiency when the number of explants is compared to the number of independent transformants. One critical aspect found in our laboratory has been the importance of the bacterial strain. In their study Nadolska-Orczyk and Orczyk (2000) found that EHA105 gave the most number of transformants. The main three methods that have been developed by Schroeder *et al.* (1994), Bean *et al.* (1997) and Grant *et al.* (1995, 1998b) have used EHA105 or AGL1 with reliable efficiencies of up to 7% but more commonly around 2% irrespective of the plasmid and the genotype. Over the last two seasons a different *Agrobacterium tumefaciens* strain has been tested in our laboratory. Comparing this strain with that of AGL1 we have obtained at least four times the number of transformants in each experiment we have carried out. These experiments used the same plasmid and selection regime on kanamycin. This result is similar for two cultivars, Bolero and Lincoln, and the garden pea breeding line 99. For Bolero, the efficiency increased from 2.77% with AGL1 to 13.58% with the new strain. The increase for Lincoln was from 0.31% to 2.19% and for the breeding line 99 from 0.21% to 1.06% (J. E. Grant, unpublished data).

2.2. DIRECT GENE TRANSFER METHODS

The only report of successful transformation of peas without using *Agrobacterium* is that of Chowrira *et al.* (1995, 1996). This group electroporated 3–4 week old peas of cultivar Sparkle at the apical node. The pea was then left to grow out from this electroporated node, flower and produce seed. Assessment of the plant tissue (R0) subsequent to the electroporated node could be analysed for expression of the introduced gene(s). Chowrira *et al.* (1996) found that patterns of expression of the *uidA* gene varied considerably in R_0 tissue. From the progeny seed transformation was assessed. 14% of R_0 shoots that were electroporated showed GUS expression and from 120 of the progeny, 20 (26%) showed GUS expression and transformation was confirmed by Southern analysis. The electroporation step requires specific, critical conditions, and although the method does not require a tissue culture regeneration protocol, selection among the progeny is very time consuming.

3. Characters transferred

Genetic engineering allows the transfer of genes from diverse sources into crop plants. Transformation of peas has now been available for some years and reports of useful genes that have been transferred are increasing.

3.1. PEST AND DISEASE RESISTANCE

3.1.1. Resistance to Bruchid beetles

One of the first experiments of transgenic peas containing other than marker genes was carried out by Schroeder *et al.* (1994, 1995, 1999) and Shade *et al.* (1994) who introduced the bean α-amylase gene for expression in pea seed. Bruchid beetle larvae infestation of legumes can cause major economic losses of grains throughout the world. The cowpea weevil and the azuki bean weevil damage stored seeds while the pea weevil attacks the growing crop.

Shade *et al.* (1994) introduced the bean α-amylase gene under the control of the bean phytohemaglutinin-L gene promoter using the method of Schroeder *et al.* (1993). The seeds used in testing were T_2 generation seeds from T_1 seeds containing the highest level of inhibitor and were derived from four primary transgenics. Measured concentrations of the protein in the T_2 seeds ranged from undetectable to 1.27%. In testing approximately 400 seeds for resistance to Azuki bean weevil these authors found that in nearly every seed where the bean α-amylase protein was detected biochemically all infesting larvae died. There were four seeds where larvae development was slowed by 3–5 weeks. In the other seed where there was no detectable protein, development of the larvae was the same as in non-transgenic control seeds.

When a further 387 of these T_2 seed were tested for resistance to cowpea weevil some were immune, some hindered growth and development of larvae to various degrees and some were as susceptible as the non-transgenic controls. The percent mortality and developmental delays in weevil larvae were highly correlated with the measured level of bean α-amylase protein in the seed.

Schroeder *et al.* (1995) also tested these peas containing the bean α-amylase gene under control of the bean phytohemaglutinin promoter for resistance to pea weevil. Of the 18 independent transformants they obtained, resistance to pea weevil was closely studied in five of them. They investigated the T_2 generation of seed. All plants had a similar infestation of the larvae and there was a wide range of responses in weevil development and mortality. As for cowpea weevil, the results correlated with the level of bean α-amylase protein in the seed. Emergence of adult weevils was gradually reduced with the increase of total soluble protein from 1.0% to 3.5%. To test this further, resistance was tested in the line that had the highest level of bean α-amylase protein (3.5%) in the T_5 generation when the gene was presumed to be homozygous. In these seeds no adult weevils emerged, while in the control adult weevils emerged from 87% of seed.

Morton *et al.* (2000b) reported the results of field trials for resistance to pea weevil. The trials were carried out over two years. In year 1 the field trial was at one site and used the cultivar Greenfeast, which had been used in the laboratory studies. In year 2 the field trial was at three sites with cultivar Laura transformed with 2 α-amylase inhibitors – αA1-1 and αA1-2. Under field conditions transformed Greenfeast peas showed that pea weevil larvae had entered 80% of the seeds but only 7% of these emerged adult weevils while the control plants emerged 98% adults. In year 2, with cultivar Laura, both transgenic lines containing the αA1-1 gave complete protection against pea weevil emergence at all 3 sites. The transgenic line containing the αA1-2 gene gave partial protection, in that it delayed the development of adults by about one month. The difference between the incomplete protection provided by transgenic Greenfeast and the complete protection provided by transgenic Laura was the level of transgene expression. The transgenic Greenfeast had 50%–70% less α-amylase inhibitor protein than the transgenic Laura lines.

These results indicated the bean α-amylase gene could be used for the control of bruchid beetles in peas. However, the presence of the protein in the seed crop may have an effect on the nutritional value of the genetically modified seed. Pusztai *et al.* (1999) conducted a short term nutritional study on rats to determine whether these transgenic peas expressing the bean α-amylase protein had an adverse effect on starch digestibility and utilization. This group found that the nutritional value of diets containing transgenic or non-transgenic peas was similar and concluded that peas transgenic for the bean α-amylase gene and expressing the protein at levels around 3% could be used in the diet of farm animals at the recommended commercial levels. Much more testing would be required to determine suitability of such transgenic peas for human consumption.

3.1.2. Alfalfa Mosaic Virus (AMV) resistance

Grant *et al.* (1998a) and Timmerman-Vaughan *et al.* (2001) have reported the details on their glasshouse tests and their field trial for peas engineered with the coat protein gene from the alfalfa mosaic virus. There is no known effective host-derived resistance to AMV in *Pisum* germplasm. The effect of AMV infection in pea crops is to reduce yield and seed size.

The peas in these experiments were selected on kanamycin using the method of Grant *et al.* (1995, 1998b). Two constructs were used, one containing the omega translation enhancer sequence from TMV inserted between the 35S promoter and the coat protein coding region and one that did not. Both constructs produced the coat protein in transgenic peas, and from these transformants, five lines were selected for glasshouse studies and four of these five lines were included in the field study of the resistance. Four of the lines had the TMV omega sequence and one did not. Two of the five lines had multiple copies, of the gene. No obvious correlation was observed between transgene banding pattern and transgene protein expression in any of the lines. Silencing was observed in the progeny of the two T_0 lines that contained the multiple copies, in that some of these progeny failed to accumulate detectable amounts of the coat protein. The single copy lines produced progeny that stably expressed detectable coat protein.

The coat protein gene was derived from AMV strain NZ1 and testing for resistance was carried out with AMV strains 425, NZ34 and NZ1 (Lincoln) which differed from the coat protein gene used for transformation by 8, 5 and 4 amino acid residues respectively. The NZ1 strain from which the coat protein gene was derived was unavailable for testing. The glasshouse testing was carried out with AMV 425 and AMV NZ1 (Lincoln). Timmerman-Vaughan *et al.* (2001) found that only plants that accumulated detectable amounts of the transgene coat protein product were resistant. In addition, some of the progeny plants that accumulated the coat protein product were susceptible to AMV infection indicating that presence of the transgene alone was not sufficient to confer resistance. The field test had three blocks of transgenic and non-transgenic progeny that had been tested in the glasshouse. The field testing was carried out with NZ1 (Lincoln) and NZ34. Three blocks of randomly arranged transgenic and non-transgenic lines were inoculated with no AMV, AMV NZ1 (Lincoln) or AMV NZ34. At flowering it was obvious that inoculations with each of the AMV strains had a large effect on the size and vigour of the non-transgenic control lines while in the uninoculated block all the plants were large and the transgenic lines were indistinguishable from the non-transgenic lines in size, vigour and yield.

These authors concluded that coat protein product accumulation was required for resistance and that the resistance was not strain specific since three different strains were used for testing and resistant plants were obtained from all of these strains. The resistance was described as partial as susceptible plants were found amongst progeny that accumulated AMV coat protein in both glasshouse and field trials. In addition these authors suggested that a sensible protocol for incorporating this technology into a breeding programme would be to produce T_0 lines, screen for coat protein accumulation with Western blots, then screen high expressors to select single copy lines with Southern blots. These individuals and/or their progeny could then be used in a breeding programme.

3.1.3. Pea Seed-borne Mosaic Virus (PSbMV) resistance

PSbMV is a potyvirus that infects pea seed and causes characteristic 'tennis ball' marking on the seed and causes yield loss. Marked seed is downgraded in the market place. There is natural resistance to PSbMV in *Pisum* and the resistance is caused by a recessive gene mapped to chromosome 6 (Timmerman *et al.*, 1993).

Grant and his coworkers (unpublished data) using their cotyledon transformation method, produced a large number of independently transformed pea lines using 10 cultivars and breeding lines containing two constructs in the sense orientation (similar to the constructs used for producing AMV resistant lines) and one construct with a non-translatable version of the PSbMV coat protein. None of the progeny of the primary transgenic lines showed resistance to PSbMV. When the primary transgenic lines were tested some resistance to PSbMV was observed but this resistance was not inherited.

Jones *et al.* (1998) approached resistance to PSbMV via the replicase sequence (NIb). They used four constructs nos: NIb, 2x35S:NIb, 2x35S:NIbIb, and 2x35S:NibbI all with PPT as the selectable marker and were inserted into AGL1. The 2x35S:NIbIb, and 2x35S:NibbI constructs were designed to be non-translatable versions with an added direct or inverted repeat respectively. The cultivar Puget was transformed with these constructs using the method of Bean *et al.* (1997). Thirty-five transformants were obtained from 2300 seeds (1.5%) and all four constructs were represented. However, three of these lines were considered to be chimeric after examining inheritance in the T_1 progeny. Resistance to PSbMV was not seen in any line tested at 10 days after infection, but after 30 days three lines had 'recovered' from the infection and this was related to presence of the transgene and NIb transcript. There was no correlation between the level of NIb transgene RNA accumulation and the ability to recover as, of the six lines with detectable NIb transcript, three showed the recovery phenotype and three did not. Line #2, the line that showed the highest NIb mRNA accumulation, was tested using different pathotypes. The recovery phenotype was observed with the most similar pathotype. With the more dissimilar pathotype, recovery was observed in some but not all experiments. The infected plants that recovered from the initial infection yielded a similar quantity of seed and reached a similar height to plants that had not been infected.

3.1.4. Pea Enation Mosaic Virus (PEMV) resistance

PEMV is the only member of the enamovirus group and in pea causes periodic crop losses, particularly in the northern parts of the USA. Resistance to PEMV can be achieved by breeding into pea cultivars a resistance gene mapped to linkage group 3 (G. M. Timmerman-Vaughan, personal communication).

Chowrira *et al.* (1998) electroporated the coat protein from PEMV into cultivar Sparkle terminal buds and obtained 23 individual R_1 plants of which three were shown to have the coat protein gene. The coat protein gene product was detected in the two plants that had a single copy of the gene and not in the third plant that had two copies of the coat protein gene. In the R_2 plants that were challenged with the virus none of the lines were completely resistant but some appeared to recover and some showed delayed development of symptoms. Chowrira *et al.* (1998) suggested that expression of the coat protein gene product was required for the resistance phenotypes that they obtained, but the small number of initial plants obtained in this study limited the conclusions that could be drawn from this work.

4. Altering quality traits

4.1. IMPROVING NUTRITIONAL QUALITY

Schroeder *et al.* (1994, 1995) have had a long interest in the nutritional quality of pea seeds. Pea seed is a rich source of protein but lacks sulphur-containing amino acids essential for a good diet. There is little variation within *Pisum* for altering this sulphur deficient state so Schroeder *et al.* (1994, 1995) have introduced the sunflower albumin gene *saf8*. The albumin contains 24% cysteine plus methionine. They have obtained transgenic plants which in the T_3 generation produced a level of SAF8 protein with 0.5–0.6% of total soluble protein in heterozygotes and 1% of total soluble protein in homozygotes. This represents an increase of 10% more carbon bonded sulphur than occurs in non-transgenic peas.

5. Peas as model systems

5.1. PEAS AS BIOREACTORS

Perrin *et al.* (2000) used field peas to investigate their suitability as bioreactors to produce antibodies. Well-established agricultural practices and processing systems, large seed, self-pollinating breeding system, and a range of pea genotypes with easily distinguished seed characters that are not used for human food were considered attractive features for testing field peas' suitability for molecular pharming. The transgene, a single chain Fv fragment antibody, was expressed in the seed and targeted to the endoplasmic reticulum. This transgene was stably inherited and the antibody was active in the seed stored at room temperature for at least two months without a decrease in antibody levels. Perrin *et al.* (2000) used the method of Bean *et al.* (1997) in combination with *Agrobacterium* strain AGL1, cultivar Puget, and phosphinothricin selection. They obtained nine independent transgenic lines six of which expressed functional single chain Fv antibodies. Although Perrin *et al.* (2000) showed levels of expression equivalent to those expressed in cereals and tobacco, the low efficiency of pea transformation achieved here is a disadvantage.

5.2. PEA AS A MODEL FOR NODULE DEVELOPMENT

Wisniewski and Brewin (2000) constructed transgenic pea lines to investigate the role of diamine oxidases in pea nodule development. Diamine oxidases (DAO) occur at high concentrations in legume species and are loosely associated with plant cell walls. Their function is not well understood and may have many functions associated with cell wall changes and permeability. The enzyme transcript is abundant in all infected tissue of pea nodules. These authors generated transgenic lines using two constructs in sense and antisense orientation using the transformation method of Bean *et al.* (1997). They chose two stable transformed lines from each construct. For all the lines nodulation ability was normal. In the lines transformed with the antisense construct the enzyme activity of DAO was unchanged from normal non-transgenic peas. The DAO activity was greatly reduced in nodules of the lines transformed with the sense constructs. These sense constructs showed co-suppression as the activity of both the native DAO homologue and the transgene was reduced. One of the conclusions from this work is that DAO does not affect nodulation ability in pea, but its critical role in nodulation is most likely in the regulation of diamine levels in host tissue.

6. Conclusions and future prospects

In this chapter several methods for routine pea transformation are described. While not all have been shown to be highly efficient, genotype independent or technically simple, they all provide workable transformation systems. Grant and coworkers (unpublished data) have demonstrated a significant increase in transformation efficiency by using a different *Agrobacterium* strain. These authors have also demonstrated that co-transformation is achievable to eliminate the marker and/or reporter genes.

The transformation systems of Grant *et al.* (1995, 1998b) and Schroeder *et al.* (1993) require actively growing pea plants from which to obtain explants for transformation while Bean *et al.* (1997) use freshly germinated seed. This latter method requires screening of progeny of primary transgenics for chimeras. Morton *et al.* (2000b) reported that the method of Schroeder *et al.* (1993) could be used with freshly germinated dry deed. The method of Grant *et al.* (1995, 1998b) is technically less complex in terms of explant preparation and tissue culture.

The three *Agrobacterium* methods and the method using *in planta* electroporation have all produced transgenic pea plants with introduced genes of agricultural interest. Further development of such transgenics requires appropriate testing for commercial release and clearance of any valid patent claims.

The use of peas as bioreactors for producing antibodies (Perrin *et al.*, 2000) is an example of the opportunities available in molecular pharming. Field peas, e.g. maple peas, which are not used for human consumption, are attractive targets as agricultural and processing systems are in place and distinct morphological markers on the seed are well characterized.

Acknowledgements

Thanks to Angela Templeton for critical reading of the manuscript and to my co-workers at Crop and Food Research for discussion.

References

Bean S J, Gooding P S, Mullineaux P M and Davies D R (1997) A simple system for pea transformation. *Plant Cell Rep.*, **16**: 513–519.

Birch R G (1997) Plant transformation: Problems and strategies for practical application. *Annu. Rev. Plant Physiol. and Plant Mol. Biol.*, **48**: 297–326.

Chowrira G M, Akella V, Fuerst P E and Lurquin P F (1996) Transgenic grain legumes obtained by *in planta* electroporation-mediated gene transfer. *Mol. Biotech*, **5**: 85–96.

Chowrira G M, Akella V and Lurquin P F (1995) Electroporation-mediated gene transfer into intact nodal meristem *in planta*. *Mol. Biotech.*, **3**: 17–23.

Chowrira G M, Cavileer T D, Gupta S K, Lurquin P F and Berger P H (1998) Coat protein-mediated resistance to pea enation mosaic virus in transgenic *Pisum sativum* L. *Transgenic Res.*, **7**: 265–271.

Davies D R, Hamilton J and Mullineaux P (1993) Transformation of peas. *Plant Cell Rep.*, **12**: 180–183.

De Kathen A and Jacobsen H J (1990) *Agrobacterium tumefaciens*-mediated transformation of *Pisum sativum* L. using binary and cointegrate vectors. *Plant Cell Rep.*, **9**: 276–279.

Filippone E and Lurquin P F (1989) Stable transformation of pea tissues after co-cultivation with *Agrobacterium tumefaciens* strains. *Pisum NewsLetter*, **21**: 16–18.

Gane A J (1985) The pea crop – agricultural progress past present and future. In: *The Pea Crop* (Eds Hebblethwaite P D, Heath M C and Dawkins T C K). Butterworths, London, 3–16.

Gelvin S B (2000) Agrobacterium and plant genes involved in T-DNA transfer and integration. *Annu. Rev. Plant Physiol. and Plant Mol. Biol.*, **51**: 223–256.

Grant J E and Cooper P A (1995) Rooting in transgenic peas. *International Plant Propagators Society Proceedings*, **45**: 362–365.

Grant J E, Cooper P A, McAra A E and Frew T J (1995) Transformation of peas (*Pisum sativum* L.) using immature cotyledons. *Plant Cell Rep.*, **15**: 254–258.

Grant J E, Cooper P A, Pither-Joyce M D, Fifield W and Timmerman-Vaughan G M (1998a) Partial resistance to alfalfa mosaic virus in transgenic pea (*Pisum sativum* L.). In: *Proc. 3rd European Conference on Grain Legumes*, Vallodolid, Spain, 372–373.

Grant J E, Cooper P A, Gilpin B J, Hoglund S J, Reader J K, Pither-Joyce M D and Timmerman-Vaughan G M (1998b) Kanamycin is effective for selecting transformed peas. *Plant Sci.*, **139**: 159–164.

Jones A L, Johansen I E, Bean S J, Bach I and Maule A J (1998) Specificity of resistance to pea seed-borne mosaic potyvirus in transgenic peas expressing the viral replicase (NIb) gene. *J. Gen. Virol.*, **79**: 3129–3137.

Komari T, Hiei Y, Saito Y, Murai N and Kumashiro T (1996) Vectors carrying two separate T-DNAs for co-transformation of higher plants mediated by *Agrobacterium tumefaciens* and segregation of transformants free from selection markers. *Plant J.*, **10**: 165–174.

Lulsdorf M M, Rempel H, Jackson J A, Baliski D S and Hobbs S L A (1991) Optimising the production of transformed pea (*Pisum sativum* L.) callus using disarmed *Agrobacterium tumefaciens* strains. *Plant Cell Rep.*, **9**: 479–483.

Miller J, Shilvanth G, Williamson B and Ramsey G (1999) Efficient genetic transformation of grain legumes for improved fungal resistance. In: *Scottish Crop Research Institute Annual Report*.

Morton R L, Gollasch S, Schroeder H E, Bateman K S and Higgins T J V (2000a) Gene technology in peas for improved field performance and enhanced grain quality. In: *Handbook of Transgenic Food Plants* (Eds Khachatourians G and Nip W H), Marcel Dekker, New York.

Morton R L, Schroeder H E, Bateman K S, Chrispeels M J, Armstrong E and Higgins T J V (2000b) Bean alpha-amylase inhibitor 1 in transgenic peas (*Pisum sativum*) provides complete protection from pea weevil (*Bruchus pisorum*) under field conditions. *Proc. Natl. Acad. Sci. USA*, **97**: 3820–3825.

Nadolska-Orczyk A and Orczyk W (2000) Study of the factors influencing *Agrobacterium*-mediated transformation of pea (*Pisum sativum* L.). *Mol. Breed.*, **6**: 185–194.

Nauerby B, Madsen M, Christiansen J and Wyndaele R (1991) A rapid and efficient regeneration system for pea (*Pisum sativum*) suitable for transformation. *Plant Cell Rep.*, **9**: 676–679.

Perrin Y, Vaquero C, Gerrard I, Sack M, Drossard J, Stoger E, Christou P and Fischer R (2000) Transgenic pea seeds as bioreactors for the production of a single-chain Fv fragment (scFV) antibody used in cancer diagnosis and therapy. *Mol. Breed.*, **6**: 345–352.

Polowick P L, Quandt J and Mahon J D (2000) The ability of pea transformation technology to transfer genes into peas adapted to western Canadian growing conditions. *Plant Sci.*, **153**:161–170.

Puonti-Kaerlas J, Eriksson T and Engström P (1990) Production of transgenic pea (*Pisum sativum* L.) plants by *Agrobacterium tumefaciens*-mediated gene transfer. *Theor. Appl. Genet.*, **80**: 246–252.

Puonti-Kaerlas J, Eriksson T and Engström P (1992) Inheritance of a bacterial hygromycin phosphotransferase gene in the progeny of primary transgenic pea plants. *Theor. Appl. Genet.*, **84**: 443–450.

Pusztai A, Bardocz G G S, Alonso R, Chrispeels M J, Schroeder H E, Tabe L M and Higgins T J V (1999) Expression of the insecticidal bean alpha-amylase inhibitor transgene has minimal detrimental effect on the nutritional value of peas fed to rats at 30% of the diet. *Journal of Nutrition*, **129**: 1597–1603.

Schaerer S and Pilet P E (1991) Roots, explants and protoplasts from pea transformed with strains of *Agrobacterium tumefaciens* and *rhizogenes*. *Plant Sci.*, **78**: 247–258.

Schroeder H E, Gollasch S, Moore A, Tabe L M, Craig S, Hardie D C, Chrispeels M J, Spencer D and Higgins T J V (1995) Bean α-amylase inhibitor confers resistance to the pea weevil (*Bruchus pisorum*) in transgenic peas (*Pisum sativum* L.). *Plant Physiol.*, **107**: 1233–1239.

Schroeder H E, Gollasch S, Tabe L M and Higgins T J V (1994) Recent advances in gene transfer to peas. *Pisum Genetics*, **26**: 1–5.

Schroeder H, Gollasch S, Tabe L, Moore A, Spencer D and Higgins T (1995) The expression and stability of transgenes in peas (*Pisum sativum* L.). In: *Proc. 2nd European Conference on Grain Legumes*, Copenhagen, Denmark, 422.

Schroeder H E, Schotz A H, Wardley-Richardson T, Spencer D and Higgins T J V (1993) Transformation and regeneration of two cultivars of pea (*Pisum sativum* L.). *Plant Physiol.*, **101**: 751–757.

Shade R E, Schroeder H E, Pueyo J J, Tabe L M, Murdock L L, Higgins T J V and Chrispeels M J (1994). Transgenic pea seeds expressing the α-amylase inhibitor of the common bean are resistant to bruchid beetles. *Bio/Technology*, **12**: 793–796.

Timmerman G M, Frew T J, Miller A L, Weedon N F and Jermyn W A (1993) Linkage mapping of *sbm-1*; a gene conferring resistance to pea seed-borne mosaic virus, using molecular markers in *Pisum sativum* L. *Theor. Appl. Genet.*, **85**: 609–615.

Timmerman-Vaughan G M, Pither-Joyce M D, Cooper P A, Russell A, Goulden D S Butler R and Grant J E (2001) Partial resistance of transgenic peas to alfalfa mosaic virus under greenhouse and field conditions. *Crop Science* (in press).

Wisniewski J P and Brewin N J (2000) Construction of transgenic pea lines with modified expression of diamine oxidase and modified nodulation responses with exogenous putrescine. *Mol. Plant-Microbe Interact.*, **13**: 922–928.

GENETIC TRANSFORMATION OF COMMON BEAN VIA PARTICLE BOMBARDMENT

FRANCISCO J. L. ARAGÃO
Embrapa Recursos Genéticos e Biotecnología, Laboratório de Introdução e Expressão de Genes, SAIN Parque Rural, C.P.O. 2372, Brasília, DF, 70770–900, Brazil
e-mail: aragao@cenargen.embrapa.br

Abstract

Success in bean (*Phaseolus vulgaris* L.) transformation requires an understanding of both tissue culture and transformations systems. Several techniques have been proposed in order to develop an efficient technology to transform bean. The susceptibility of bean to *Agrobacterium* has been demonstrated and electroporation or polyethylenglycol (PEG)-mediated transformation has been evaluated. Using the particle bombardment process, transgenic bean plants containing agronomically important traits from several varieties have been obtained. Some aspects influencing the achievement of transgenic *P. vulgaris* by particle bombardment, limitations and future prospects are discussed.

1. Introduction

In Latin America, Africa and India, beans are among the most important grain legume for direct consumption. Beans have been a very important source of protein and calories for these areas of the world. The world annual production is about 8.5 million metric tons. Production has been carried out in a wide range of cropping systems, diverse environments, primarily on small farms with little purchased input, in association with other crops. Despite its nutritional importance, productivity has been declining in some regions. The main limiting factors are: poor agronomic practices, diseases, insects, nutritional deficiencies, soils, climate constraints, lack of improved cultivars and weed competition. Consequently, there is considerable interest in the introduction of agronomically useful traits into beans by breeding and genetic engineering.

With the advent of the recombinant DNA technology, genes can be cloned from any organism, independent of the kingdom. Then, the genes can be introduced and expressed in crop plants. Thus, the source of variability available for crop improvement was expanded to all living organisms. Indeed, this technology, associated with sexual breeding

P. K. Jaiwal and R. P. Singh (eds.), Applied Genetics of Leguminosae Biotechnology, 35–45.

methods, may accelerate the production of plants with useful traits. Considerable advances have been achieved in the technologies to transform plants, mainly using the particle bombardment process. This chapter will discuss the technologies developed to transform *Phaseolus vulgaris* and prospects for development of new systems.

2. Gene transfer to *Phaseolus vulgaris*

Several methods have been developed for inserting genetic information into plant cells, such as *Agrobacterium*-mediated system, direct DNA uptake into protoplasts and particle bombardment. Due the advances in the methodologies for gene delivery and plant tissue culture, it was assumed that plant transformation would become a routine for most important crops. Unfortunately, legumes, in particular bean (*P. vulgaris*) provided some of the greatest challenges to transformation efforts.

Early efforts to transform bean demonstrated its susceptibility to *Agrobacterium*, and some transgenic tissues, such as callus, leaves, meristems, cotyledon and hypocotyl, have been achieved (Lippincott *et al.*, 1968; McClean *et al.*, 1991; Franklin *et al.*, 1993; Becker *et al.*, 1994; Lewis and Bliss, 1994; Brasileiro *et al.*, 1996; Nagl *et al.*, 1997). Mariotti *et al.* (1989) reported the production of transgenic bean plants through the utilization of the *Agrobacterium* system. However, there was no molecular evidence for genetic transformation or progeny analysis. Transient gene expression using either electroporation or PEG-mediated protoplast transformation was demonstrated (Crepy *et al.*, 1986; Bustos, 1991; Leon *et al.*, 1991; Giovinazzo *et al.*, 1993). Dillen *et al.* (1995) have demonstrated the applicability of electroporation of intact tissue to introduce and express the *gus* gene in bean embryonic axes.

Nevertheless, no transgenic bean plants were obtained due to the difficulties in regenerating plants. During the last two decades, efforts to achieve an efficient methodology for bean transformation were obstructed due to the lack of an efficient tissue culture system to regenerate bean plants from transformed cells. Numerous attempts have been made to regenerate bean plants from several types of isolated cells and tissues. Although no satisfactory results have been achieved, some methodologies have described shoot organogenesis (through multiple shoot induction) of the apical and axillary meristems from bean embryonic axis (McClean and Grafton, 1989; Malik and Saxena, 1992; Mohamed *et al.*, 1992, 1993). This background information, in association with the particle bombardment process, allowed us to recover transformed bean plants.

3. Particle bombardment-mediated transformation

The particle bombardment process was initially proposed by the group of Dr. John Sanford, at Cornell University, with the objective to introduce genetic material into plant genome (Sanford *et al.*, 1987; Klein *et al.*, 1987). Since the 80s, the universality of applications of particle bombardment has been evaluated, demonstrating to be an effective and simple process for the introduction and expression of genes in bacteria, protozoa, fungi, algae, insects, mammals, plants and isolated organelles, such as chloroplasts and mitochondria

(Daniell *et al.*, 1991; Carrer *et al.*, 1993; Klein and Fitzpatrick-McElligott, 1993; Sanford *et al.*, 1993; Vainstein *et al.*, 1994; Bogo *et al.*, 1996; Rech *et al.*, 1996). The basis of the particle bombardment process is the acceleration of DNA-coated microprojectiles (mainly particles of tungsten or gold, 0.2 to 1.5 μm in diameter) at high speed (about 1500 km/h) towards living cells. After penetration in the cell, the DNA dissociates from the projectiles and integrates into the chromosome. Several devices (particle guns) have been constructed in order to accelerate these microprojectiles. All these systems are based on the generation of a shock wave with enough energy to move the microprojectiles. The shock wave can be generated through: a chemical explosion (Sanford *et al.*, 1987), discharge of gas helium (Sanford *et al.*, 1991; Finer *et al.*, 1992; Takeuchi *et al.*, 1992), vaporization of a drop of water through the electric discharge with high voltage and low capacitance (McCabe *et al.*, 1988; Christou, 1993b) or low voltage and high capacitance (Rech *et al.*, 1991), and discharge of compressed air (Morikawa *et al.*, 1989).

The particle bombardment allowed the bombardment of intact plant cells turning possible the transformation of plants for which a regeneration system was not available. The status of *Phaseolus* tissue culture was recently well reviewed by Nagl *et al.* (1997). Although several authors have claimed the achievement of *P. vulgaris* regeneration, so far, only a cytokinin-induced shoot organogenesis was obtained. Consequently, the apical region of embryonic axes became an obvious target for the development of a system based on the bombardment of meristematic cells. The association of shoot organogenesis and particle bombardment was first utilized to produce transgenic soybean (McCabe *et al.*, 1988).

The particle bombardment was also utilized to mechanically inoculate virus in bean tissues and study plant-pathogen interactions (Gilbertson *et al.*, 1991; Aragão *et al.*, 1995).

4. Transformation based on the bombardment of meristematic cells

The first experiments showed the applicability of the particle bombardment to introduce and transiently express genes into apical meristematic cells (Genga *et al.*, 1991; Aragão *et al.*, 1992, 1993). However, one question has arisen: how deep the particle should penetrate in order to reach the cells that could generate transgenic plants. The different parts of the meristem have been divided in layers (L1, L2 and L3) (Satina *et al.*, 1940). The layer L1 is the most external and forms the epidermis of the differentiated regions. The layers L2 and L3 divide preferentially in the anticlinal and periclinal planes to form the organs. Several studies have demonstrated that the differentiated *de novo* shoots are originated from the sub-epidermal layers (L2 and L3) of the apical meristem. However, the L1 layer could participate in their formation (McClean and Grafton, 1989; Franklin *et al.*, 1991; Malik and Saxena, 1992; Mohamed *et al.*, 1992). These shoots are formed in the peripheral regions of the apical meristem (Aragão and Rech, 1997). The bombardment of *P. vulgaris* meristematic cells showed that it was possible to efficiently reach these layers (Fig. 1), demonstrating that it would be possible to achieve transgenic plants (Aragão *et al.*, 1993).

The utilization of the apical meristematic region to clonally multiply plants was first demonstrated by Morel (1960), studying *Cymbidium* micropropagation. Since then, significant advances in the development of several micropropagation techniques have been

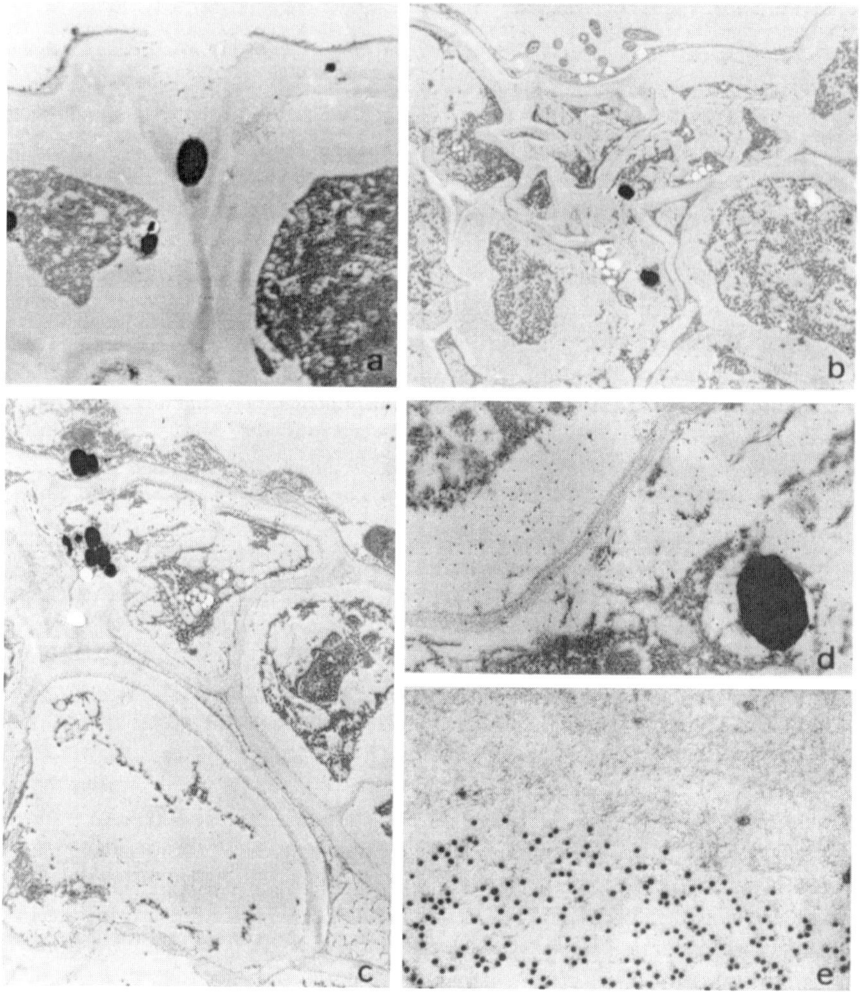

Figure 1. Transmission electron micrographs of bombarded bean tissues. (a) Disrupted cell wall (4,300 ×). (b) Clump of particle penetrating the L1 and L2 cell layers (2,700 ×). (c) Single particle penetration into the L2 and L3 cell layers (2,000 ×). (d) Immunolabeling, showing GUS activity (7,200 ×). (e) Detail of d, showing colloidal gold immunoreacted with GUS-specific antibody (70,000 ×) (Aragão et al., 1993).

reached. The induction of shoot morphogenesis in bean meristems has been achieved with success, through the culture of mature embryos in cytokinins, such as kinetin, zeatin and N^6-benzyl aminopurine (BAP) (Martins and Sondal, 1984; McClean and Grafton, 1989; Kartha *et al.*, 1991; Franklin *et al.*, 1991; Malik and Saxena, 1992; Mohamed *et al.*, 1992, 1993; Aragão and Rech, 1997). Compounds that present cytokinin-like effects, such as thidiazuron (TDZ) and N(2-chloro-4-pyridyl)-N′-phenylurea (CPPU), have also been evaluated (Mohamed *et al.*, 1992).

Early efforts to achieve transgenic bean through the particle bombardment process showed the possibility to introduce and express genes in any type of tissue (Genga *et al.*, 1991; Aragão *et al.*, 1992, 1993). Russel *et al.* (1993) were able to achieve transgenic navy bean (cv. Seafarer) plants using an electrical particle acceleration device. It was the first report of *P. vulgaris* transformation, presenting molecular evidences and progeny. However, the frequency of transgenic plants obtained was much lower (0.03%) and variety-limited. In addition, their described tissue culture protocol was time-consuming, involving several temperature treatments and medium transfers of the bombarded embryos before recovery of transgenic shoots. Kim and Minamikawa (1996) achieved transformation by bombarding embryonic axes and obtained stably transformed bean plants (cv. Goldstar).

We developed an efficient and reproducible system to achieve routinely transgenic bean plants. The bean transformation system was also based upon the development of a tissue culture protocol of multiple shoot induction, shoot elongation and rooting (Fig. 2). The average frequency of transformation (the total number of putative transgenic plants divided by the total number of bombarded embryonic axes) was 0.9% (Aragão *et al.*, 1996; Aragão and Rech, 1997). In addition, we have been able to transform several varieties of *P. vulgaris*, including those that were considered recalcitrant to transformation in previous study. Molecular analysis and progeny test of several generations of transgenic lines revealed the presence of a small number of integrated copies of the foreign genes and segregation in a Medelian fashion in most of them. This is extremely important in order to accelerate the introduction of these plants in a breeding program as well as the production of transgenic commercial varieties.

The particle bombardment has been considered a universal process to introduce macromolecules into any living cell, not limited by genotype or variety (Sanford, 1990; Christou, 1993a). However, the morphology of the explants utilized during the bombardment process may greatly influence the successful recuperation achievement of transgenic bean plants (Aragão and Rech, 1997). In some cultivars, the embryonic axes revealed the apical meristematic region partially exposed, whereas only the central region could be visualized (Fig. 3). The number of meristematic cells, which could be reached by the microparticle coated-DNA, will be drastically reduced. Consequently, the efficiency of transformation could also be reduced. Several studies have shown that *de novo* shoot differentiation in embryos of bean cultivated on cytokinins appeared in the peripheral layers of the meristematic ring (McClean and Grafton, 1989; Franklin *et al.*, 1991; Malik and Saxena, 1992; Aragão and Rech, 1997). Thus, based on these concepts, cultivars with a non-exposed apical meristematic region are not suitable for particle bombardment transformation, considering that the removal of the leaf primordia is not practical.

5. Introduction of useful traits into *Phaseolus vulgaris*

Since the first report of transformation of *Phaseolus vulgaris*, in 1993, few groups have transformed bean to introduce useful traits. Difficulties that still exist in obtaining transgenic plants might account for this fact.

Figure 2. Bean transformation. (a) Embryonic axis expressing the β-glucuronidase (GUS) gene (uidA) 24 h after bombardment. (b) Transgenic shoot from a multiple shooting apical meristem. (c) Leaf from an acclimatized plant expressing uidA gene. (d) Microspores showing (GUS) segregation. (e) Pods and (f) Seeds from transgenic plants (below) and control (above). (g) R1 generation of transgenic plants growing in the greenhouse (Aragão et al., *1996).*

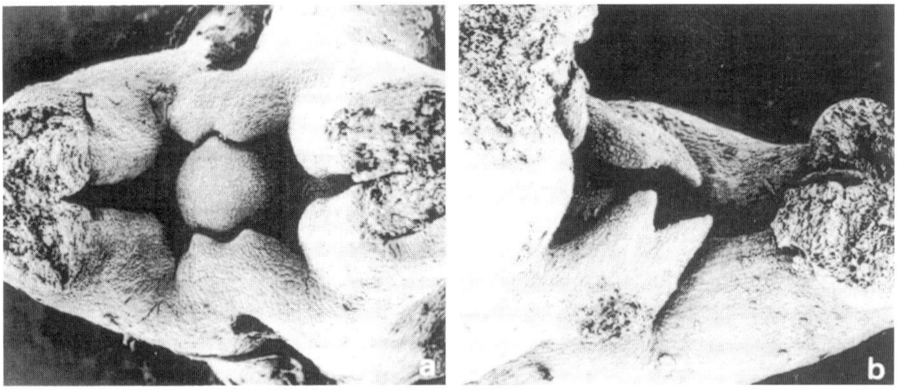

Figure 3. Scanning electron micrographs of the shoot apical meristem regions of the bean cultivars (a) Carioca and (b) Jalo (190X) (Aragão et al., *1993).*

Our group first introduced an agronomically important gene in *Phaseolus*. Mature embryos were transformed in order to transient express a methionine-rich albumin gene from Brazil nut (*be2s1* gene), which could be detected by Western blot and ELISA, twenty-four hours later. However, no stable transformants were recovered (Aragão *et al.*, 1992). Recently, we stable transformed bean with the *be2s1* gene in order to improve the methionine content in the seeds. The transgene was stable and correctly expressed in homozygous R_2 to R_6 generations. In two of the five transgenic lines, the methionine content was significantly increased 14% and 23% over the non-transgenic plants (Aragão *et al.*, 1999).

Russel *et al.* (1993) introduced the *bar* gene that confers resistance to the herbicide phosphinothricin and the coat protein gene from the bean golden mosaic geminivirus (BGMV) in an attempt to achieve virus-resistant plants. The introduced *bar* gene showed it confers strong resistance in transgenic bean to the herbicide in the greenhouse. Bean plants containing the BGMV coat protein gene did not show resistance to the virus (D. Maxwell, University of Wisconsin, personal communication).

The 5′ regulatory sequences from several genes have been transient and stably studied in bean tissues transformed using the particle bombardment. Regulatory regions from the seeds specific promoter from Brazil nut 2S gene were studied in bean embryos (Grossi de Sá *et al.*, 1994; Vincentz *et al.*, 1997). Transgenic bean plants were produced containing the β-glucuronidase gene (*uidA*) under the control of the canavalin gene promoter (Con A) from jack bean. Both the organ and maturation stage-specific promoter regulation was studied in seeds of transgenic plants (Kim and Minamikawa, 1996, 1997).

In order to achieve bean plants resistant to BGMV, the genes *Rep-TrAP-REn* and *BC1* from the virus were cloned in antisense orientation under the control of the CaMV 35S promoter and used to transform bean. Two transgenic lines had both delayed and attenuated viral symptoms (Aragão *et al.*, 1998). The transgenic bean lines are being used in our bean-breeding program in order to evaluate the gene expression in different genetic backgrounds in both greenhouse and field conditions.

6. Limitations and future prospects

The possibility of combining the *Agrobacterium* and particle bombardment systems has been recently proposed. It would associate the high efficiency of the *Agrobacterium* system to integrate the T-DNA in a susceptible host, and the capacity to introduce genetic information into any living cell by the particle bombardment process. This new technology allowed the transformation of tobacco, sunflower, banana and grape (Bidney *et al.*, 1992; Malone-Schoneberg *et al.*, 1994; May *et al.*, 1995; Scorza *et al.*, 1995). Our studies have demonstrated that it is possible to combine *Agrobacterium*/particle bombardment systems to transform bean apical meristems (Brasileiro *et al.*, 1996). The micro-wounds caused by the particle bombardment can enhance the frequency of *Agrobacterium*-mediated transformation. However, we were unable to recover transgenic bean plants. Further investigation will be necessary in order to confirm the practical utility of the combined techniques for bean transformation.

Several species within the genera *Phaseolus*, such as *P. coccineus*, *P. acutifolius* and *P. angularis* (reviewed by Nagl *et al.*, 1997), have been regenerated in *strictu sensu*, e.g. plants regenerated *de novo* from undifferentiated cells originated from differentiated cells. The extrapolation of these technologies to *P. vulgaris* would be a breakthrough regarding the development of a transformation system to accelerate the generation of new transgenic commercial varieties. The ideal system would be the regeneration of fertile mature plants through somatic embryogenesis or organogenesis, preferentially from mature embryonic axes, which has been considered the most adequate tissue for transformation by *Agrobacterium*, intact tissue electroporation and particle bombardment.

Another important constraint in the transformation system based on the bombardment of meristematic tissue of embryonic axes is the difficulty of having an efficient selection of transformed cells. In this case, only few cells from the meristem are transformed, while the others, including those which are in contact with the culture medium and the selective agent, are not transformed. Antibiotics, such as kanamycin, have been used as selective agent, associated to neomycin phosphotransferase (*nptII*) gene (Aragão *et al.*, 1996). However, this selection is not efficient. Recently, we have developed a novel system to select transgenic meristematic cells based on the use of imazapyr, a herbicidal molecule capable of systemically translocating and concentrating in the apical meristematic region of the plant. This selectable marker system, combined with an improved multiple shoot induction protocol, resulted in a significant increase in the recovery of fertile, transgenic material compared with standard soybean transformation protocols (Aragão *et al.*, 2000). We are now extending these findings to improve the *P. vulgaris* transformation system.

In our experience, we have seen that for each 20 transgenic plants obtained, only one has the potential to be introduced in a breeding program to generate a commercial variety. Consequently, a bean transformation technology that would allow selection of transgenic meristematic cells and regeneration of fertile plants should accelerate the development of transgenic bean varieties with improved agricultural characteristics.

Studies of the behavior of transgenic bean plants in field conditions need to be conducted. In these studies, the interactions would be appraised of the transgenic plants with other plants of the agricultural and natural environment. Moreover, the stability of foreign genes and factors related to their interaction with the complex physiology of these plants submitted to natural stress in tropical areas would be evaluated.

Acknowledgements

This work was supported by Empresa Brasileira de Pesquisa Agropecuária (EMBRAPA) and Conselho Nacional de Desenvolvimento Científico e Tecnológico (CNPq), Brazil, grant No. 300107/97–8.

References

Aragão F J L, Barros L M G, Brasileiro A C M, Ribeiro S G, Smith F D, Sanford J C, Faria J C and Rech E L (1996) Inheritance of foreign genes in transgenic bean (*Phaseolus vulgaris* L.) co-transformed via particle bombardment. *Theor. Appl. Genet.*, **93**: 142–150.

Aragão F J L, Barros L M G, Sousa M V, Grossi-de-Sá M F, Almeida E R P, Gander E S and Rech E L (1999) Expression of a methionine-rich storage albumin from Brazil nut (*Bertholletia excelsa* H.B.K., Lecythidaceae) in transgenic bean plants (*Phaseolus vulgaris* L., Fabaceae). *Gen. Mol. Biol.*, **22**: 445–449.

Aragão F J L, Brasileiro A C M, Ribeiro S G, Faria J C and Rech E L (1995) Inoculation of bean and soybean with cloned bean golden mosaic virus (BGMV) DNA using particle acceleration. *Fitopatol Bras.*, **20**: 642–644.

Aragão F J L, Grossi-de-Sá M F, Davey M R, Brasileiro A C M, Faria J C and Rech E L (1993) Factors influencing transient gene expression in bean (*Phaseolus vulgaris* L.) using an electrical particle acceleration device. *Plant Cell Rep.*, **12**: 483–490.

Aragão F J L and Rech E L (1997) Morphological factors influencing recovery of transgenic bean plants (*Phaseolus vulgaris* L.) of a carioca cultivar. *Intl. J. Plant Sci.*, **158**: 157–163.

Aragão F J L, Ribeiro S G, Barros L M G, Brasileiro A C M, Maxwell D P, Rech E L and Faria J C (1998) Transgenic beans (*Phaseolus vulgaris* L.) engineered to express viral antisense RNAs show delayed and attenuated symptoms to bean golden mosaic geminivirus. *Mol. Breed.*, **4**: 491–499.

Aragão F J L, Sá M F G, Almeida E R, Gander E S and Rech E L (1992) Particle bombardment-mediated transient expression of a Brazil nut methionine-rich albumin in bean (*Phaseolus vulgaris* L.). *Plant Mol. Biol.*, **20**: 357–359.

Aragão F J L, Sarokin L, Vianna G R and Rech E L (2000) Selection of transgenic meristematic cells utilizing a herbicidal molecule results in the recovery of fertile transgenic soybean (*Glycine max* (L.) Merrill) plants at high frequency. *Theor. Appl. Genet.*, **101**: 1–6.

Becker J, Vogel T, Iqbal J and Nagl W (1994) *Agrobacterium*-mediated transformation of *Phaseolus vulgaris* L. Adaptation of some conditions. *Annu. Rep. Bean Improv. Coop. USA.*, **37**: 127–128.

Bidney D L, Scelonge C, Martich J, Burrus M, Sims L and Huffman G (1992) Microprojectile bombardment of plant tissues increases transformation frequency by *Agrobacterium tumefaciens. Plant Mol. Biol.*, **18**: 301–313.

Bogo M R, Vainstein M H, Aragão F J L, Rech E L and Schrank A (1996) High frequency gene conversion among benomyl resistant transformants in the entomopathogenic *Mtarhizium anisopliae. FEMS Microbiol. Lett.*, **142**: 123–127.

Brasileiro A C M, Aragão F J L, Rossi S, Dusi D M A, Barros L M G and Rech E L (1996) Susceptibility of common and tepari bean to *Agrobacterium* spp. strains and improvement of *Agrobacterium*-mediated transformation using microprojectile bombardment. *J. Amer. Soc. Hort. Sci.*, **121**: 810–815.

Bustos M (1991) Transgenic gene expression in *Phaseolus vulgaris* by direct gene transfer to protoplasts. *Plant Mol. Biol. Rep.*, **9**: 322–332.

Carrer H, Hockenberry T N, Svab Z and Maliga P (1993) Kanamycin resistance as a selectable marker for plastid transformation in tobacco. *Mol. Gen. Genet.*, **241**: 49–56.

Christou P (1993a) Philosophy and practice of variety-independent gene transfer into recalcitrant crops. *In Vitro Cell. Dev. Biol. Plant*, **29**: 119–124.

Christou P (1993b) Particle gun mediated transformation. *Curr. Opin. Biotech.*, **4**: 135–141.

Crepy L, Barros L M G and Valente V R N (1986) Callus production from leaf protoplasts of various cultivars of bean (*Phaseolus vulgaris* L.). *Plant Cell Rep.*, **5**: 124–126.

Table 2. Studies on growth responses of explants of various genotypes of Cajanus cajan *L. Millsp*

Variety	Basal medium	Growth regulators concentration	Explant(s)	Morphogenetic potential	Reference
T-21	MS	2,4-D (0.1 ppm) + Kin (1 ppm) + IAA (0.01 ppm)	Hypocotyl of γ-irradiated seeds	Callus	Rao and Narayanaswami, 1975
	White	Kin (1 ppm) + IAA (0.01 ppm) + CH (400 ppm) + inositol (100 ppm)	Callus	Shoot buds	
Prabhat	B_5	BAP (5×10^{-6} M) + GA_3 (5×10^{-6} M)	Leaves, stem, epicotyl and root	Callus	Mehta and Mohan Ram, 1980
		BAP (10^{-5} M) + NAA (10^{-6} M)	Cotyledon	Plantlets	
—	B_5	BAP (5×10^{-6} M) + GA_3 (10^{-6} M) + IAA (10^{-6} M)	Anthers	Multicelled bodies	Mohan Ram *et al.*, 1982
		BAP (10^{-5} M) + 2,4-D (2×10^{-5} M) + PVP (1000 ppm)	Multicelled bodies	Light green friable callus	
ICP-7035,	Blaydes	2,4-D (2 mg/L) + Kin (0.5 mg/L) +	Leaf, epicotyl, root,	Multiple shoots	Kumar *et al.*, 1984
ICP-2836,	B_5	PVP (100 mg/L) + BAP (5 μM)	cotyledons	via callus	
ICP-7186,	MS				
BDN-2	SN				
T-21	LS				
ICPL-161	MS				
T-21, T-148	V-47	NAA (3 mg/L) + BAP (0.8 mg/L) + Mannitol 8%	Mesophyll protoplast	Callus	Kulkarni and Krishnamurthy, 1989
ICPL-87, Rayagada, local, 82208, Kandulta	Liq. KM 8 p modified	2,4-D (0.4 mg/L) + NAA (1.0 mg/L) + BAP (0.5 mg/L) + Kin (0.5 mg/L)	Leaves protoplast	Cell clusters	Sarangi *et al.*, 1992
	Liq. Caboche's	2,4-D (0.4 mg/L) + Kin (0.5 mg/L) + Mannitol (0.35 M)	Cell clusters	Protocolonies	
ICPL-87, Rayagada, local, 82208, Kandulta	Liq. Caboche's	2,4-D (0.4 mg/L) + Kin (0.5 mg/L) + Mannitol (Man) (0.25 M)	Protocolonies	Protocalli	Sarangi *et al.*, 1992

Table 2. (Continued)

Variety	Basal medium	Growth regulators concentration	Explant(s)	Morphogenetic potential	Reference
	Solid Caboche's	BAP (1.0 mg/L) + Kin (0.5 mg/L) + AdS (40 mg/L) + Man (0.25 M)	Protocalli	Green callus	
ICPL-161	MS	BAP (10×10^{-6} M) or BAP (10×10^{-6} M) + IAA (0.5×10^{-6} M) or BAP (10×10^{-6} M) + IAA–alanine (0.5×10^{-6} M) or BAP (10×10^{-6} M) + IAA–aspartic acid (0.5×10^{-6} M) or BAP (10×10^{-6} M) + IAA–glycine (0.5×10^{-6} M)	Leaf disc	Callus	Eapen and George, 1993
		BAP (5.0×10^{-6} M) + IAA (0.5×10^{-6} M)	Callus	Multiple shoots	
Gaut-89-8, Gault-88-29, BP 86-34, SPMA-4, CC 11295, CC 2376	MS	BAP (2 mg/L)	Cotyledonary node	Multiple shoots	Shiva Prakash *et al.*, 1994
ICPL-161	MS	BAP (5.0 mg/L) + IAA (0.1 mg/L)	Epicotyl	Callus	George and Eapen, 1994
		BAP (10.0–15.0 mg/L)	Whole seeds	Multiple buds	
		BAP (1.0 mg/L) + IAA (0.1 mg/L)	Mature cotyledons	Plantlets	
ICPL-161	MS	BAP (1.0–5.0 mg/L) + IAA (0.1 mg/L)	Leaves	Plantlets	George and Eapen, 1994
		BAP (1.0–5.0 mg/L) + IAA (0.1 mg/L)	Immature embryonal axes	Small protuberance	
—	MS and B_5	2,4-D (1.0, 2.0 mg/L) + IAA (1.0 mg/L) + Kin (0.5 mg/L) + CW (70.0 mg/L) or	Immature hybrid embryos	Hybrid embryo rescued	Swamy *et al.*, 1994

Table 2. (Continued)

Variety	Basal medium	Growth regulators concentration	Explant(s)	Morphogenetic potential	Reference
		2,4-D (1.0 mg/L) + Kin (0.25 mg/L) + CW (100.0 mg/L)			
ICPL-161	L_6 salts + B_5 (vit.) 1% sorbitol, 0.1% Charcoal, activated, 6% sucrose, CH (200 mg/L)	2,4-D (5.0 mg/L) or NAA (5.0 mg/L) or Picloram (5.0 mg/L)	Immature cotyledons and embryonal axes	Somatic embryos	George and Eapen, 1994
T-15-15, Gaut-82-90, Banda-palera, NP-(WR) 15	MS, EC_6, LS White, B_5	BAP (22.2×10^{-6} M) + Kin (2.3×10^{-6} M) + AdS (271.0×10^{-6} M); BAP (2.22×10^{-6} M) + Kin (0.23×10^{-6} M) + Ads (27.1×10^{-6} M)	Distal cotyledon	Somatic embryos	Patel *et al.*, 1994
	C (MS + B_5)	IBA (2.4×10^{-6} M) + GA_3 (2.9×10^{-6} M)	Mature embryo	Plantlets	
ICPL-87	MS	NAA (50 mg/L) + BAP (1 mg/L)	Leaflet segments, Root discs	Somatic embryos	Mallikarjuna *et al.*, 1996
		NAA (5 mg/L) + BAP (1 mg/L)	Epicotyl/ hypocotyl	Somatic embryos	
		NAA (5–50 mg/L) + BAP (1.0 mg/L)	Cotyledons	Somatic embryos	
ICPL-87	MS	BAP (10 mg/L) for 15 days; BAP (1 mg/L)	Somatic embryos developed on cotyledon	Plants	Mallikarjuna *et al.*, 1996
Pusa-609, Pusa-852, Pusa-855, Pusa-856, H-86-5	MS	TDZ (10×10^{-6} M)	Leaf, cotyledon	Embryogenic calli	Sreenivasu *et al.*, 1998
	$^1/_2$ MS	—	Embryogenic calli	Plantlets	
ICPL 89021	Modified MS ($^1/_2$ MS macro + micro	BAP (0.1–1.0 mg/L), NAA (0.1–0.5 mg/L), Boric acid (25 mg/L)	Uninucleate microspores	Callus with embryoids	Kaur and Bhalla, 1998

Table 2. (Continued)

Variety	Basal medium	Growth regulators concentration	Explant(s)	Morphogenetic potential	Reference
	salts and vitamins of Nitsch and Nitsch)				
	MS	BAP (0.1 mg/L) + NAA (0.1 mg/L)	Embryoid	Plantlet	
T-15, Gaut-82-90, BDN-1, BDN-2, ICP-7182, KPL-87, IPCL-37119, TV-1	MS or B_5 or modified B_5 or EC_6 or LS or White's	BAP (22.2 μM) + Kin (2.3 μM) + AdS (271.0 μM)	Distal cotyledons segments	Shoot buds	Mohan and Krishnamurthy, 1998
		BAP (2.22 μM +) Kin (0.23 × μM) + AdS (27.1 μM)	Shoot buds	Multiple shoots	
Hydrabad C	MS	BAP (1.0–5.0 mg/L), Kin (1.0–5.0 mg/L)	Cot. node, epicotyl, hypocotyl, cotyledon, leaf	Multiple shoots	Geetha *et al.*, 1998
ICPL-161, ICPL-88039, UPAS-120	MS	TDZ (1.0 or 2.0 mg/L)	Primary leaf	Multiple shoots	Eapen *et al.*, 1998
Vamban-1	MS	BAP (13.31 μM)	Cot. node and shoot tip in contact with medium	Multiple shoots	Franklin *et al.*, 1998
			Shoot tip in contact with medium	Multiple shoots	
			Excised cot. node	Multiple shoots	
			Seeds with or without seed coats	Multiple shoots	
VB N1, VB N2, SA1, Co-5	4 × MS micro-nutrient + B_5 vitamins	BAP (8.86 μM) + NAA (1.07 μM)	Mature embryonal axes	Multiple shoots	Franklin *et al.*, 1999
Bahar	MS salts + B_5	BAP (0.5–2.0 mg/L) vitamins	Decapitated embryonic axes	Multiple shoots	Rathore *et al.*, 1999

Table 2. (Continued)

Variety	Basal medium	Growth regulators concentration	Explant (s)	Morphogenetic potential	Reference
Bahar	Modified MS	BAP (0.1 to 4.0 mg/L) + IAA (0.1 mg/L)	Primary leaf disc	Multiple shoots	Rathore *et al.*, 2000
Vamban-1	MS	2,4-D (2.26–11.30 μM)	Leaf	Callus	Anbazhagan and Ganapathi, 1999
	Liq. MS	2,4-D (1.12–6.78 μM)	Callus	Somatic embryos	
	MS	2,4-D (0.0 μM)	Somatic embryos	Plantlets	

Cot. node: cotyledonary node; CW: coconut water; see text for other abbreviations.

3.1. REGENERATION OF PLANT THROUGH ORGANOGENESIS

3.1.1. Direct shoot organogenesis

Seed culture. Induction of direct multiple shoot initials at axillary bud region of cotyledonary node of young seedlings raised on B_5 (Gamborg *et al.*, 1968) or MS (Murashige and Skoog, 1962) basal medium supplemented with N^6-benzylaminopurine (BAP) (Mehta and Mohan Ram, 1980; George and Eapen, 1994; Shiva Prakash *et al.*, 1994; Franklin *et al.*, 1998) or with thidiazuron (TDZ) (Singh *et al.*, 2002b) has been obtained. TDZ at 0.05 μM and BAP at 10.0 to 13.3 μM were found to be optimal for shoot organogenesis. The multiple shoot initials elongated when transferred either on low BAP containing medium or hormone-free medium. Multiple shoots were induced not only at cotyledonary node but also at shoot tip when seedling explants with cotyledonary node and shoot tip were placed horizontally on the BAP-containing medium and also when the shoot tip of the seedling explants alone touched the medium (Franklin *et al.*, 1998). The latter response was not observed by Shiva Prakash *et al.* (1994). Presence of BAP and light conditions were found to be critical during initial 48 h after seed culture for multiple shoot formation. The presence of gibberellic acid (GA_3, 0.5 mg/L) in MS + BAP (10.0 μM) during first two days of seed culture completely inhibited formation of multiple shoots (Shiva Prakash and Bhalla Sarin, 1995). Franklin *et al.* (1998) also studied the effect of the presence/absence of seedcoat on multiple shoot formation by culturing seeds with or without seed coat on MS basal medium containing BAP. Higher number of multiple shoot initials developed at cotyledonary node region of seeds with seed coat than the decoated seeds.

Explant culture. Direct shoot organogenesis without an intervening callus phase has been achieved from various seedling explants. Multiple shoot buds regeneration has been reported from entire mature cotyledons (Mehta and Mohan Ram, 1980; George and Eapen, 1994; Geetha *et al.*, 1998) or distal halves of cotyledons (Mohan and Krishnamurthy, 1998); cotyledonary nodes (Franklin *et al.*, 1998; Geetha *et al.*, 1998; Singh *et al.*, 2002); epicotyl and hypocotyl segments (Geetha *et al.*, 1998); primary leaf explants (Eapen *et al.*, 1998; Geetha *et al.*, 1998; Singh *et al.*, 2002) and shoot apices (Singh *et al.*, unpublished). However, Franklin *et al.* (1998) failed to induce shoots from shoot tip explants. The cotyledons and cotyledonary node explants produced lesser number of shoot buds than those regenerated from the proximal end of cotyledons at the nodal region of the seedling on BA enriched medium (Mehta and Mohan Ram, 1980). It appears that the presence of embryonal axis stimulates the formation of shoot buds on cotyledons. The specific nature of stimulus is not yet known. The formation of shoot buds at the distal end of cotyledons was also observed when whole cotyledons were cultured (George and Eapen, 1994). The type of explant, choice of cytokinin and its dosage influenced the frequency of shoot formation. Among various seedling explants, the highest frequency of shoot bud regeneration was observed from cotyledonary node explants (Geetha *et al.*, 1998). BAP alone was found to be suitable for both multiple shoot bud induction and proliferation without any auxin supplementation. It appears that the explants contain endogenous auxin. Thidiazuron, *N*-phenyl-*N*2–1,2,3-thiadiazol-5-ylurea (1, 2 mg/L) alone evoked *de novo* high frequency shoot regeneration from primary leaf segments and presence of indole-3-acetic acid (IAA, 0.1 mg/L) did not alter the average number of shoot buds per culture (Eapen *et al.*, 1998). Among cytokinins, BAP is the most effective for multiple shoot formation (Mehta and Mohan Ram, 1980; Shiva Prakash *et al.*, 1994; Geetha *et al.*, 1998). However, the efficient regeneration can be optimized by balancing the auxin-cytokinin level by exogenous supply (Shiva Prakash *et al.*, 1994). Seed derived entire or decapitated embryonal axes and cotyledons produced multiple shoots on MS medium supplemented with BAP and naphthalene acetic acid (NAA) (Franklin *et al.*, 1998; Rathore and Chand, 1999) or IAA (George and Eapen, 1994), respectively. The multiple shoots obtained on various concentrations of BAP failed to elongate on same medium. Subculture to fresh medium with lower level of BAP and NAA enhanced the multiple shoots and shoot elongation (Geetha *et al.*, 1998; Mohan and Krishnamurthy, 1998). Elongation of shoot buds has also been achieved on MS medium without growth regulator supplements or on half strength-MS medium containing GA_3 (Mehta and Mohan Ram, 1980; Mohan and Krishnamurthy, 1998) or MS medium containing GA_3 and IAA (Eapen *et al.*, 1998). Topical supplementation of indole acetic acid has been suggested by Shiva Prakash *et al.* (1994) to elongate the shoot buds produced on cotyledonary node explants. The shoot buds regenerated from embryonic axes incubated continuously in dark on MS medium containing BAP and NAA elongated well on the same medium, this may be due to etiolation; however, shoot primordia induced under light-dark conditions needed to be transferred to basal medium for elongation (Franklin *et al.*, 1998).

3.1.2. Organogenesis from callus

Regeneration of plants from the callus may result in genetic variability. Callus cultures on periodic subculture undergoes genetic erosions, mutations and change in ploidy number, etc. Though highly undesirable for maintenance of clones, nevertheless, it is a novel and rich source of genetic variability. The adventitious regeneration directly from organs generally gives a limited number of propagules. This may be increased several fold by an intervening callus phase.

Shoot organogenesis has been reported from callus derived from hypocotyls obtained from gamma (5 kR) irradiated seeds (Shama Rao and Narayanaswamy, 1975) or non-irradiated seeds (Xu *et al.*, 1984); cotyledons (Kumar *et al.*, 1983); primary leaves (Kumar *et al.*, 1983; Eapen and George, 1993; George and Eapen, 1994; Rathore *et al.*, 2000), and roots of seedlings (George and Eapen, 1994). Primary leaf segments produced hard nodular and compact calli from cut ends. IAA (0.1 mg/L) and BAP (1 mg/L) in MS medium supported maximum growth of callus with multiple shoot initials within 4 weeks. Later the growth was significantly retarded due to exudation of phenolics from the explants. Therefore monthly transfer of shoot-bearing callus to fresh medium was essential to recover well developed shoots (George and Eapen, 1994). Callus retained its regeneration potential even after 2 years (George and Eapen, 1994). On the contrary, callus subcultured beyond 3 weeks failed to regenerate (Kumar *et al.*, 1984). Closely related species of *Cajanus cajan*, *Atylosia cajanifolia*, *A. albicans* and *A. sericea* have also been regenerated via calli derived from cotyledons (Kumar *et al.*, 1984). Cultures of epicotyl segments of seedlings yielded only hard, compact and non-morphogenic callus (George and Eapen, 1994).

3.2. SOMATIC EMBRYOGENESIS

Proper choice of explant, genotype and hormone concentration are critical for induction of somatic embryogenesis (Griga, 1999). In pigeonpea, somatic embryogenesis has been induced either directly or indirectly through callus formation from a variety of explants. Direct induction of somatic embryos has been obtained at cotyledonary node of the seedlings raised on MS medium containing high concentrations of TDZ (10.0 μM or 20.0 μM) (Singh *et al.*, 2002b) or from abaxial or adaxial surface of distal halves of cotyledons on a high cytokinin MS or B_5 medium containing BA (22.2 μM), kinetin (2.3 μM) and adenine suphate (AdS, 271.0 μM) (Patel *et al.*, 1994). Reduction of cytokinins favoured maturation of somatic embryos. Complete withrawal of TDZ (Singh *et al.*, 2002b) or cytokinins (Patel *et al.*, 1994) from medium and addition of GA_3 (Singh *et al.*, 2002b) or GA_3 (2.9 μM) with indole butyric acid (2.4 μM) (Patel *et al.*, 1994), respectively, proved effective in germination of embryos. About 70% of the embryos converted into plantlets (Patel *et al.*, 1994). Patel *et al.* (1994) also found that the induction of somatic embryos was dependent on genotype and basal medium used. On B_5 or modified B_5 medium, the embryos become vitreous, whereas on EC_6 (Maheswaran and William, 1984), LS (Linsmaier and Skoog, 1965) and White's medium, they turned chlorotic and failed to develop beyond the heart or early cotyledonary stage. Of the four cultivars (T-15-15, Gaut 82-92, NR (WR) 15, Bandapalera), the highest frequency (100%) of responding cultivars and the highest average number of somatic embryos per responding culture (97%) were obtained with NR (WR) 15 and Bandapalera, respectively.

Globular somatic embryos were induced on embryogenic calli initiated from immature cotyledons and embryonic axes on L_6 salts (Kumar *et al.*, 1988) supplemented with B_5 vitamins, 5 mg/L of 2,4-dichlorophenoxyacetic acid (2,4-D) or picloram or NAA (George and Eapen, 1994). The type and concentration of auxins for optimal embryogenic response varied with explant. Picloram was the most effective for embryogenesis from cotyledon explants while 2,4-D was the most effective for embryonal axes (George and Eapen, 1994). The explants with globular embryos were transferred to MS medium with 3% mannitol and then to the same medium containing zeatin (0.1 mg/L), GA_3 (1 mg/L), ABA (0.25 mg/L) and $AgNO_3$ (10 mg/L) for maturation and germination of embryos. Five normal mature embryos were obtained from cotyledons and one grew up to 1.5 cm to produce a small plant. However, plantlets capable for transfer to field could not be obtained from the somatic embryos (George and Eapen, 1994). Mallikarjuna *et al.* (1996) reported the induction of callus and subsequent somatic embryogenesis from immature leaf segments, root discs, epicotyl and hypocotyl segments, as well as from mature cotyledons on MS medium supplemented with different concentrations of NAA and BAP. However, the conversion of somatic embryos to plants was obtained only from the cotyledons. Cotyledons produced embryogenic callus and somatic embryos on NAA (50.0 mg/L) and BAP (1.0 mg/L). Further development of embryos was observed on medium with 10.0 mg/L of BAP for 15 days and then to 1 mg/L of BAP. Three per cent cotyledons showed direct regeneration of embryos into plants. A notable feature of pigeonpea somatic embryos is the presence of the suspensor but it is not known if the suspensor had any specific role in viability or conversion of somatic embryos. Sreenivasu *et al.* (1998) induced pale-green, compact and nodular callus (embryogenic calli) from mature cotyledon and leaf explants (on MS medium containing 10.0 μM TDZ) which on subsequent subculture form the cluster of globular embryos. Subsequent withrawal of TDZ from the induction medium resulted in the maturation and growth of the embryos into plantlets on MS basal medium. The conversion of embryos into plantlets was 70%.

Anbazhagan and Ganapathi (1999) produced greenish white friable calli from leaf explants on MS semi-solid medium containing 6.7 μM 2,4-D. The maximum frequency of somatic embryogenesis was observed when this callus was transferred to MS liquid medium containing 4.5 μM 2,4-D. The somatic embryos germinated into plantlets upon transfer to auxin-free MS basal medium. About 5–6% of embryos converted into plantlets.

3.3. HISTOLOGICAL STUDIES ON REGENERATING EXPLANTS

Histological evidence for *de novo* shoot regeneration from the cut ends of leaf explants (Eapen *et al.*, 1998), entire cotyledons (Mehta and Mohan Ram, 1980), cotyledonary segments (Mohan and Krishnamurthy, 1998) and embryonal axes (Franklin *et al.*, 1999) have been provided. Histology of regenerating cotyledonary node explants after cutting its surface layers (to completely remove the pre-existing shoot initials) indicates the formation of new shoot initials from the partially differentiated cells just below the excised meristem (Shiva Prakash *et al.*, 1994).

Histological analysis of embryogenic calli revealed the presence of globular and torpedo-shaped somatic embryos (Mallikarjuna *et al.*, 1996; Sreenivasu *et al.*, 1998).

and *C. cajanifolius* (*A. cajanifolia*) are those wild species which have been successfully crossed with different pigeonpea cultivars (Reddy, 1981; Reddy and De, 1983; Kumar *et al.*, 1985; Pundir and Singh, 1985). In these interspecific *Cajanus* crosses, the fertility of pollens, pod set and number of seeds per pod of the F_1 plants has been lower than that found in parental material (Kumar L *et al.*, 1958; Reddy and De *et al.*, 1983; Dundas *et al.*, 1987). The pod set and seed's content after interspecific cross-pollination have been shown to increase after application of GA_3 or kinetin (Kumar P. S. *et al.*, 1985; Dhanju and Gill, 1985).

Embryo culture techniques have also been used to rescue immature hybrid embryos from interspecific crosses, i.e. pigeonpea × *C. scarabaeoides* (Dhanju and Gill, 1985), × *C. platycarpus* (Dhanju *et al.*, 1985), pigeonpea × *C. acutifolius*, × *C. cajanifolius* and × *C. scarabaeoides* (Moss *et al.*, 1988). Difficulty in producing hybrids between pigeonpea and wild species has been encountered especially with *C. platycarpus* and *C. crassus* var. *crassus* (*A. volubilis*) although Dhanju and Gill (1985) claimed to have successfully crossed *C. platycarpus* with pigeonpea.

It should be noted that the majority of the known wild species in *Cajanus* have not been crossed with pigeonpea (Dundas, 1990). *Cajanus* interspecific hybridization work through immature hybrid embryo rescue techniques should be taken up not only to get a better understanding of inter-relationships between *Cajanus* species but also to introgress desirable genes from wild relatives. An efficient method for raising young pigeonpea embryos (<11 days) may be a great asset for future studies with incompatible crosses.

3.8. SOMACLONAL VARIATION

In vitro raised plants showed wide range of genetic variations called somaclonal variation. Such potentially useful variations have been identified in many crop plants including important legumes such as early maturity and salt tolerance (Gulati and Jaiwal, 1993) and chlorophyll and morphological variants (Mathews *et al.*, 1986) in *Vigna radiata*, resistance to *Fusarium oxysporium* sp *medicaginis* in *Medicago sativa* (Hartmann *et al.*, 1984), leaf variegation and change in growth habit in *Glycine max* (Freylag *et al.*, 1989). Pigeonpea plants regenerated from cotyledon explants exhibited a spectrum of alterations in floral morphology and architecture, plant height, seed mass and damage due to insect pest, *H. armigera* (Chintapalli *et al.*, 1997). Such variations for agronomically important traits like resistance to diseases and herbicides, new plant type, resistant to abiotic stresses, high yield potential, etc. in pigeonpea are yet to be determined.

4. Factors affecting regeneration of pigeonpea

4.1. EXPLANTS

A variety of explants excised from immature or mature seeds, seedlings and mature plants have been used to initiate *in vitro* cultures (Table 2). However, mature seeds and seedlings have been the favourite source of explants. Immature explants have been excised from the immature seeds removed from the pods collected one week before the harvest of field grown plants (George and Eapen, 1994). The explants have also been excised from sterile distilled water-soaked mature seeds (Patel *et al.*, 1994; Mohan and Krishnamurthy, 1998; Franklin *et al.*, 1999) or seedlings raised on moist filter paper/moist sterile cotton (Kumar *et al.*, 1983; Mallikarjuna *et al.*, 1996; Geetha *et al.*, 1998; Anbazhagan and Ganapathi, 1999) or MS basal medium (Sreenivasu *et al.*, 1998) or MS or B_5 basal medium containing BAP (Mehta and Mohan Ram, 1990; Shiv Prakash *et al.*, 1994; George and Eapen, 1994; Franklin *et al.*, 1998) after various days of germination. Every explant of pigeonpea can be induced to develop variable amount of callus. The presence of embryonic axis stimulated the production of buds in cotyledon cultures (Mehta and Mohan Ram, 1980). Patel *et al.* (1994) eliminated the proliferation of axillary meristems of cotyledons by removing the proximal half of the cotyledons to obtain somatic embryos. In contrast to this, the development of multiple shoots from the distal half of cotyledon has also been reported by Mohan and Krishnamurthy (1998). Cotyledonary node from the germinating seeds with seed coats produces more number of shoots per explant as compared to the decoated seeds cultured on the same conditions (Franklin *et al.*, 1998). However, in our recent observations, the number of multiple shoots and the percentage of cultures showing somatic embryogenesis were reduced when seeds with seed coats were cultured on MS basal medium supplemented with TDZ (Singh *et al.*, 2002b).

4.2. BASAL MEDIUM

The most suitable and widely used medium for pigeonpea regeneration, either through embryogenesis or direct or indirect organogenesis is MS. However, B_5 basal medium has also been used for shoot regeneration (Mehta and Mohan Ram, 1980). Reduction of MS inorganic salts to half has favoured regeneration of complete plantlets from mature somatic embryos (Sreenivasu *et al.*, 1998) and rooting of shoots (Franklin *et al.*, 1998; Eapen *et al.*, 1998). In another study, out of the different basal media tried, the induction of maximum number of shoot buds from distal half of cotyledons and calli was observed on EC_6 medium (Mohan and Krishnamurthy, 1998) and on Blaydes medium (Kumar *et al.*, 1984), respectively. MS and B_5 media have been used for direct induction of somatic embryos, other media (MS, B_5, LS, EC_6 and White's) have induced formation of callus as well as embryo (Patel *et al.*, 1994).

4.3. GROWTH REGULATORS

Various explants have shown difference in growth regulator's requirement for organogenesis and somatic embryogenesis. Different cytokinins either alone or in combinations have been used to induce organogenesis and embryogenesis in pigeonpea (Table 2). Direct organogenesis was obtained from diverse explants cultured on cytokinin containing medium. Out of different cytokinins, BAP was found to be the most effective (Shiva Prakash *et al.*, 1994; Singh, 2002a). However, TDZ at low concentrations have been optimal for the induction of multiple shoots in the seed cultures of pigeonpea (Singh *et al.*, 2002b). A combination of three cytokinins, BAP, kinetin and adenine sulphate was used for *de novo* organogenesis from the distal half of cotyledonary explants (Mohan and Krishnamurthy, 1998). Addition of low concentration of IAA to BAP medium showed development of shoot buds from calli derived from mature cotyledons, leaf and root segments (George and Eapen, 1994). The combination of BAP with NAA increased the number of multiple shoots as well as shoot elongation, which is again dramatically enhanced when GA_3 was added (Geetha *et al.*, 1998).

In most plant species presence of high auxin concentration followed by its gradual withdrawal from culture media has been reported to be necessary for somatic embryogenesis. Among the auxins, 2,4-D, NAA, and picloram have been tested for the induction of somatic embryos from immature cotyledon explants. 2,4-D induced callus and suspension cultures have been shown to produce well-developed somatic embryos (Anbazhagan and Ganapathi, 1999). In contrast to this, only cytokinins, BAP, kinetin and adenine sulphate in combination induced somatic embryogenesis (Patel *et al.*, 1994). TDZ alone also induced somatic embryos directly (Singh *et al.*, 2002b) or through embryogenic calli (Sreenivasu *et al.*, 1998) and subsequent withdrawal of TDZ from the medium resulted in the maturation and growth of the embryos into plantlets (Sreenivasu *et al.*, 1998; Singh *et al.*, 2002b).

4.4. OTHER COMPOUNDS

Several compounds such as tyrosine, polyvinyl pyrrolidone (PVP), cysteine and bovine serum albumin (BSA) have been used to overcome the problem of darkening of tissue *in vitro*. Incorporation of PVP (1000 ppm) with B_5 containing NAA eliminated excessive production of phenolic compounds, reduced callusing of roots and stimulated plant growth in pigeonpea (Mehta and Mohan Ram, 1980). Addition of activated charcoal to the culture medium enhanced the response of androgenesis and absorbed the inhibitory phenolic substances (Kaur and Bhalla, 1998). L-proline was used in MS medium containing BAP and NAA for regeneration of multiple shoots from immature embryonal axes (Franklin *et al.*, 1999). Mannitol and casein hydrolysates (CH) have been used as adjuvants with L_6 salts for induction of somatic embryos (George and Eapen, 1994).

4.5. GENOTYPE

A large number of genotypes have been evaluated for their morphogenic response (Table 2). Efficiency of the shoot or embryo induction from diverse explants varied with genotype. Kumar *et al.* (1984) reported cultivars ICP-7035 and ICP-6917 to be prolific shoot producers. Variety VBN2 was reported to be more efficient for shoot regeneration (Franklin *et al.*, 1999, 2000) while Mohan and Krishnamurthy (1998) reported regeneration of shoots only from explants of Gaut-80-90 and T-15-15 among the eight cultivars that were tested. Interestingly the genotypes reported to have poor somatic embryos induction (Patel *et al.*, 1994) were found to produce maximum number of shoots in the study by Mohan and Krishnamurthy (1998). However, some studies have reported genotype independent morphogenic response (Shiva Prakash *et al.*, 1994; Eapen *et al.*, 1998; Sreenivasu *et al.*, 1998).

4.6. LIGHT

The effect of light on regeneration has been studied by Shiva Prakash and Bhalla-Sarin (1995). Total dark conditions completely inhibited the formation of multiple shoots in presence of BAP. Light has a weak effect on regeneration in the first 24 h after sowing but it is more effective between 24–48 h after sowing for multiple shoot formation. In contrast, the explants incubated under continuous dark produce small whitish shoots after 20 days, which elongated within 8 days in the same medium. The calli induced from the explants incubated under light-dark conditions became dark brown and produced small protuberances which later formed shoots, whereas, culture under continuous light conditions could not produce any shoots (Franklin *et al.*, 1999).

4.7. RADIATIONS

Although numerous investigations on the effect of ionizing radiation in mutation breeding of crop plants have been reported, such reports on explanted tissue grown *in vitro* are rare (Devereux, 1973). In *Cajanus cajan*, Rao and Narayananswamy (1975) studied the effect of gamma irradiation on cell proliferation and regeneration in explanted tissues. Hypocotyls of seedlings obtained from irradiated tissue at 5 kR produce normal shoots and abundant callus.

5. Genetic transformation

Although genetic transformation of grain legumes has been difficult and challenging till now, significant progress has been made in the recovery of transformed plants via *Agrobacterium* especially in soybean and pea or via particle bombardment as in case of soybean and bean (*Phaseolus vulgaris*) (Christou, 1997). Genetic transformation of pigeonpea using any of the transformation system is still far from routine (Christou, 1997). Attempts have been made to transform pigeonpea by co-cultivating cotyledonary node explants with non-oncogenic strains of *Agrobacterium tumefaciens* (Shiva Prakash and Bhalla Sarin, 1995). The explants showed sign of hypersensitivity when exposed to non-oncogenic strains, which was overcome by topical supplement of auxin during infection. As auxin application adversely affected regeneration, it was concluded that the cotyledonary node regeneration system is unsuitable for transformation using *Agrobacterium* aimed at obtaining transgenic pigeonpea. Rathore and Chand (1997) evaluated the susceptibility of five pigeonpea genotypes (UPAS-120, Bahar, Pant A-106, H82-1 and ICPL-151) to three wild strains of *A. tumefaciens*, A_{281} (succinamopine), T37 (nopaline) and A_6 (octopine). The strain A_{281} induced tumor formation in all five cultivars, among which Bahar and Pant A-106 cultivars showed the highest response. Similar genotype variation in compatible reactions between host-plant and *Agrobacterium* spp has also been reported in other legumes, e.g. *Cicer arietinum* (Islam and Riazuddin, 1994) and pea (Hobbs *et al.*, 1989).

Ramana *et al.* (1997) obtained putative transformants via organogenesis from immature embryonal axis co-cultivated with *Agrobacterium tumefaciens* strain LBA4404. The transgenic nature of regenerants was confirmed by GUS expression and by rooting them on medium containing 50 mg/L kanamycin. The data on molecular analysis of putative transformants is not available. Recently, Geetha *et al.* (1999) reported the production of pigeonpea transgenic plants using *Agrobacterium* strain LBA4404 (Hoekema *et al.*, 1983) harbouring a binary plasmid pBI121 (Jefferson *et al.*, 1987). Shoot apices and cotyledonary nodes of 5-day old seedlings were used as explants for co-cultivation with *A. tumefaciens*. Before incubation with bacterial suspension, explants were precultured for 2 days on a regeneration medium. Selection of transformed shoot *via* direct organogenesis was achieved on MS medium supplemented with BAP (2 mg/L), kanamycin (50 mg/L) and cefotaxime (300 mg/L). Histochemical GUS expression and PCR analysis of plants regenerated on selection medium have been presented as proof of transformation. Integration of foreign genes was ascertained by Southern analysis, but transmission of genes to progeny has not been demonstrated (Geetha *et al.*, 1999). Cotyledonary node explants were reported to have higher transformation efficiency than shoot apices. Shoot apices of 16 h water soaked seeds on co-culture with A. tumefaciens strain EHA105 containing binary vector pCAMBIA-αai (carrying α-amylase inhibitor gene with 5′ and 3′ control regions from bean phyto-haemagglutinin gene, 35S*gus-gfp* nos fusion as reporter gene and 35S *hpt* polyA as a selectable marker gene) showed transient GUS expression. The various parameters affecting gene transfer, such as bacterial concentration, inoculation time, co-cultivation period and presence/absence of acetosyringone in co-cultivation medium were optimized using transient GUS activity. Shoot apices co-cultured with *Agrobacterium* using optimal conditions regenerated shoots on regeneration medium containing hygromycin (7.5 mg/L). The shoots recovered on selection medium turned blue when stained with X-gluc (Singh *et al.*, 2002c). Chandra and Pental (1998) have used particle bombardment to transfer genes into apical meristems of pigeonpea. Conditions for efficient gene delivery through bombardment were optimized using transient GUS activity in apical meristem cells. But transgenic plants could not be recovered so far, from bombarded meristems. Thus development of an efficient regeneration system and optimization of parameters for transformation by *Agrobacterium* or by particle bombardment of DNA coated microprojectiles can lead to the production of transgenic pigeonpea plants carrying agronomically important traits.

6. Conclusions and future prospects

Pigeonpea plants in several genotypes have been regenerated through direct or indirect organogenesis or embryogenesis using diverse explants on different media. Initial morphogenic response in terms of induction of organized structure is very high. However, conversion of buds or embryos to shoots seems to be very low. The regeneration frequency in most of the existing protocols is very low, inefficient, genotype specific and also depends on the type of explant, media composition and culture conditions. Despite many reports, no specific nutrient and hormonal regimes for organogenesis or embryogenesis have been identified. Regeneration of plants from protoplasts and rescued embryo cultures has been achieved with limited success. In general, pigeonpea explants are highly responsive to cytokinin especially BAP containing media. Among explants, cotyledonary node is the best explant for direct shoot organogenesis. The single report available on genetic transformation via. *A. tumefaciens* used cotyledonary node explants for the development of T_0 transgenic plants. The inheritance of transgenes to T_0 progenies have not been studied. The protocols of plant regeneration from calli through shoot organogenesis or embryogenesis have not yet been tried in developing transgenics using any of the gene transfer techniques. An optimization of various factors affecting regeneration and gene transfer will help in developing an efficient protocol which can routinely be used for the production of transgenic plants with desired agronomic traits.

Acknowledgements

Authors are grateful to CSIR, New Delhi for financial support.

References

Akinola J O, Whiteman P C and Wallis E S (1975) The agronomy of pigeonpea (*Cajanus cajan* L. Millsp.). Rev. Sr. No. 1 CAB, Bureau of Pastures and Field Crops Review Series 1/1975.

Anbazhagan V R and Ganapathi A (1999) Somatic embryogenesis in cell suspension cultures of pigeonpea (*Cajanus cajan* L. Millsp.). *Plant Cell Tiss. Org. Cult.*, **56**: 179–184.

Anonymous (1999) Agricultural Statistics at a Glance, Directorate of Economics and Statistics, Department of Agriculture and Co-operation, Ministry of Agriculture, Government of India.

Bajaj Y P S, Singh H and Gosal S S (1980) Haploid embryogenesis in anther cultures of pigeonpea. *Theor. Appl. Genet.*, **58**: 157–159.

Butterfass T and Kohlenbach H W (1979) Monosomics of diploid *Nicotiana sylvestris* produced at will by androgenesis. *Naturwissenschaften.*, **66**: 162.

Chandra A and Pental D (1998) Optimising parameters for particle bombardment mediated transformation of pigeonpea (*Cajanus cajan* L. Millsp.). *Natl. Symp. on Commercial Aspects of Plant Tissue Culture*, Meeting of the PTCA (India) Feb. 25–27, 1998. New Delhi. Abstr. (GE09), 95.

Chintapalli Prasanna L, Moss J P, Sharma Kiran K and Bhall J K (1997) *In vitro* culture provides additional variation for pigeonpea (*Cajanus cajan* L. Millsp.). Crop improvement. *In Vitro Cell. Dev. Biol. Plant.*, **33**: 30–37.

Christou P (1997) Biotechnology applied to grain legumes. *Field Crop Res.*, **53**: 83–97.

de Wet J M J (1989) Inaugural Address. In: Collaboration on Genetic Resources: Proceedings of a joint ICRISAT/NBPGR (ICAR) workshop on Germplasm Exploration and Evaluation in India. Nov. 14–15 ICRISAT Centre, Patancheru, A.P. India, 5–6.

Deka P C, Mahera A K, Pathak N N, Sen S K (1977) Isolation and fusion studies on protoplasts from pollen tetrads. *Experientia*, **33**: 182–184.

Devereux M (1973) *In vitro* culture and mutation breeding. In: Induced mutations vegetatively propagated plants. *Proc. of a Panel I.A.E.A.*, Vienna, 1972, 41–51.

Dhanju M S and Gill B S (1985) Intergeneric hybridization between *Cajanus cajan* and *Atylosia platycarpa. Ann. Biol.*, **1**: 229–231.

Dias C A R, Lal S S and Yadav C P (1980) *International Workshop on Pigeonpea* **2**: 337–340.

Dundas I S, Britten E J, Byth D E and Gordon G H (1987) Meiotic behavior of hybrids of pigeonpea and two Australian native *Atylosia* species. *J. Heredity*, **78**: 261–265.

Dundas I S (1990) Pigeonpea: Cytology and cytogenetics – perspectives and prospects. In: *The Pigeonpea* (Eds Nene Y L, Susan D H and Sheila V K), CAB International, Wallingford, UK, 117–136.

Eapen S and George L (1993) Plant regeneration from leaf discs of peanut and pigeonpea: influence of benzyladenine, indole-acetic acid and indole-acetic acidamino acid conjugates. *Plant Cell Tiss. Org. Cult.*, **35**: 223–227.

Eapen S, Suchita T and George L (1998) Thiadiazuron-induced shoot regeneration in pigeonpea (*Cajanus cajan* L. Millsp.). *Plant Cell Tiss. Org. Cult.*, **53**: 217–220.

Ekeke G I and Shode F O (1985) The reversion of sickled cells by *Cajanus cajan. Planta Medica*, 504–507.

Faris D G and Singh U (1990) Pigeonpea: nutrition and products. In: *The Pigeonpea* (Eds Nene Y L, Susan D H and Sheila V K), CAB International, Wallingford, UK, 1–14.

Fillippone E (1993) To improve resistance against diseases and pests. *Grain legumes*, **2**: 20–21.

Franklin G, Jayachandran R and Ignacimuthu S (1999) Factors affecting regeneration of pigeonpea (*Cajanus cajan* L. Millsp.) from mature embryonal axes. *Plant Growth Reg.*, **90**: 1–6.

Franklin G, Jayachandran R, Melchias G and Ignacimuthu S (1998) Multiple shoot induction and regeneration of pigeonpea (*Cajanus cajan* L. Millsp.) cv. Vamban 1 from apical and axillary meristems. *Curr. Sci.*, **74**: 36–37.

Freylag A H, Rao-Arell A P, Anard S C, Wrather J A and Owens L D (1989) Somaclonal variation in soybean plants regenerated from tissue culture. *Plant Cell Rep.*, **8**: 199–202.

Geetha N, Venkatachalm P and Laxmi Sita G (1999) *Agrobacterium*-mediated genetic transformation of pigeonpea (*Cajanus cajan* L. Millsp.) and development of transgenic plants via direct organogenesis. *Plant Biotech.*, **16**: 213–218.

Geetha N, Venkatachalam P, Rakesh V and Laksmi Sita G (1998) High frequency induction of multiple shoots and plant generation from seedling explants of pigeonpea (*Cajanus cajan* L. Millsp.). *Curr. Sci.*, **75**: 1036–1041.

George L and Eapen S (1994) Organogenesis and embryogenesis from diverse explants in pigeonpea (*Cajanus cajan* L. Millsp.). *Plant Cell Rep.*, **13**: 417–420.

Gill L S and Husaini S W H (1986) Cytological observations in leguminosae from southern Nigeria. *Wildenowia*, **15**: 521–527.

Griga M (1999) Somatic embryogenesis in grain legumes. *Advances in Regulation of Plant Growth and Development*, 233–250.

Gulati A and Jaiwal P K (1993) *In vitro* selection of salt resistance *Vigna radiata* (L.) Wilczek plants by adventitious shoot formation from cultured cotyledon explants. *J. Plant Physiol.*, **142**: 99–102.

Hartmann C L, McCoy T J, Knous T R (1984) Selection of alfalfa (*Medicago sativa*) cell lines and regeneration of plant resistance to the toxin (S) produced by *Fusarium oxysporium* f. sp. *Medicaginins*. *Plant Sci. Lett.*, **43**: 183–194.

Hobbs Shaun L A, Jennifer A, Jackson and Mohan John D (1989) A binary vector strategy based on separation of *vir*- and T-region of the *Agrobacterium tumefaciens* Ti-plasmid. *Nature*, **303**: 179–180.

Huysmans A A C (1970) Storage of food grains – problems and prospects. *Bulletin of Grain Technology*, **8**: 92–97.

Islam R and Riazuddin S (1994) Influence of genotype and age seedlings on chickpea response to *Agrobacterium tumefaciens*. *Intl. Chickpea and Pigeonpea News Lett.*, **1**: 23–25.

Jefferson R A, Kavanagh T A and Bevan M (1987) GUS fusions: beta-glucuronidase as a sensitive and versatile gene fusion marker in higher plants. *EMBO J.*, **6**: 3901–3907.

Kaur P and Bhalla J K (1998) Regeneration of haploid plants from microspore culture of pigeonpea (*Canajus cajan* L. Millsp.). *Indian J. Exp. Biol.*, **36**: 736–738.

Khan S and Ghosh P D (1986) Pigeonpea. In: *Perspectives in Cytology and Genetics. Vol. 5* (Eds Manna G K and Sinha U), Rashtravani Printers, New Delhi, India, 181.

Kulkarni D D and Krishnamurthy K V (1989) Isolation and culture of protoplasts of pigeonpea (*Cajanus cajan* L. Millsp). *Indian J. Exp. Biol.*, **27**: 939–942.

Kumar A S, Gamborg O L and Nabors M W (1988) Plant regeneration from cell suspension cultures of *Vigna aconitifolia*. *Plant Cell Rep.*, **7**: 138–141.

Kumar A S, Reddy T P and Reddy G M (1983) Plantlet regeneration from different callus cultures of pigeonpea (*Cajanus cajan* L.). *Plant Sci. Lett.*, **32**: 271–278.

Kumar A S, Reddy T P and Reddy G M (1984) Multiple shoots from cultured explants of pigeonpea and *Atylosia* species. *SABRAO J.*, **16**: 101–105.

Kumar A S, Reddy T P and Reddy G M (1985) Genetic analysis of certain *in vitro* and *in vivo* parameters in pigeonpea (*Cajanus cajan* L. Millsp.). *Theor. Appl. Genet.*, **70**: 151–156.

Kumar L S S, Thombre M V and D'Cruz R (1958) Cytological studies of an intergeneric hybrid of (*Cajanus cajan* L. Millsp.) and *Atylosia lineata* W&A. *Proc. Indian Acad. Sci.*, Section B, **47**: 252–262.

Kumar P S, Subramanyam N C and Faris D G (1985). Intergeneric hybridization in pigeonpea. I. Effect of hormone treatments. *Field Crop Res.*, **10**: 315–322.

Lackey J A (1980) Chromosome numbers in the *Phaseoleae* (*Fabaceae*; *Faboidae*) and their relation to taxonomy. *Am. J. Bot.*, **67**: 595–602.

Linsmaier E M and Skoog F (1965) Organic growth factor requirements of tobacco tissue cultures. *Physiol. Plant.*, **18**: 100–127.

Maheswaran G and William E G (1984) Direct somatic embryoid formation in immature embryos of *Trifolium repens*, *T. pratense* and *Medicago sativa* and speed clonal propagation of *T. pratense*. *Ann. Bot.*, **54**: 201–211.

Mallikarjuna N, Reena M J T, Sastri D C and Moss J P (1996) Somatic embryogenesis in pigeonpea (*Cajanus cajan* L. Millsp.) *Indian J. Exp. Biol.*, **34**: 282–284.

Mathews V H M, Rao P S, Bhatia C R (1986) Somaclonal variation in cotyledonary plants of mungbean. *Z. Pflanzenzuech*, **96**: 169–173.

Mattingly C F and Collins G B (1974) The use of anther derived haploids in *Nicotiana* III. Isolation of nullisomics from monosomic lines. *Chromosoma*, **46**: 29–36.

Mehta V and Mohan Ram H Y (1980) Regeneration of plantlets from cotyledons of *Cajanus cajan*. *Indian J. Exp. Biol.*, **18**: 800–802.

Miller C D, Branthoover B, Sekiguchi N, Dening H and Bauer A (1956) Vitamin values of foods used in Hawaii. *Hawaii Agriculture Experiment Station Technical Bulletin*, **30**: 303–313.
Mohan M L and Krishnamurthy K V (1998) Plant regeneration in pigeonpea (*Cajanus cajan* L. Millsp.) by organogenesis. *Plant Cell Rep.*, **17**: 705–710.
Mohan Ram H Y, Usha Mehta, Ramanuja Rao IV and Narasimham M (1982) Haploid induction in legumes. *Proc. 5th Intl. Cong. Plant Tissue and Cell Culture*, Japan, 32.
Moss J P, Singh A K, Sastri D C and Dundas I S (1988) Wide hybridization in legumes at ICRISAT. In: Biotechnology in Tropical Crop Improvement: *Proceedings of the International Biotechnology Workshop*, ICRISAT, India, 87–95.
Nene Y L and Sheila V K (1990) Pigeonpea: Geography and importance. In: *The Pigeonpea* (Eds Nene Y L, Susan D H and Sheila V K), CAB International, Wallingford, UK, 1–14.
Patel D B, Barve D M, Nagar N and Mehta A R (1994) Regeneration of pigeonpea, *Cajanus cajan* through somatic embryogenesis. *Indian J. Exp. Biol.*, **32**: 740–744.
Pundir R P S and Singh R B (1985) Crossability relationships among *Cajanus*, *Atylosia* and *Rhynchosia* species and detection of crossing barriers. *Euphytica*, **34**: 303–308.
Pundir R P S and Singh R B (1987). Possibility of genetic improvement of pigeonpea (*Cajanus cajan* L. Millisp.) utilizing wild gene sources. *Euphytica*, **36**: 33–37.
Ramaiah S M and Skinner D Z (1997) Particle bombardment: A simple and efficient method of alfalfa (*Medicago sativa* L.) pollen transformation. *Curr. Sci.*, **73**: 674–682.
Ramana R V, Venu C H, Jayasree T and Sadanandan A (1997) Genotype-dependent embryogenesis, organogenesis and *Agrobacterium* mediated transformation in pigeonpea (*Cajanus cajan* L. Millsp.). *Congress on In Vitro Biology* Abtr. P 1132.
Ramanandan P (1981) The wild gene pool of *Cajanus* at ICRISAT, present and future. In: *Proceedings of the International Workshop on Pigeonpeas, Vol. 2*. ICRISAT, Patancheru, India, 29–38.
Rathore R S and Chand L (1997) *In vitro* transformation of pigeonpea genotypes by wild strains of *Agrobacterium tumefaciens*. *Intl. Chickpea* and *Pigeonpea Newslett.*, **4**: 38–39.
Rathore R S and Chand L (1999) Plantlet regeneration from decapitated embryonic axes of pigeonpea (*Cajanus cajan* L. Millsp.). *Indian J. Exp. Biol.*, **37**: 496–498.
Rather R S, Chand L, Singh N K and Garg G K (2000) A new method for enhancement of regeneration efficiency in pigeonpea (*Cajanus cajan* L. Millsp.). *J. Plant Biol.*, **27**: 81–84.
Razzaaque M A, Dias S C, Shaheen Akhtar and Sayeed M (1986) Economic and chemurgic prospects of *Cajanus cajan*. *Bano Biggyan Patrica*, **15**: 16–29.
Reddy L J (1981) Pachytene analysis in *Atylosia sericea* and *Cajanus cajan* × *A. sericea* hybrid. *Cytologia*, **46**: 567–577.
Reddy L J and De D N (1983) Cytomorphological studies in *Cajanus cajan* × *Atylosia lineata*. *Indian J. Genet. and Plant Breed.*, **42**: 96–103.
Reddy L J, Green J M, Singh U, Bisen S S and Jambunathan R (1979) Seed protein studies on *Cajanus cajan*, *Atylosia* spp and some hybrid derivatives. In: *Proceedings, Symposium on Seed Protein Improvement in Cereals and Grain Legumes*. Intl. Energy Agency, Vienna, Austria, **2**: 105–117.
Sarangi B K, Kuckuk N and Gleba Yu Y (1992) Isolation and culture of protoplasts of pigeonpea (*Cajanus cajan* L. Millsp.). *Plant Cell Rep.*, **11**: 462–465.
Shama Rao H K and Narayanaswamy S (1975) Effect of gamma irradiation on cell proliferation and regeneration in explanted tissues of pigeonpea (*Cajanus cajan* L. Millsp.). *Radiation Botany*, **15**: 301–305.
Shiva Prakash N and Neera Bhalla Sarin (1995) Critical factors affecting regeneration and transformation of pigeonpea (*Cajanus cajan* L. Millsp.) Quality-Session IV for Transformation. 2nd European Conference on Grain Legumes, Copenhagen.
Shiva Prakash N, Pental D and Neera Bhalla Sarin (1994) Regeneration of pigeonpea from cotyledonary node via multiple shoot formation. *Plant Cell Rep.*, **13**: 623–637.
Shobana Sangawan P S, Nainawatee H S and Lal B M (1976) Chemical composition of some improved varieties of pulses. *Food Sci. Tech.*, **13**: 49–51.
Singh N D, Sahoo L, Sonia and Jaiwal P K (2002a) *In vitro* shoot organogenesis and plant regeneration from cotyledonary node and leaf explants of pigeonpea (*Cajanus cajan* L. Millsp.). *Physiol. Mol. Biol. Plant*, **8**: 133–140.
Singh N D, Sahoo L, Bhalla-Sarin N and Jaiwal P K (2002b) Dose and exposure time dependent morphoregulatory role of TDZ: organogenesis and somatic embryogenesis in pigeonpea (*Cajanus cajan* L. Millsp.). *Plant Sci.* (in press).
Singh N D, Sahoo L, Bhalla-Sarin N and Jaiwal P K (2002c) *In vitro* plant regeneration and recovery of primary transformants from shoot apices inoculated with *Agrobacterium*. *In Vitro Cell Dev. Biol. Plant* (in press).
Sreenivasu K, Malik S K, Kumar P A and Sharma R P (1998) Plant regeneration via somatic embryogenesis in pigeonpea (*Cajanus cajan* L. Millsp.). *Plant Cell Rep.*, **17**: 294–297.

van der Maesen L J G (1986) *Cajanus* DC and *Atylosia* W&A. (*Leguminosae*). Agricultural University Wageningen Papers 85-4. Agricultural University, Wageningen, The Netherlands, 225.

van der Maesen L J G (1990) Pigeonpea: Origin, history, evolution and taxonomy. In: *The Pigeonpea* (Eds Nene Y L, Susan D H and Sheila V K), CAB International, UK, 15–46.

Xu Z H, Yang L J, Wei Z M and Garo M X (1984) Plant regeneration in tissue culture of four leguminous species. *Acta Biol. Exp. Sin.*, **17**: 483–486.

IN VITRO REGENERATION AND GENETIC TRANSFORMATION OF CHICKPEA

SONIA[1], RANA P. SINGH[1], K. K. SHARMA[2] AND PAWAN K. JAIWAL[1,*]

[1]*Department of Biosciences, M.D. University*
Rohtak – 124 001, India
**e-mail: pkjaiwal@yahoo.com*
[2]*Genetic Resources and Enhancement Program, Genetic Transformation Laboratory, International Crops Research Institute for the Semi-Arid Tropics (ICRISAT), Patancheru, Andhra Pradesh–502 324, India*

Abstract

Chickpea is one of the major grain legume crops of the world and of the Indian subcontinent. Production of chickpea has not made any significant increase in the last few decades as vulnerability of the crop to various diseases and pests seriously hamper the crop yield. Attempts to introgress desirable traits from wild relatives into cultivated chickpea have had limited success due to sexual incompatibilities and high degree of autogamy. Genetic engineering techniques to incorporate useful traits have not made any significant progress for want of an efficient regeneration protocol. Although chickpea has been successfully regenerated *in vitro* using diverse explants on different media and somatic embryogenesis has been reported in various genotypes of chickpea, rooting of shoots and conversion of embryos to plants is still problematic besides the genotype dependency of the reported methods. Regeneration of plants from protoplast and anther has not been given due attention. Even though the genetic transformation of chickpea has been reported where the primary transformants were confirmed by molecular analysis, the methods were genotype dependent and produced low frequencies of transformed tissue or plantlets. This review presents the current status of research encompassing various aspects of shoot regeneration in tissue culture and genetic transformation of chickpea. Critical analysis and directions for future strategy of research on genetic manipulation of chickpea are also presented.

1. Introduction

Chickpea, *Cicer arietinum* L., is a grain legume crop that is of prime importance in India and all over the world. It is the third most important pulse crop after dry beans

P. K. Jaiwal and R. P. Singh (eds.), Applied Genetics of Leguminosae Biotechnology, 69–87.

(*Phaseolus vulgaris* L.) and dry peas (*Pisum sativum* L.). It is a diploid species with 2n = 16 chromosomes that is self-pollinating crop with natural cross-pollination ranging between zero and one per cent. The center of origin of chickpea is believed to be in southern Caucasus and northern Persia (van der Measen, 1972, 1987) or south-eastern Turkey (Ladizinsky, 1975). For convenience, chickpea has been classified into two main types, namely desi (characterized by small size, angular shape and colored seeds with a high percentage of fibre) and kabuli (characterized by large size ram-head shape and beige colored seeds with low percentage of fibre). The desi type accounts for about 85% of the world production, the remainder being kabuli (Singh *et al.*, 1985).

Chickpea is traditionally grown in many parts of the world – Asia, Africa, Europe, and North and South America – but the bulk of it is produced and consumed in South Asia and, increasingly, the Middle East and some Mediterranean countries (Jodha and Subba Rao, 1987). Chickpea seeds are consumed in various forms all round the globe, as they are free from any antinutritional factors and are rich in phosphorus and calcium and digestible proteins. It can grow well despite the low inputs under edaphic and arid environments, which make it an important component of the cropping systems of subsistence farmers in the Indian subcontinent, West Asia and North Africa. It can manage to squeeze out 70% of its nitrogen requirement from symbiotic nitrogen fixation and is an attractive option in intercropping. In spite of being an economically important crop, the production of chickpea has remained low due to several biotic and abiotic constraints. Drought and cold stresses cause considerable losses in chickpea yield. Chickpea crop standing in field is vulnerable to many insect pests and diseases. Lepidopteran pod borer (*Heliothis armigera*) larvae feed on the leaves, flowers and pods in addition to the various fungi – *Ascochyta rabiei* (ascochyta blight), *Fusarium oxysporum* (fusarium wilt), *Rhizoctonia bataticola* (dry root rot), *Botrytis cinerea* (botrytis gray mould), to name a few – cause crop destruction to a vast extent (Jaiwal *et al.*, 2001). Use of chemical insecticides and fungicides to control them adversely affect the seed quality, pollute the environment and also result in the evolution of resistant biotypes of insects.

Over 20,000 accessions of chickpea (*Cicer arietinum* L.) are now held in the gene banks of International Crops Research Institute for the Semi-Arid Tropics (ICRISAT, in India) and the International Centre for Agricultural Research in the Dry Areas (ICARDA, in Syria) (Singh, 1997). Chickpea lines exhibiting desirable traits like resistance to drought, cold and severe diseases like ascochyta blight, fusarium wilt, leaf miner, etc. have been identified and have been employed in various breeding programs to develop elite cultivars resistant to biotic and abiotic stresses. However, major success has not been achieved in this area because resistant lines breakdown with time or succumb to new races of pathogens (Singh and Reddy, 1991).

Wild relatives of chickpea possess genes for resistance towards fungal disease and cold stress. Classical breeding attempts by utilizing both cultivated and wild germplasms to introgress inherent resistance in cultivated chickpea have been carried out but could not lead to any breakthrough in the scenario.

During the last decade, biotechnology has made impressive advancements. Genetic transformation of crop plants is its valuable aspect, leading towards the building of an organized and healthy agriculture system free from the use of polluting insecticides and fungicides. Exploring the possibilities of transferring pest resistance genes in pea

(Shade *et al.*, 1994; Morton *et al.*, 2000) has opened new vistas for transferring agronomically important traits into other crop plants to develop elite cultivars (Birch, 1997). The technology emphatically supports the transfer, integration and statement of foreign genes in heterologous organisms. Agronomically important genes to improve the crop yield qualitatively as well as quantitatively have been isolated and cloned from various sources. These include genes for fungal resistance – chitinases and glucanases; viral resistance – coat protein and ribosome inhibiting proteins; pest resistance – amylase inhibitors, proteinase inhibitors and Bt toxin genes and genes for nutritional improvement – 2S albumin gene. Biotechnology encompassing genetic engineering offers avenues to overcome the barriers in achieving maximum chickpea production. This review highlights the present status of research conducted on various aspects of chickpea, particularly *in vitro* regeneration of plants and genetic transformation.

2. *In vitro* regeneration

The prerequisite for the *in vitro* manipulation for improvement of a plant species is the ability of an efficient and reproducible plant regeneration system. Towards the development of an *in vitro* regeneration protocol in chickpea several methods have been reported by various workers.

2.1. DIRECT SHOOT ORGANOGENESIS

High frequency multiple shoot regeneration from mature seeds cultured on 1 to 20 μM thidiazuron (TDZ) containing medium has been reported (Malik and Saxena, 1992). TDZ, a substituted phenylurea derivative was effective in inducing nearly 50 shoots per explant from the nodal and basal regions of the primary epicotyl, as compared to the other purine based cytokinins. Similarly, N^6-benzylaminopurine (BAP) at a concentration of 5 μM has been shown to induce as much as 20 shoots from seeds devoid of radicle tip (Polisetty *et al.*, 1997). *In vitro* culture of seeds obviates the complexities regarding selection and preparation of explants. The technique proves useful when large-scale multiplication of a species is required. However, zygotic embryo-based regeneration systems have met with limited success in transformation studies.

An increasing number of protocols for regeneration from different seedling explants, viz. cotyledons, shoot tips, embryonic axes and hypocotyls of chickpea, have been reported. Immature cotyledons formed multiple shoots when cultured on B_5 medium containing zeatin (13.7 μM) and indole-3-acetic acid (IAA, 0.2 μM) (Shri and Davis, 1992), while multiple shoots have been regenerated from cotyledonary node explants cultured on medium containing BAP (0.2–1 mg/L) either alone or in combination with picloram (0.012 mg/L). Twelve to fourteen shoots per explant differentiated directly without the intervention of callus phase (Subhadra *et al.*, 1998). Embryo axes with apical meristem gave rise to multiple shoots when cultured on MS medium supplemented with kinetin (1.0 mg/L) (Fontana *et al.*, 1993) and also on medium containing MS macrosalts, 4X MS microsalts, B_5 vitamins, BAP (3 mg/L) and naphthalene acetic acid (NAA, 0.004 mg/L) (Kar *et al.*, 1996, 1997). Shoot tip explant differentiated multiple shoots when cultivated on MS medium containing kinetin (0.5 mg/L) and IAA (2.0 mg/L) (Bajaj and Dhanju, 1979) or on MS medium supplemented with NAA (2.0 mg/L) and indole-3-butyric acid (IBA) (2.0 mg/L) (Chandra *et al.*, 1993; Polisetty *et al.*, 1996, 1998). Hypocotyls were reported to produce maximum number of shoots on MS medium supplemented with IAA (1.0 mg/L) and BAP (5.0 mg/L) (Bansal and Bansal, 1995), while epicotyl explants resulted in the induction of multiple shoots when cultured on MS medium containing 1.0 mg/L BAP, 1.0 mg/L kinetin and 0.5 mg/L IAA (Sudha Vani and Reddy, 1996). Production of multiple shoot regeneration from shoot tips, stem nodes and cotyledonary nodes of chickpea were obtained on MS medium supplemented with BAP (8.8 μM) and IBA (0.04 μM) where a maximum number of shoots were produced from cotyledonary node explants (Sharma *et al.*, 1998).

Almost every seedling explant derived from embryonal axis of chickpea produces multiple shoots upon *in vitro* culture. However, maximum frequency of regeneration and highest number of shoots per explant has been obtained from embryonic axes that have been used for the genetic transformation in chickpea (Fontana *et al.*, 1993; Kar *et al.*, 1996, 1997; Krishnamurthy *et al.*, 2000).

2.2. ORGANOGENESIS VIA CALLUS

Various seedling explants of chickpea are reported to be capable of forming multiple shoots via an intervening callus phase. Callus derived from the explant may differentiate multiple shoots on callus induction medium or may require a subsequent subculture on shoot induction medium. Whole cotyledons and shoot tip explants cultured on B_5 medium supplemented with BAP (1.5 mg/L) and NAA (1.0 mg/L) formed callus which later on developed multiple shoots on the same medium (Prakash *et al.*, 1992a). Shoot organogenesis was observed in hypocotyl-derived callus on B_5 medium supplemented with BAP (1.0 mg/L) (Vani and Reddy, 1996). Multiple shoots have also been induced from cotyledonary node explants cultivated on B_5 medium containing BAP (0.5 to 1.0 mg/L) or NAA (0.5 mg/L) (Khan and Ghosh, 1984). Hypocotyls produced callus on B_5 medium containing 2% sucrose, BAP (1.0 mg/L), kinetin (1.0 mg/L) and IAA (0.5 mg/L), which subsequently developed shoots upon transfer to MS medium containing 2% sucrose, kinetin (2.0 mg/L) and NAA (2.0 mg/L) (Neelam *et al.*, 1986). Barna and Wakhlu (1994) demonstrated multiple shoot induction from immature leaflet derived callus of chickpea on MS medium supplemented with 2,4-D (25 μM) or with 2,4-D (5 μM) and BAP (10 μM) that regenerated multiple shoots upon transfer to MS medium containing BAP (10 μM).

2.3. SOMATIC EMBRYOGENESIS

Somatic embryogenesis refers to the *in vitro* development of embryo like structures from somatic cells of explants under the influence of plant growth regulators. Two distinct pathways leading to somatic embryogenesis have been observed in different crops (Brown *et al.*, 1995). These include direct somatic embryogenesis that results in the formation of somatic embryos without an intervening callus phase while indirect somatic embryogenesis is usually a two step process where the explants give rise to callus that can then be induced to form somatic embryos on the same or different media conditions.

2.3.1. *Direct somatic embryogenesis*

Somatic embryogenesis has been induced directly as well as through an intervening callus phase in chickpea. Intact mature seeds directly form somatic embryos when cultured in medium containing MS salts, B_5 vitamins, 10 μM TDZ and 1 mM proline (Murthy *et al.*, 1996). Immature cotyledons have proven to be very responsive for the induction of somatic embryogenesis by using different media formulations where somatic embryos were produced either on B_5 medium containing 7.8 μM 2,4,5-trichlorophenoxyacetic-acid (2,4,5-T) and 4.4 μM BAP (Ramana *et al.*, 1996) or on B_5 medium containing 2.0 mg/L zeatin (Islam, 1995). Hita *et al.* (1997) reported somatic embryogenesis from immature cotyledons on B_5 medium containing 13.7 μM zeatin, 1.5% sucrose and 0.2 μM IAA that elongated on B_5 medium supplemented with 14.4 μM GA and 1.0 μM IBA. Somatic embryos intermixed with shoot buds were formed when immature cotyledons were cultured on B_5 medium containing zeatin (14 μM) and IAA (5 μM), out of which a few germinated upon transfer to B_5 medium supplemented with 1 to 2 mg/L BA and 0. 5 mg/L IAA (Mallikarjuna *et al.*, 1993). Immature cotyledons and embryo axes taken 14 to 21 days after pollination and cultured on medium containing MS salts, B_5 vitamins, zeatin (1.3 or 5.2 mg/L) and IAA (0 or 35 mg/L), under dim light (150 m^{-2} s^{-1}), developed embryoids directly on the edges of immature cotyledons or petiole stumps in 2 weeks' time. These further regenerated into morphogenetically normal plantlets when subcultured on medium containing half strength MS salts, B_5 vitamins, 2.0 mg/L NAA and 0.5 g/L activated charcoal. These plantlets were successfully established in vermiculite and soil potting mix. Immature cotyledons and embryo axes explants gave rise to somatic embryos on MS medium supplemented with 2,4,5-T (3.0 mg/L) that developed into complete plantlets upon transfer to half strength MS medium containing 1.0 mg/L zeatin (Sagare *et al.*, 1993). Somatic embryos have also been induced directly from mature embryo axes explants cultured on semi-solid or liquid MS medium containing 2,4,5-T (3.0 mg/L) which were later transferred to half strength MS medium containing abscisic acid (1.0 mg/L) for maturation and then on half strength MS supplemented with zeatin (1.0 mg/L) for conversion to plantlets (Suhasini *et al.*, 1994).

2.3.2. *Somatic embryogenesis via callus*

Indirect somatic embryogenesis preceded by a callus phase has been reported in chickpea. Mature leaflets of chickpea gave rise to embryogenic callus and somatic embryos on MS medium containing 2,4-dichlorophenoxy acetic acid (2,4-D, 0.5 mg/L) and BAP (0.5 mg/L) in dark, which develop into complete plantlets upon transfer to MS medium containing 0.1 mg/L IAA and 3.0 mg/L BAP (Rao and Chopra, 1989). Mature leaflets cultured on MS medium supplemented with 2,4-D (1.25 mg/L) and kinetin (0.25 mg/L) developed somatic embryos in dark, which were subsequently transferred to light on medium containing MS salts, B_5 vitamins, IBA (0.125 mg/L) and BAP (0.25 mg/L) for maturation. Subsequent transfer to B_5 medium containing BAP (0.25 mg/L) for conversion and finally to B_5 basal medium for plantlet formation led to complete recovery of plants (Dineshkumar *et al.*, 1994). The same was repeated by Kumar *et al.* (1995) by using three different cultivars.

Root, leaf, epicotyl, cotyledons and cotyledonary nodes taken from 7-d old seedlings were cultured on MS medium containing 2,4-D and picloram (0.1 to 1.0 mg/L) either singly or in combination with kinetin (0.5 to 1.0 mg/L) to develop somatic embryos via a callus phase. However, such embryos developed only up to torpedo stage and failed to develop further (Shankar and Mohan Ram, 1993). Embryogenic cell suspension cultures can also be obtained on B_5 medium containing BA (1.5 mg/L) and NAA (1.0 mg/L) that produce plantlets at very low frequency upon transfer to semi-solid media (Prakash *et al.*, 1994).

In a comparative study, zygotic embryos isolated from immature pods 3 to 15 days after pollination were compared with somatic embryos obtained from immature cotyledonary segments and embryo axes induced on MS containing 3.0 mg/L 2,4,5-T (Suhasini *et al.*, 1997; Sagare *et al.*, 1999). The suspensor of the zygotic embryos of chickpea was found to consist of six tiers of biseriate filamentous cells whereas somatic embryos were raised on a broad multiseriate suspensor like stalk. The size of a globular stage somatic embryo was much larger than the zygotic embryo at a similar stage of development. The cells that composed the apical dome were flat in zygotic embryos in comparison to the curved cells of the somatic embryos. Germinated seedlings and converted somatic embryos showed similarity in organization, differing only in the size of cotyledons.

Histology of somatic embryo initiation and development from different chickpea explants has also been studied (Sagare *et al.*, 1995, 1999). In this work, cotyledonary segments and embryo axes from immature and mature embryos were cultured on MS medium supplemented with 3.0 mg/L 2,4,5-T either liquid or semi-solid medium. Young leaflets were cultured on MS medium supplemented with 2,4-D (1.25 mg/L) and kinetin (0.25 mg/L) or 2,4-D (3.0 mg/L) and BAP (0.1 mg/L). Histological examination of explants fixed at different embryo development stages, revealed that somatic embryos arose indirectly with an intervening callus phase from immature cotyledons and immature embryo axes, directly from mature embryo axes and by both routes from young leaflets.

A direct relationship of somatic embryogenesis with the evolution of ethylene and methane in three chickpea cultivars has been shown (Chandra *et al.*, 1998; Guru *et al.*, 1999). In BG-362, which was more embryogenic than the other genotypes, a high ethylene : methane ratio (5.8 : 1) on day one after inoculation on induction medium and a lower ethylene: methane ratio of 2.89 : 1 on maturation medium was found. In contrast, in BG-372 with the least embryogenic potential, a low ethylene: methane ratio of 4 : 1 on the maturation medium was found. However, the inclusion of polyamines, i.e. putrescine (0.1, 0.2 or 0.5 mM) and spermidine, and of the ethylene inhibitors like ethrel (5 or 10 ppm), aminooxyacetic acid (10 or 20 μM), salicylic acid (25, 50 or 75 μM) in the induction medium resulted in reduction in somatic embryogenesis (Patil *et al.*, 1998).

The frequency of somatic embryogenesis and subsequent conversion of embryos to fertile plantlets is dependent on a range of factors that include genotype, explant, macronutrients, plant growth hormones, carbon source, light, etc. Like other legumes, in chickpea young meristematic tissues such as immature cotyledons, embryos and leaf explants readily undergo somatic embryogenesis on media containing chlorophenoxy-acetic acids. However, poor germination and low conversion frequency into plantlets accompanied with somaclonal variations delimit the vast horizons of this immensely useful technique. The future research should be directed towards fundamental aspects uncovering the underlying relationships of the process with the endogenous hormone levels in order to increase the efficiency of obtaining fertile plantlets.

2.4. ANTHER CULTURE

In vitro culture of anthers or pollen produce haploid plants, which are useful for production of homozygous lines and for the detection of recessive mutants. In chickpea, the lone attempt of anther culture was reported by Gosal and Bajaj (1979); only callus was obtained and ploidy level of the callus was not clearly mentioned.

2.5. PROTOPLAST CULTURE

Isolated plant protoplasts serve as starting material for cell cloning and development of mutant lines, somatic hybrids and transgenic plants (Bhojwani and Razdan, 1996). Isolated protoplasts of chickpea derived from hypocotyls produced microcalli on V47 medium supplemented with NAA (1.5 mg/L) and BAP (0.5 mg/L) but failed to undergo differentiation and organogenesis to produce plantlets (Sagare and Krishnamurthy, 1991).

2.6. *IN VITRO* ROOTING OF SHOOTS

Although multiple shoot regeneration via organogenesis and somatic embryogenesis from meristematic tissues of chickpea has become a routine, rooting of *in vitro* regenerated shoots does not occur easily and various workers have suggested different media formulations to obtain complete plantlets of chickpea *in vitro*. Inclusion of auxins in nutrient media has been widely practiced to attain rooting of regenerated shoots. *In vitro* formed shoots were cultured on B_5 medium containing NAA (1.0 mg/L), agar (8.0 g/L), prior to their transfer to B_5 containing NAA (20.0 mg/L), agar (12.0 g/L) and activated charcoal (0.2%) to obtain complete plantlets (Prakash *et al.*, 1992a). A different method has been suggested by Malik and Saxena (1992) where the shoots were cultured on basal medium containing MS macro and micronutrients, B_5 vitamins, 3% sucrose and 0.25% gelrite for 1–3 weeks and before transfer to basal medium containing NAA (2.5 μM). However, Khan and Ghosh (1984) achieved rooting of shoots on a simple B_5 medium containing NAA (0.5 mg/L). Another auxin, IBA is frequently incorporated in media for rooting of shoots (Bansal and Bansal, 1995; Kar *et al.*, 1996). More recently, the *in vitro* formed shoots were found to produce healthy multiple roots with high frequencies when cultured on filter paper bridges in liquid MS medium containing up to 10 μM IBA (B. Jayanand and K. K. Sharma, unpublished data).

Reducing the amount of salts in medium can also facilitate rooting of *in vitro* regenerated shoots of chickpea. Ninety to 100% of shoots formed roots on one-fourth-strength MS containing 0.75% sucrose and 0.8% agar-agar (Polisetty *et al.*, 1996). Root induction and proliferation from chickpea shoots have been reported on half-strength MS containing IAA (1.0 mg/L) alone or in combination with kinetin (0.01 mg/L) (Subhadra *et al.*, 1998) or on half-strength MS medium containing 1% sucrose and IBA (2.0 μM) (Sharma and Amla, 1998).

Roots arising directly from the base of the shoot without intermediary callus have well-defined vasculature connected with that of shoot and thus these plantlets have higher survival frequency in pots. Roots that are fragile or succulent or arise through callus make the plant difficult to establish in soil.

2.7. ESTABLISHMENT OF THE PLANTLET

Success of any regeneration protocol depends upon the establishment of the *in vitro* raised plantlets in soil. Several methods have been suggested to adapt the plantlets to soil. Prior to transfer to soil, hardening of plantlets is done. The plantlets were kept on one-fourth-strength liquid MS medium devoid of sucrose for 4–5 days and then transferred to soilrite (Dineshkumar *et al.*, 1994) or on sterile vermiculite for 1 week and transferred to a mixture of soil, farmyard manure and sand mixed in 3 : 1 : 1 ratio (Dineshkumar *et al.*, 1994) or to pots containing garden soil (Barna and Wakhlu, 1993). Plantlets raised via somatic embryogenesis were established in the soil of greenhouse containing mixture of peat, vermiculite and soil in 1 : 1 : 1 ratio by volume (Sagare *et al.*, 1993). Shoots regenerated from shoot tip explants formed roots on one-fourth-strength Hoagland's solution for 20 days and then established in vermiculite (Polisetty *et al.*, 1996). Plantlets derived from callus obtained from immature leaflets were established in pure vermiculite (Barna and Wakhlu, 1994).

In general, shoots up to 2–3 cm in length with well-developed roots arising directly from the base of the shoots acclimatize with ease to the field conditions and have higher frequency of establishment. However to date, the transplantation record of rooted shoots to glasshouse have not been very successful which has led some workers to resort to time-consuming and inefficient grafting of the *in vitro* produced shoots on to the stocks of seedlings (Krishnamurthy *et al.*, 2000).

2.8. WIDE HYBRIDIZATION

Many agronomically important traits like resistance to ascochyta blight, fusarium wilt, leaf miner, nematode cysts, seeds beetle and cold can be found in wild *Cicer* species namely – *C. reticulatum*, *C. echinospermum*, *C. judaicum*, *C. pinnatifidum*, *C. bijugum* and *C. cuneatum* (Malhotra, 1987). Various workers have studied the isozyme polymorphism patterns of *Cicer* species to determine phylogeny of the cultivated chickpea (Ahmed *et al.*, 1992; Tayyar and Waines, 1996). As revealed by these studies *C. arietnum*, *C. reticulatum* and *C. echinospermum*, form one cluster while *C. judaicum*, *C. pinnatifidum* and *C. bijugum* formed another cluster. *Cicer chorassanicum* was grouped with *C. yamashitae*, *C. anatolicum* and *C. songoricum* whereas *C. cuneatum* formed an independent group and showed largest genetic distance from *C. arietinum*. *Cicer reticulatum* has been proposed to be the putative progenitor of *C. arietinum* and has also been classified as its subspecies (Ladizinsky and Adler, 1976a, b).

Classical breeding methods comprising of interspecific hybridization offers a means of broadening the gene pool of a particular plant species by introgressing desirable traits of a high degree of autogamy. However, very little success has been achieved in gene introgression from difficult to cross species. To improve the yield and yield related traits in F_2, F_3 and F_4 generations of *C. arietinum* × *C. reticulatum* and *C. arietinum* × *C. echinospermum*, F_1 with 28–158% hybrid vigor and F_2 population with numerous transgressive segregants for high yield were produced (Jaiswal and Singh, 1989; Singh and Ocampo, 1993). Four crosses *C. arietinum* × *C. echinospermum* and *C. arietinum* × *C. reticulatum* and their reciprocals produced transgressive segregants for high yield and other desirable traits (Singh and Ocampo, 1997). A high level of heterosis was observed in F_1 and nine F_7 lines out-yielded the cultigen parent ILC 482 (a high yielding variety) by up to 39%. Moreover, these lines were free from any undesirable traits from the wild species, which contributed towards the recovery of superior lines.

Amongst the non-crossable species of *Cicer* that cannot be crossed with *C. arietinum*, *C. bijugum*, *C. judaicum* and *C. pinnatifidum* possess several agronomically important traits. The introgression of desirable characters into chickpea from these wild relatives by employing embryo rescue and *in vitro* culture will prove beneficial for overcoming post-fertilization barriers and raising hybrids. A few attempts to utilize tissue culture techniques have been made to recover the hybrids in some cases with limited success (Mallikarjuna, 1999). For example, Singh and Singh (1989) reported crossing of *C. arietinum* with *C. cuneatum* where the ovaries harboring hybrid embryos were cultured *in vitro* to recover plantlets. However, the hybrid plants failed to flower. Naik *et al.* (1994) reported hybrid of *C. arietinum* × *C. pinnatifidum* employing embryo rescue technique.

3. Genetic transformation

Although grain legumes have been notoriously known to be recalcitrant towards *in vitro* regeneration and genetic transformation, an increasing number of protocols are becoming available for gene transfer to legume species and transgenic plants expressing desirable foreign genes have been recovered in soybean (Stewart *et al.*, 1996), pea (Shade *et al.*, 1994; Morton *et al.*, 2000) and adzuki bean (Ishimoto *et al.*, 1996).

Chickpea has been shown to be susceptible to *Agrobacterium rhizogenes* as well as wild *Agrobacterium tumefaciens* and infection results in the formation of hairy roots and crown galls respectively on the infected explant. Srinivasan *et al.* (1991) inoculated leaf and stem explants with wild and disarmed strains of *Agrobacterium tumefaciens*. The tumors induced were capable of phytohormone independent growth. Transgenic calluses obtained proliferated on medium containing kanamycin (100 mg/L), but failed to regenerate. Formation of tumors and hairy roots following infection with wild *Agrobacterium tumefaciens* or *Agrobacterium rhizogenes* respectively have been reported in various genotypes of chickpea viz. 6153 (Islam *et al.*, 1992, 1993); C-235, H75-35 (Prakash *et al.*, 1992b); red chickpea, canitez-87 and MB-10 (Altinkut *et al.*, 1997a, b); in four genotypes (Seifkes-Boer *et al.*, 1995). Transgenic calluses have also been obtained from leaf and epicotyl explants of 12-d old seedlings, inoculated with *Agrobacterium tumefaciens* strain Ag10/ pKiwi105. The calli were GUS positive and proliferated on medium containing 100 mg/L kanamycin (Gray *et al.*, 1993).

Disarmed *Agrobacterium tumefaciens* strains and particle bombardment mediated gene transfer have resulted in raising putative transgenic plants of chickpea. Fontana *et al.* (1993) employed *Agrobacterium tumefaciens* strain LBA4404 harboring pBI121 carrying *nptII* and *uidA* genes to inoculate 24-h old embryonic axes deprived of shoot apex. Shoots regenerated from epicotylar end were selected on kanamycin (50 mg/L) and the shoots derived from the axillary bud primordia of cotyledonary node failed to survive on kanamycin. This indicated the need of wounding for agroinfection, which could not be substituted by acetosyringone. Integration of transgenes was confirmed by stable *uidA* expression in the leaves and roots of plantlets and Southern blot analysis for *nptII*.

In contrast to this, cotyledonary node explants from overnight-germinated soaked seedlings were co-cultivated with *Agrobacterium tumefaciens* strain LBA4404 containing pBI121 carrying the *nptII* and *gus* genes (Kar *et al.*, 1996). The explants were first cultured on medium containing 25 mg/L of kanamycin and the shoots thus obtained were repeatedly subcultured onto kanamycin 50 mg/L containing medium. This eliminated all the chimeric shoots and only the green shoots recovered were subsequently rooted and transplanted. The plants showed uniform GUS staining in the leaf and roots and Southern analysis for *nptII* revealed the integration of multiple copies of T-DNA in most of the plants. Since, the explant contains axillary bud primordia of cotyledonary node responsible for regeneration from pre-existing meristem and *de novo* regeneration from the epicotylar end also cannot be ruled out, but the authors are silent about the origin of transformed shoots. Moreover, as reported by Fontana *et al.* (1993) transformed shoots could only be obtained from the epicotylar end of the explant. Further, analysis regarding the inheritance of transgenes to progenies was not reported. Similarly, the transgenic plants of chickpea were obtained by co-cultivating the shoot primordia of mature embryos with *A. tumefaciens* carrying pBI121 (Altinkut *et al.*, 1997b). The shoots raised on kanamycin (100 mg/L) containing medium were found to be positive for Southern blot analysis for *nptII* and *gus* genes.

Various factors affecting gene delivery into zygotic embryos via particle bombardment has been studied on the basis of statement of *nptII* and *uidA* genes (Husnain *et al.*, 1997). Plasmid DNA at a concentration of 12 μg per mg of tungsten particles when accelerated with an inflow of helium gas at a pressure of 60 kg/cm^3 through a distance of 24 cm in chamber maintained at a negative pressure of 71.1 cm of mercury resulted in optimal transient *gus* expression. CaMV35S promoter was found to be better than Actin and Win6 promoters. Zygotic embryos were also infected with *A. tumefaciens* and the resultant plants were found to be positive for the statement of *uidA* and *nptII* genes.

Transgenic chickpea plantlets expressing the *cry1Ac* gene for resistance towards the field pest pod borer (*Heliothis armigera*) have been reported (Kar *et al.*, 1997). Embryo axes deprived of root and shoot meristems were co-transformed with *cryIAc* and *nptII* genes using a Bio-Rad Biolistic 1000/HE particle gun. Putative transformed shoots were obtained on medium containing kanamycin (50 mg/L). Molecular analysis by Southern and Northern blots revealed the presence of *cryIAc* and its statement was confirmed by the inhibition of larval development of *Heliothis armigera* on feeding transgenic shoots. The co-transformation frequency of *nptII* and *cry1Ac* was estimated to be 45.8%.

Inheritance of transgenes to T_1 progeny was shown by PCR. However, the workers have not carried out the insect bioassays with the pods or seeds of chickpea. The presence and statement of transgenes and resistance level in subsequent progenies were not reported.

Following a similar protocol Krishnamurthy *et al.* (2000) used embryo axes deprived of root and shoot meristems for *Agrobacterium* mediated transformation. The two strains used were C58C1/pGV2260 and EHA101/pIBGUS carrying NOS-NPTII and CaMV35S-PAT as plant selectable markers. Explants co-cultivated with the *Agrobacterium* were cultured on selection medium containing either kanamycin (100 mg/L) or PPT (10 mg/L). The shoots obtained were subcultured on same fresh medium for 3–4 months to eliminate escapes and chimeric shoots the explants regenerated 2 shoots from the nodal region and 2 shoots from the shoot apex region of embryo axes after 10–15 days of culture. In all, 16 shoots were obtained from 4000 explants giving transformation efficiency of 0.4%. Due to problems with efficient rooting the shoots were grafted onto germinated seedlings of wild type chickpea, prior to their hardening and transfer to soil. Genomic analysis was carried out for only 4 out of the 16 plants, which revealed the integration of single or multiple copies of T-DNA. The plantlets transferred to soil had reduced vigor and fertility, which the authors attributed to the suboptimal glasshouse conditions. Out of 36 plants growing in glasshouse only 5 plants flowered and set seeds. T_1 progeny as analyzed by PCR was found to be positive for *nptII* but not for histochemical GUS assay. The study carried out reports the transformation of chickpea via *Agrobacterium*, but lacks an efficient regeneration protocol imperative for the success of any transformation experiment, resulting in the loss of transformed shoots.

An overall view of the scenario clearly conveys that an efficient, genotype independent and reproducible protocol for the chickpea transformation has not been developed to date. The major lacunae in various protocols reported are the lack of an efficient regeneration system compatible with the transformation techniques. This results in low transformation frequencies and stable transmission and statement of transgenes to subsequent generations of plants. Any transformation experiment depends upon the interplay of a number of factors comprising of *in vitro* regeneration, selection of transformed shoots, rooting, establishment and fertility of transgenic plants and finally the statement of transgenes in T_0 plants and their progenies. In the light of all the reports, it is evident that future research should be conducted keeping in mind the above-mentioned prerequisites for developing transformation protocols and their subsequent deployment for incorporating agronomically important traits.

4. Conclusions and future prospects

Chickpea, *Cicer arietinum* L., has been regenerated *in vitro* through direct or indirect organogenesis and also via somatic embryogenesis in various genotypes by culturing diverse explants on different media. Most of the workers have been successful in achieving efficient shoot organogenesis in terms of frequency of regeneration and number of shoots per explant. However, all these systems rely on the use of embryonal explants or meristematic tissues associated with embryo axis. Further, there are limited reports of reliable rooting and establishment of *in vitro* regenerated shoots to soil. Similarly, the

Table 1. Direct or callus-mediated shoot regeneration in various genotypes of chickpea

Cultivar	Basal medium[1]	Growth regulators	Explant	Morpho-genetic response	Reference
DIRECT					
—	MS	IAA (2 mg/L), Kinetin (0.5 mg/L)	Apical meristem	Plantlet	Bajaj and Dhanju, 1979
Kabuli	MS salts + B_5 vitamins	TDZ (5 μM)	Seed	Multiple shoots	Malik and Saxena, 1992
ICC640	B_5	Zeatin (13.7 μM), IAA (0.2 μM)	Immature cotyledon	Multiple shoots	Shri and Davis, 1992
Pusa-256	MS	NAA (2 mg/L), IBA (2 mg/L)	Shoot tip	Multiple shoots with roots	Chandra *et al.*, 1993
Local ecotype	MS	Kinetin (1.5 mg/L)	Embryonic axis devoid of shoot apical meristem	Plant	Fontana *et al.*, 1993
Annigeri, ICCV6	B_5, DKW-C	BAP (4.4 μM)	Cotyledonary node, meristems tip	Multiple shoots, plant	Brandt and Hess, 1994
JG-317, JG-1265	MS	IAA (1.0 mg/L) BAP (5.0 mg/L)	Hypocotyl	Plant	Bansal and Bansal, 1995
UC5, Narayen, Amethyst, Tyson, Barwon	MS	Zeatin (1–5.2 mg/L), IAA (3 mg/L)	Immature cotyledon, embryo axis	Multiple shoots, plant	Adkins *et al.*, 1995
BG-362, BG-329, BG-267, BG-256, C-235	MS	NAA (2 mg/L) IBA (2 mg/L)	Shoot tip	Plantlet	Polisetty *et al.*, 1996
ICCV-1, ICCV-6 and a desi variety	MS macro + 4X MS micro + B_5 vitamins	BAP (3.0 mg/L) NAA (0.004 mg/L)	Embryonic axis without root, and shoot meristem	Plant	Kar *et al.*, 1996
Local	MS	TD2 (10 μm)	Seed	Multiple shoots	Murthy *et al.*, 1996
JG-62, C-235, PGC1	B_5	BAP (1.0 mg/L) Kinetin (1.0 mg/L) IAA (0.5 mg/L)	Epicotyl	Multiple shoots	Sudha Vani and Reddy, 1996

Table 1. (Continued)

Cultivar	Basal medium[1]	Growth regulators	Explant	Morphogenetic response	Reference
BG–362, BG-329, BG-267, BG-256, C-235	MS salts + B_5 vitamins	Different concentrations of BAP	Embryo axis-based	Multiple shoots, plants	Polisetty *et al.*, 1997
ICC435M	—	BAP (0.2–1.0 mg/L), Picloram (0.012 mg/L)	Cotyledonary node	Plant	Subhadra *et al.*, 1998
—	MS	BAP (8.8 μM) IBA (0.04 μM)	Shoot tip, stem node, cotyledonary node	Plant	Sharma *et al.*, 1998
VIA CALLUS					
B-108	B_5	NAA (0.5 mg/L)	Cotyledonary node	Plantlet	Khan and Ghosh, 1984
ICCC-4, H-208, Annigeri	B_5	BAP (1.0 mg/L) Kinetin (1 mg/L) IAA (0.5 mg/L)	Hypocotyl, shoot tip	Multiple shoots	Neelam *et al.*, 1986
C-235, H-208	B_5	BAP (1.5 mg/L) NAA (1.0 mg/L)	Seed, shoot tip, cotyledon	Multiple shoots	Prakash *et al.*, 1992
C-235	MS	2,4-D (25 μM) or 2,4-D (5 μM) + BAP (10 μM)	Immature leaflet	Multiple shoots	Barna and Wakhlu, 1993
JG-62	B_5	BAP (1.0 mg/L)	Hypocotyl	Plantlet	Vani and Reddy, 1996

[1]MS: Murashige and Skoog, 1962; B_5: Gamborg *et al.*, 1968; DKW-C: McGranahan GH *et al.*, 1987.

frequency of conversion and germination of somatic embryos has remained very low and restricted to few genotypes. Anther and protoplast culture in chickpea are still in their infancy. Due to these impediments efforts towards genetic transformation of chickpea have not made any significant progress. In the few reports published by independent research groups, transgenic chickpea plants have been obtained by both particle bombardment as well as *Agrobacterium*-mediated transformation. However, their transgenic nature has not been reliably established. By comparing the different protocols, it is evident that most of the workers have employed cotyledonary node explants for *Agrobacterium*-mediated transformation and used *uidA* as reporter gene and *nptII* as selectable marker gene under the control of NOS and CaMV 35S promoters respectively. But still none of the available protocols can be employed for routine gene transfer experiments because of the low transformation frequency in addition to the genotype specificity. Chickpea being an economically important legume crop, calls for serious targeted efforts to evolve regeneration and transformation protocols aiming to engineer it

Table 2. Somatic embryogenesis in various genotypes of chickpea

Cultivar	Basal medium	Growth regulators	Explant	Morphogenetic response	Reference
DIRECT					
ICCV6	B_5	Zeatin (14 μM) IAA (5 μM)	Immature cotyledon	Germinated embryos	Mallikarjuna *et al.*, 1993
PG-12, C-235	MS	2,4,5-T (3.0 mg/L)	Immature cotyledon	Plant	Sagare *et al.*, 1993
PG-5, PG-12, C-235	MS	2,4,5-T (3.0 mg/L)	Mature embryo axis	Plant	Sagare *et al.*, 1993
Kabuli	MS salts + B_5 vitamins	TDZ (10 μM) + L-Proline (1 μM)	Mature seed	Somatic embryo	Murthy *et al.*, 1996
PG-12	MS	2,4,5-T (3.0 mg/L)	Mature embryo axis	Plant	Suhasini *et al.*, 1994
—	B_5	Zeatin (13.7 μM), IAA (0.2 μM)	Immature cotyledons	Plant	Hita *et al.*, 1997
VIA CALLUS					
BG-256	MS	2,4-D (0.5 mg/L)	Leaflet	Multiple shoots/ somatic embryos	Rao and Chopra, 1989
JG-62, Gaurav, Annigeri, BG-267, C-235	MS	2,4-D (25 μM)	Immature leaflet	Plant	Barna and Wakhlu, 1993
—	MS	2,4-D (0.1–1.0 mg/L), Picloram (0.1–1.0 mg/L), Kinetin (0.5–1.0 mg/L)	Root, leaf, epicotyl, cotyledon, cotyledonary node	Somatic embryo	Shankar and Mohan Ram, 1993
—	B_5	BAP (1.5 mg/L) NAA (1.0 mg/L)	—	Somatic embryo	Prakash *et al.*, 1994
C-235	MS	2,4-D (1.25 mg/L), Kinetin (0.25 mg/L)	Leaf	Plant	Dineshkumar *et al.*, 1994
BG-256	MS	2,4-D (0.25 mg/L), picloram (0.25 mg/L), BAP (0.1 mg/L)	Leaflet	Somatic embryo	Dineshkumar *et al.*, 1995
C-235, JG-262, P-144, P-209	MS	2,4-D (1.25 mg/L), Kinetin (0.25 mg/L)	Leaf	Plant	Kumar *et al.*, 1995
ICC4918	B_5	2,4,5-T (7.8 μM), BAP (4.4 μM)	Immature cotyledon	Somatic embryo	Ramana *et al.*, 1996

Table 3. Current status of chickpea transformation

Cultivar	Explant	Mode of transformation[1]	Status	Gene statement[2]	Inheritance	Reference
Pusa-256	Leaf, stem	Wild AT (pTVK291)	Callus	+ve *nptII* (S)	NA	Srinivasan *et al.*, 1991
6153	Seedlings	Wild AT	Tumors	NA	NA	Islam and Riazuddin, 1992
C-235, H75-35	Seedlings	AR	Hairy roots, tumors	NA	NA	Prakash *et al.*, 1992
Amethyst, Tyson, Barwon	Leaf, epicotyl	AT	Callus	NA	NA	Gray *et al.*, 1993
Local ecotype	Embryo axes deprived of apical meristem	AT (pBI121)	Plants	+ve *uidA*, *nptII* (S)	NA	Fontana *et al.*, 1993
—	—	AT	Tumors	NA	NA	Islam *et al.*, 1994
ICC4918	Immature cotyledon	AT (Bin19 GUSInt)	Somatic embryos	+ve *uidA* histology	NA	Ramana *et al.*, 1996
ICCV-1, ICCV-6 and a desi variety	Embryo axes without root, and shoot meristem	AT (BI121)	Plants	+ve *uidA*, *nptII* (S)	NA	Kar *et al.*, 1996
Red chickpea, Canitez 87, MB10	Shoot primordia	AT, AR	Plantlets, hairy roots	+ve *nptII* and *uidA* (S)	NA	Altinkut *et al.*, 1997
Red chickpea, Canitez 87, MB10	Seedlings, callus derived plantlets	AR	Hairy roots	NA	NA	Altinkut *et al.*, 1997
6153, CM72	Zygotic embryos	Biolistics, AT	Plants	NA	NA	Husnain *et al.*, 1997
ICCV-1 and ICCV-6	Embryo axes without root and shoot meristem	Biolistics (pcry1Ac BOS1)	Plants	+ve *cry1Ac* and *nptII* (S)	+ve PCR (T_1)	Kar *et al.*, 1997
PG1, PG12, Chafa, Turkey	Embryo axes	AT (pGV 2260 and pIBGUS)	Plantlets	+ve *uidA* and *nptII* (S)	+ve PCR *nptII* (T_1)	Krishnamurthy *et al.*, 2000

[1]AT: *Agrobacterium tumefaciens*; AR: *Agrobacterium rhizogenes*; biolistics: particle bombardment, [2]NA:Data not shown; S: Southern blot analysis.

with agronomically desirable genes for pest resistance, fungal resistance and nutritional improvement, in order to develop superior germplasm contributing towards increased yields of chickpea.

Acknowledgement

The senior author is grateful to CSIR, New Delhi for a Senior Research Fellowship.

References

Adkins A L, Godwin I D and Adkins S W (1995) An efficient *in vitro* regeneration system for Australian-grown chickpea (*Cicer arietinum*) cultivars. *Aust. J. Bot.*, **43**: 491–497.

Ahmed F, Gaur P M and Slinkard A E (1992) Isozyme polymorphism and phylogenetic interpretations in the genus *Cicer* L. as revealed by polyacrylamide gel electrophoresis of seed storage proteins. *Theor. Appl. Genet.*, **84**: 688–692.

Altinkut A, Bajrovic K and Gozukirmizi N (1997a) Regeneration and hairy root formation of chickpea using callus-derived plantlets and seedlings. *Intl. Chickpea and Pigeonpea Newslett.*, **4**: 30–31.

Altinkut A, Gozukirmiz N, Bajrovic K and Altman A (1997b) High percentage of regeneration and transformation in chickpea. *Acta Hortic.*, **447**: 319–320.

Bajaj Y P S and Dhanju M S (1979) Regeneration of plants from apical meristem tips of some grain legumes. *Curr. Sci.*, **48**: 906–907.

Bansal S and Bansal Y K (1995) An efficient multiple shoot and plantlet formation schedule in chickpea (*Cicer arietinum* L.). *J. Phytological Res.*, **8**: 31–34.

Barna K S and Wakhlu A K (1993) Somatic embryogenesis and plant regeneration from callus cultures of chickpea (*Cicer arietinum* L.). *Plant Cell Rep.*, **12**: 521–524.

Barna K S and Wakhlu A K (1994) Whole plant regeneration of *Cicer arietinum* from callus cultures via organogenesis. *Plant Cell Rep.*, **13**: 510–513.

Bhojwani S S and Razdan M K (1996) *Plant Tissue Culture: Theory and Practice*, revised edition. Elsevier, Amsterdam, 767.

Birch R G (1997) Plant transformation: problems and strategies for practical application. *Annu. Rev. Plant Physiol. Mol. Biol.*, **48**: 297–326.

Brandt E B and Hess D (1994) *In vitro* regeneration and propagation of chickpea (*Cicer arietinum* L.) from meristems tips and cotyledonary nodes. *In Vitro Cell. Dev. Biol. Plant*, **30**: 75–80.

Brown D C W, Finstad K I and Watson E M (1995) Somatic embryogenesis in herbaceous dicots. In: *In Vitro Embryogenesis in Plants* (Ed Thorpe T A), Kluwer Academic Publ., The Netherlands, 345–415.

Chandra R, Chatrath A, Polisetty R and Khetarpal S (1993) Differentiation of *in vitro* grown explants of chickpea (*Cicer arietinum* L.). *Indian J. Plant Physiol.*, **36**: 121–124.

Chandra R, Khetarpal S and Polisetty R (1998) Effect of plant growth regulators on evolution of ethylene and methane by different explants of chickpea. *Biol. Plant*, **40**: 337–343.

Dineshkumar V, Kirti P B, Sachan J K S and Chopra V L (1994) Plant regeneration via somatic embryogenesis in chickpea (*Cicer arietinum* L.). *Plant Cell Rep.*, **13**: 468–472.

Dineshkumar V, Kirti P B, Sachan J K S and Chopra V L (1995) Picloram induced somatic embryogenesis in chickpea (*Cicer arietinum* L.). *Plant Sci.*, **109**: 207–213.

Fontana G S, Santini L, Caretto S, Frugis G and Mariotti D (1993) Genetic transformation in the grain legume *Cicer arietinum* L. *Plant Cell Rep.*, **12**: 194–198.

Gamborg O L, Miller R A and Ojima K (1968) Nutrient requirements of suspension cultures of soybean root cells. *Exp. Cell Res.*, **12**: 151–158.

Gosal S S and Bajaj Y P S (1979) Establishment of callus tissue cultures, and the induction of organogenesis in some grain legumes. *Crop Improv.*, **6**: 154–160.

Gray S, Godwin T and Adkins S (1993) Development of *Agrobacterium* mediated transformation system for chickpea. *Intl. Chickpea Newslett.*, **28**: 23–24.

Guru S K, Chandra R, Raj A, Khetarpal S and Polisetty R (1999) Evolution of ethylene and methane in relation to somatic embryogenesis in chickpea. *Biol. Plant.*, **42**: 149–154.

Hita O, Lafarga C and Guerra H (1997) Somatic embryogenesis from chickpea (*Cicer arietinum* L.) immature cotyledons: the effect of zeatin, gibberellic acid and indole-3 butyric acid. *Acta Physiol. Plant*, **19**: 333–338.

Husnain T, Malik T, Riazuddin S and Gordon M P (1997) Studies on the expression of marker genes in the chickpea. *Plant Cell Tiss. Org. Cult.*, **49**: 7–16.

Ishimoto M, Sato T, Chrispeels M J and Kitamura K (1996) Bruchid resistance of transgenic adzuki bean expressing seed α-amylase inhibitor of common bean. *Entomol. Expt. Appl.*, **79**: 309–315.

Islam A K M R and Riazuddin S (1992) Transformation of chickpea tissue by *Agrobacterium tumefaciens* in *ex vitro* conditions. *Intl. Chickpea Newslett.*, **26**: 11–12.

Islam R (1995) Somatic embryogenesis from immature cotyledons of chickpea (*Cicer arietinum* L.). *Pak. J. Bot.*, **33**: 77–78.

Islam R, Malik T, Husnain T and Riazuddin S (1994) Strain and cultivar specificity in the *Agrobacterium* and chickpea interaction. *Plant Cell Rep.*, **13**: 561–563.

Jaiswal H K and Singh B D (1989) Analysis of gene effects for yield traits in crosses between *Cicer arietinum* L. and *C. reticulatum* Ladz. *Indian J. Genet.*, **49**: 9–17.

Jaiwal P K, Sonia and Upadhyaya K C (2001) Chickpea regeneration and transformation. *Curr. Sci.*, **80**: 1368–1369.

Jodha N S and Subba Rao K V (1987) Chickpea: world importance and distribution. In: *The Chickpea* (Eds Saxena M C and Singh K B), CAB International, Wallingford, UK, 1–10.

Kar S, Basu D, Das S, Ramakrishanan N A, Mukherjee P, Nayak P and Sen S K (1997) Expression of *cry1Ac* gene of *Bacillus thuringiensis* in transgenic chickpea plants inhibits development of pod borer (*Heliothis armigera*) larvae. *Transgenic Res.*, **6**: 177–185.

Kar S, Johnson T M, Nayak P and Sen S K (1996) Efficient transgenic plant regeneration through *Agrobacterium*-mediated transformation of chickpea (*Cicer arietinum* L.). *Plant Cell Rep.*, **16**: 32–37.

Khan S K and Ghosh P D (1984) Plantlet regeneration from cotyledonary nodes of chickpea. *Intl. Chickpea Newslett.*, **11**: 22–24.

Krishnamurthy K V, Suhasini K, Sagare A P, Meixner M, Kathen A de, Pickardt T and Scheider O (2000) *Agrobacterium*-mediated transformation of chickpea (*Cicer arietinum* L.) embryo axes. *Plant Cell Rep.*, **19**: 235–240.

Kumar P A, Bisaria S, Pai R A and Sharma R P (1995) Comparative shoot regeneration in different genotypes of chickpea, *Cicer arietinum* L. *Indian J. Exp. Biol.*, **33**: 77–78.

Ladizinsky G (1975) A new *Cicer* from Turkey. *Notes of the Royal Botanic Garden Edinburgh*, **34**: 201–202.

Ladizinsky G and Adler A (1976a) Genetic relationships among the annual species of *Cicer* L. *Thoer. Appl. Genet.*, **48**: 197–203.

Ladizinsky G and Adler A (1976b) The origin of chickpea *Cicer arietinum* L. *Euphytica*, **25**: 211–217.

Malhotra R S, Pundir R P S and Slinkard A E (1987) Genetic resources of chickpea. In: *The Chickpea* (Eds Saxena M C and Singh K B), CAB International, Wallingford, UK, 67–81.

Malik K A and Saxena P K (1992) Thidiazuron induces high frequency shoot regeneration in intact seedlings of pea (*Pisum sativum*), chickpea (*Cicer arietinum* L.) and lentil (*Lens culinaris*). *Aust. J. Plant Physiol*, **19**: 731–740.

Mallikarjuna N (1999) Ovule and embryo culture to obtain hybrids from interspecific incompatible pollinations in chickpea. *Euphytica*, **110**: 1–6.

Mallikarjuna N and Moss J P (1993) Regeneration from elite chickpea genotypes. *Intl Chickpea Newslett.*, **29**: 4–6.

McGranahan G H, Driver J A and Tulecke W (1987) Tissue culture *Juglans*. In: *Cell and Tissue Culture in Forestry* (Eds Bonga J M and Durzan D I), Nijhoff Publ., Dordrecht, 261–271.

Morton R L, Schroeder H E, Bateman K S, Chrispeels M J, Armstrong E and Higgins T J V (2000) Bean α-amylase inhibitor 1 in transgenic peas (*Pisum sativum*) provides complete protection from pea weevil (*Bruchus pisorum*) under field conditions. *Proc. Natl. Acad. Sci. USA*, **97**: 3820–3825.

Murashige T and Skoog F (1962) A revised medium for rapid growth and bioassays with tobacco tissue cultures. *Physiol. Plant*, **15**: 437–498.

Murthy B N S, Victor J, Singh R P, Fletcher R A and Saxena P K (1996) *In vitro* regeneration of chickpea (*Cicer arietinum* L.): stimulation of direct organogenesis and somatic embryogenesis by thidiazuron. *Plant Growth Regul.*, **19**: 233–240.

Naik S V, Katiyar S K and Moss J P (1994) A new hybrid: *C. arietinum* × *C. pinnatifidum* obtained through embryo-rescue. In: *Proc. of the Second Asia-Pacific Conference on Agricultural Biotechnology*. March 1994, Madras, India, 163, 6–10.

Neelam A, Reddy C S and Reddy G M (1986) Plantlet regeneration from callus cultures of *Cicer arietinum* L. *Intl. Chickpea Newslett.*, **14**: 12–13.

Prakash S, Chowdhury J B, Jain R K and Chowdhury V K (1992a) Factors affecting plant regeneration in chickpea, *Cicer arietinum* L. *Indian J. Exp. Biol.*, **30**: 1149–1153.

Prakash S, Chowdhury J B, Yadav N R, Jain R K and Chowdhury V K (1994) Somatic embryogenesis in suspension cultures of chickpea. *Annals of Biol.*, **10**: 7–14.

Prakash S, Yadav R C, Chowdhury J B and Sareen P K (1992b) Transformation of *Cicer arietinum* L. with *Agrobacterium rhizogenes. Intl. Chickpea Newslett.*, **26**: 13–14.

Patil P, Chandra R, Khetrapal S and Polisetty R (1998) Influence of polyamines and ethylene inhibitors on somatic embryo induction in chickpea (*Cicer arietinum* L.). *Indian J. Plant Physiol.*, **3**: 26–31.

Polisetty R, Patil P, Deveshwar J J, Khetarpal S and Chandra R (1996) Rooting and establishment of *in vitro* shoot tip explants of chickpea (*Cicer arietinum* L.). *Indian J. Exp. Biol.*, **34**: 806–809.

Polisetty R, Paul V, Deveshwar J J, Khetrapal S, Suresh K and Chandra R (1997) Multiple shoot induction by benzyladenine and complete plantlet regeneration from seed explants of chickpea (*Cicer arietinum* L.). *Plant Cell Rep.*, **16**: 565–571.

Ramana R V, Venu C H, Jayashree T and Sadanandam A (1996) Direct somatic embryogenesis and transformation in *Cicer arietinum* L. *Indian J. Exp. Biol.*, **34**: 716–718.

Rao B G and Chopra V L (1989) Regeneration in chickpea (*Cicer arietinum* L.) through somatic embryogenesis. *J. Plant Physiol.*, **134**: 637–538.

Reinert J (1958) Morphogenese und ihre Kontrolle an Gewebekulturen aus Karotten. *Naturwiss.*, **45**: 344–349.

Sagare A P and Krishnamurthy K V (1991) Protoplast regeneration in chickpea (*Cicer arietinum* L.). *Indian J. Exp. Biol.*, **29**: 930–932.

Sagare A P, Suhasini K and Krishnamurthy K V (1995) Histology of somatic embryo initiation and development in chickpea (*Cicer arietinum* L.). *Plant Sci.*, **109**: 87–93.

Sagare A P, Suhasini K and Krishnamurthy K V (1993) Plant regeneration via somatic embryogenesis in chickpea (*Cicer arietinum* L.). *Plant Cell Rep.*, **12**: 652–655.

Sagare A P, Suhasini K, Krishnamurthy K V and Kishore P B K (1999) Comparative study of the development of zygotic embryos of chickpea (*Cicer arietinum* L.). *Proc. of a Symposium Held at Hyderabad, India,* 29–31 Jan. 1997, 56–63.

Shade R E, Schroeder H E, Pueyo J J, Tabe L M, Murdock L L, Higgins T J V and Chrispeels M J (1994) Transgenic pea seeds expressing the α-amylase inhibitor of the common bean are resistant to bruchid beetles. *Bio/Technology*, **12**: 793–796.

Shankar S and Mohan Ram H Y (1993) Induction of somatic embryogenesis in the tissue cultures of chickpea (*Cicer arietinum* L.). *Proc. Indian Natl. Sci. Acad.*, **59**: 629–632.

Sharma L and Amla D V (1998) Direct shoot organogenesis in chickpea (*Cicer arietinum* L.). *Indian J. Exp. Biol.*, **36**: 605–609.

Shri P V and Davis T M (1992) Zeatin induced shoot regeneration from immature chickpea (*Cicer arietinum* L.) cotyledons. *Plant Cell Tiss. Org. Cult.*, **28**: 45–51.

Siefkes-Boer H J, Noonan M J, Bullock D V and Conner A J (1995) Hairy root transformation system in large seeded grain legumes. *Israel J. Plant Sci.*, **43**: 1–5.

Singh K B (1997) Chickpea (*Cicer arietinum* L.). *Field Crops Res.*, **53**: 161–170.

Singh K B and Ocampo B (1993) Interspecific hybridization in annual *Cicer* species. *J. Genet. Breed.*, **47**: 199–204.

Singh K B and Ocampo B (1997) Exploitation of wild *Cicer* species for yield improvement in chickpea. *Theor. Appl. Genet.*, **95**: 191–222.

Singh K B and Reddy M V (1991) Advances in disease-resistance breeding in chickpea. *Adv. Agron.*, **45**: 191–222.

Singh K B, Reddy M V and Malhotra R S (1985) Breeding kabuli chickpeas for high yield, stability and adaptation. In: *Proceedings, Faba Beans, Kabuli Chickpeas and Lentils in the 1980s* (Eds Saxena M C and Varma S), ICARDA, Allepo, 71–90.

Singh R P and Singh B D (1989) Recovery of rare interspecific hybrids of gram *C. arietinum* × *C. cuneatum* L. through tissue culture. *Curr. Sci.*, **58**: 874–876.

Srinivasan, Mohapatra T and Sharma R P (1991) *Agrobacterium*-mediated genetic transformation of chickpea (*Cicer arietinum* L.). *Indian J. Exp. Biol.*, **29**: 758–761.

Stewart N C, Adang M J, All J N, Boerma H R, Cardineau G, Tucker D and Parrot W A (1996) Genetic transformation, recovery and characterization of fertile soybean transgenic for a synthetic *Bacillus thuringiensis cryIAc* gene. *Plant Physiol.*, **112**: 121–129.

Subhadra, Vashishth R K, Chowdhury J B, Singh M and Sareen P K (1998) Multiple shoots from cotyledonary node explants of non-nodulating genotype (ICC435M) of chickpea, *Cicer arietinum* L. *Indian J. Exp. Biol.*, **36**: 1276–1279.

Sudha Vani A K and Reddy V D (1996) Morphogenesis from callus cultures of chickpea, *Cicer arietinum* L. *Indian J. Exp. Biol.*, **34**: 285–287.
Suhasini K, Sagare A P and Krishnamurthy K V (1994) Direct somatic embryogenesis from mature embryo axes in chickpea (*Cicer arietinum* L.). *Plant Sci.*, **102**: 189–194.
Suhasini K, Sagare A P, Sainkar S R and Krishnamurthy K V (1997) Comparative study of the development of zygotic and somatic embryos of chickpea (*Cicer arietinum* L.). *Plant Sci.*, **128**: 207–216.
Tayyar R I and Waines J G (1996) Genetic relationships among annual species of *Cicer* (Fabaceae) using isozyme variation. *Theor. Appl. Genet.*, **92**: 245–254.
Van der Measen L J G (1972) *Cicer* L. A monograph of the genus with special reference to chickpea (*Cicer arietinum* L.), its ecology and cultivation. Maded, Landbouw, Wageningen, 72–100.
Van der Measen L J G (1987) Origin, history and taxonomy of chickpea. In: *The Chickpea* (Eds Saxena M C and Singh K B), CAB International, Wallingford, UK, 11–34.

IN VITRO REGENERATION AND GENETIC TRANSFORMATION OF COWPEA, MUNGBEAN, URDBEAN AND AZUKI BEAN

LINGARAJ SAHOO, TWINKLE SUGLA AND PAWAN K. JAIWAL*
Plant Genetic Engineering Lab, Department of Biosciences,
M.D. University, Rohtak – 124 001, India
**e-mail: pkjaiwal@yahoo.com*

Abstract

Vigna species contribute a vital component of vegetarian diet of the resource poor population of underdeveloped and developing countries. Focus is to increase its yield by elevating and/or imparting resistance for diseases, insect pests and rescuing from natural genetic inadequacies. Absence of sufficient and satisfactory level of genetic variability within the germplasm has been the major hurdle in their improvement by conventional breeding. The immense potential of biotechnological tools to supplement breeding programmes to bring into elite germplasm of *Vigna* species is being realized. Much progress has been made in *in vitro* regeneration of sexually mature plants, in majority of these recalcitrant species, from various tissues and considerable success in generating plants from protoplast of *Vigna aconitifolia*. While attempts are being made to develop suitable transformation protocols for most of these species, commendable success has been achieved in *Vigna angularis* by generation of transgenics resistant to storage pests. An overview of the *in vitro* regeneration and critical analysis of the genetic transformation studies in major *Vigna* species is presented. The bottlenecks to overcome their recalcitrance *vis-à-vis* the advantages and limitations of all those transformation techniques applicable, along with future directions in research, are discussed.

1. Introduction

The genus *Vigna* consists of eight subgenera (Verdcourt, 1969) and about 150 species, the majority of which are found in Asia and Africa. Only seven species of *Vigna* are cultivated, five Asian of subgenus *Ceratotropis*, *Vigna radiata*, *V. mungo*, *V. aconitifolia*, *V. angularis* and *V. umbellata*, and two African of subgenus, *Vigna*, *V. unguiculata* and *V. subterranean* (Table 1). In addition, there are a few wild or semi-wild species, *V. trilobata*, *V. vexillata*, *V. glabrescens* and *V. hosei*. The members of genus *Vigna* were formally placed in the genus *Phaseolus*. These species of *Vigna* are now cultivated throughout the world especially in

P. K. Jaiwal and R. P. Singh (eds.), Applied Genetics of Leguminosae Biotechnology, 89–120.

Table 1. General information about different species of Vigna

Species name/ common name	Adaptation	Edible parts and uses	Area of cultivation	Negative factors
Vigna aconitifolia (Jacq.) Marechal *syn. Phaseolus aconitifolia* (Jacq.) Moth bean, mat bean	Highly drought resistant plants, suited for arid and semi-arid conditions	Grown as pulse, a green pod vegetable, forage and green manure	India, Arabia, Iran, Indonesia, South China, North East Africa, South West USA, Japan, Korea and China	Indeterminate, late in maturity, prone to shattering, susceptible to Yellow mosaic, root-knot nematode, photosensitive
V. angularis (Willd.) Ohwi and Ohashi *syn. Phaseolus angularis* W. F. Wright Azuki bean	Subtropical countries, grown at high altitude	Seeds, bean jam and confectionary	Japan	Sensitive to insect pests, diseases and cold weather, poor yield
V. radiata (L.) Wilczek *syn. Phaseolus radiatus* L., *Phaseolus aureus* Roxb. Mung bean, green gram	Arid, semiarid and humid regions (widely adaptable)	Sprouted seeds, boiled seeds	South East Asia	Susceptible to Yellow mosaic virus, Powdery mildew, Cercospora leaf spot, insect pests
V. mungo (L.) Hepper *syn. Phaseolus mungo* Urd, Black gram	Dry conditions (more drought resistant than mung bean)	Seeds boiled or ground	South East Asia	Susceptible to Yellow mosaic virus, Powdery mildew, Cercospora leaf spot, insect pests
V. umbellata (Thumb.) Ohwi and Ohashi *syn. Phaseolus calcaratus* (Roxb.) rice bean, red bean, pegin bean	Drought resistant	Seeds boiled, edible pods and leaves	India, China and Malaysia	Poor yield
V. trilobata syn. Phaseolus trilobatus (L.) Schreb Jungli bean	—	Edible pods, boiled seeds and leaves	India and Sri Lanka	Poor yield
V. unguiculata (L.) Walps Black-eyed bean, catyang	Semi-arid tropics	Fresh vegetable (leaves, young pods and peas) and seeds	West and Central Africa, South America, Asia	Pathogens, insect pests, parasitic weeds
V. subterranea (L.) Verdc Bambara, ground nut	Semi-arid tropics	Fresh vegetable (leaves, young pods and peas) and seeds	Western Africa	—

the tropics and subtropics of Africa, South East Asia, particularly India, Pakistan, Bangladesh, Burma, Philippines, Thailand, Indonesia, and in China, North America and Australia. This indicates that *Vigna* species are adapted to a wide range of agro climatic conditions and also grow on poor marginal soils.

V. unguiculata, *V. radiata* and *V. mungo* are the widely cultivated species of the *Vigna* group. All *Vigna* species are annual, nitrogen fixing, generally short duration plants, and therefore, are widely grown as mixed intercrop or in rotation to improve nitrogen status of soil and to break disease/pest cycles. These are primarily grown for their protein rich edible seeds and sprouts although the immature fruits and seeds are occasionally used as green vegetables and the plants are also used as fodder or green manure. The seeds are favoured due to their easy digestibility, low production of flatulence, rich in phosphorus and pro-vitamin A and relatively free from the anti-nutritional factors. Their high protein levels and high lysine/low methionine amino acid profile complement the high carbohydrate and low lysine/high methionine content of cereals to form a balanced amino acid diet.

Seeds of *V. radiata* on germination produce sprouts, rich in thiamine, riboflavin, niacin and ascorbic acid concentration, and are preferred throughout most of Asia, Europe and North America. In Japan, *V. mungo* is favoured for the production of bean sprouts and the recent growth of Thai and Australian industries are directed towards their market. *V. angularis* is grown in Japan for bean jam and confectionary.

The production of *Vigna* species has not improved significantly in spite of the best efforts of the plant breeders. Among the factors responsible for reduction and uncertainty in yield besides the abiotic salinity and drought, etc. factors, are their prominent susceptibility to several fungal, bacterial or viral pathogens and insects. The other yield limiting factors are their indeterminate growth habit, defective plant type, excessive vegetable growth, low harvest index, sensitivity to photoperiod and temperature, asynchronous flowering and shattering of pods (Table 1).

Diseases are the major constraints to yield throughout Asia (Kim, 1994). The three diseases which are responsible for considerable yield losses in these crops are (i) yellow mosaic, a viral disease transmitted by white flies; (ii–iii) fungal diseases, i.e. powdery mildew caused by *Erysiphe polygoni* and cercospora leaf spot caused by *Cercospora canescens* (Jeswani and Baldev, 1990).

Vigna species are susceptible to a wide range of insect pests, which feed on flowers, fruits and seeds. Foremost among these are the pod boring caterpillars such as *Heliothis* spp. and *Etiella* spp., the bean pod borer (*Maruca testulalis*) and pod sucking bugs such as *Nazara viridula*. The storage pest bruchids (*Callosobruchus* spp.) can cause complete damage to grains if stored uninterrupted for three months. Chemical control measures for diseases and pests are expensive and ineffective. Moreover, the increasing use of chemicals in agriculture is affecting the quality of the product as well as posing a threat to the ecosystem (Fillippone, 1993).

Strategies to improve resistance against diseases and insect pests by classical breeding has been slow due to the lack of a desirable and satisfactory level of variability in germplasm. Sources of resistance for the diseases and insects have been identified in few wild or closely related species, but these were not included in breeding programmes due to sexual incompatibility with the cultivated species. The other alternative is to transfer desirable genes from alien sources in order to increase the production of these crops to combat the shortage in food supplies, especially the vegetable protein. The success of any gene transfer technique depends upon efficient, reproducible and rapid *in vitro* regeneration system. However, grain legumes in general and species of *Vigna* in particular has remained recalcitrant to regeneration *in vitro* (Christou, 1994, 1995; Jaiwal and Gulati, 1995; Nagl *et al.*, 1997). Recently, some progress has been made in both regeneration and transformation of different *Vigna* species (Sharma *et al.*, 1998; Das *et al.*, 1998; Avenido and Hattori, 1999; Sahoo *et al.*, 2000; Jaiwal *et al.*, 1998, 2001; Yamada *et al.*, 2001; Tivarekar and Eapen, 2001). The present review provides an overview of tissue culture and genetic transformation undertaken for the improvement of *Vigna* species and also outlines the future perspectives.

2. *In vitro* regeneration

Studies on *in vitro* culture of *Vigna* spp. are summarized in Table 2. The main areas in which *in vitro* studies on *Vigna* have been centered are: (1) establishment of callus cultures, (2) regeneration of plants (via direct shoot organogenesis or organogenesis from callus), (3) somatic embryogenesis, (4) anther culture, (5) wide hybridization, (6) protoplast culture and somatic hybridization, (7) cell suspension culture, and (8) genetic transformation.

2.1. *VIGNA UNGUICULATA*

2.1.1. Direct shoot organogenesis

In an attempt to develop cryopreservation methods for germplasm storage, Kartha *et al.* (1981) described a method for cowpea regeneration from shoot apical meristem using a simple procedure, which was also found to be applicable to other legumes including soybean, peanut, chickpea and bean. Whereas most other legumes required the presence of a cytokinin, and in some cases an auxin, for efficient regeneration, cowpea meristems did not require exogenously supplied hormones for maximum response. Since plant cells are not generally known to proliferate in the absence of exogenously supplied phytohormones, autonomous regenerability of cowpea meristems was attributed to endogenous hormones synthesized in the cowpea explant by primordial leaves. Interestingly, at low level of exogenously supplied cytokinin N^6-benzylaminopurine (BAP) at 0.1–0.001 μM), meristem differentiated into whole plants at higher frequencies (90–100%). However, by increasing the levels of BAP to 0.5 μM, plant regeneration was reduced to 21%. The authors concluded that this response strengthened the assumption that endogenous hormones in cowpea meristems were adequate to effect plant regeneration. Unlike soybean and bean meristems, cowpea did not regenerate multiple buds. However, recently multiple shoots have been developed from shoot apical meristems of *V. unguiculata* on B_5 medium supplemented with BAP (5 μM) (Sahoo *et al.*, 2000).

Muthukumar *et al.* (1996) cultured mature de-embryonated cotyledons with intact proximal end on B_5 basal medium containing BAP (8 μM); 36% of cultures produced multiple shoots with an average of 6 shoots per explant.

Cowpea mature cotyledons and hypocotyls were induced to form multiple shoots within 45 days from the wounded region of the cotyledons and upper part of primary

Table 2. Studies on growth responses of explants of various genotypes of Vigna mungo

Variety	Basal medium	Growth regulators (μM)	Explant	Morphogenic responses	References
T-9, No. 55	MS	BAP 44.4	Cotyledon with cotyledonary node	Multiple buds	Gill *et al.*, 1987
	MS	Picloram 4.14 + Zeatin 4.5 + IAA 0.57		Callus	
	MS liquid	Picloram 2.48–0.24 + Zeatin 4.5 + IAA 0.57	Cotyledonary node	Somatic embryoids, callus	
LBG	MS salts + B_5 vitamins	2,4-D 13.5 + NAA 10.8 + Kin 2.32	Hypocotyl	Callus	Geetha *et al.*, 1997
	MS liquid	2,4-D 6.75 + BAP 2.22	Hypocotyl callus	Somatic embryos in cell suspension culture	
LBG	MS salts + B_5 vitamins	NAA 16.1 + BAP 2.2	Cotyledonary node, Hypocotyl, Epicotyl, Axillary bud and Immature leaf	Callus	Geetha *et al.*, 1997
	MS	NAA 2.6 + BAP 6.6	Callus of cotyledonary node, Hypocotyl, Epicotyl, Axillary bud and Immature leaf	Shoots	
Vamban-1	MS salts + B_5 vitamins	BAP 13.3	Cotyledonary node with both cotyledons	Multiple shoots	Ignacimuthu *et al.*, 1997
	MS (1/2 inorganic + full organic)	IBA 0.49	Regenerated shoots	Rooting and *in vitro* fruiting	
T-9, Pusa-1 Pusa-2	MS	BAP 8.8 or TDZ 2.27	Seedling (without hypocotyl and root)	Multiple shoots	Das *et al.*, 1998
	MS $^1/_2$ strength	NAA 0.54	Stem with apex from axillary shoots	Callus, shoot buds and roots	
PS-1, Tm-9, PU-19, PU-35	B_5	Kinetin 9.29	Cotyledonary node	Multiple shoots	Sen and Mukherjee, 1998
	MS salts + B_5 vitamins	TDZ 0.045	Hypocotyl Cotyledon	Callus Callus and shoots	Sen and Mukherjee, 1998

Table 2. (Continued)

Variety	Basal medium	Growth regulators (μM)	Explant	Morphogenic responses	References
Vamban-1	MS salts + B_5 vitamins	BAP 13.31 + NAA 0.161 + Proline 12	Cotyledon and Embryonal axis	Multiple shoots	Ignacimuthu and Franklin, 1999
T-9, PU-19, PU-30	MS	BAP 4.4	Cotyledonary node	Multiple shoots	Hinger and Pareek, 1999
—	MS	BAP 8.87 + IAA 2.85	Primary leaves	Callus	Ignacimuthu and Arockiasamy, 1999
	MS	BAP 13.3 + TDZ 2.27 + IAA 0.57	Primary leaf callus	Shoots	—
	MS	GA_3 2.89 and/or BAP 0.44	Differentiated callus	Somatic embryos	—
VBN-1	MS salts + B_5 vitamins	BAP 13.31	Cotyledonary node	Multiple shoots	Franklin and Ignacimuthu, 2000

hypocotyl and cotyledons in different media containing a high cytokinin concentration. The induced explants on transfer to the same media for 2 days in light developed numerous shoot buds (from the upper part of the hypocotyl and wounded part of the cotyledons), which subsequently developed into plants. These buds emerged *de novo*, as they had no apparent vascular connection with the parent tissue. The developed protocol was found applicable to five other cowpea genotypes. Addition of putrescine to high cytokinin medium enhanced the shoot forming response of the explants (Pellegrineshi, 1997).

2.1.2. *Organogenesis from callus*

Embryonic axes and cotyledons (excised from immature seeds) cultured on B_5 medium containing 1 mg/L 2,4-dichlorophenoxyacetic acid (2,4-D) produced light yellow to cream colored calli with highest fresh weight. Casein hydrolysate (4 g/L) and glutamine (2 g/L) greatly promoted callus growth. The effect of various factors on callus growth and polyphenol accumulation was also studied. Polyphenol content increased with 2,4-D concentration above 1 mg/L. Glutamine followed by casein hydrolysate and coconut water (100 ml/L) showed lowest polyphenol accumulation (Singh *et al.*, 1982).

Pandey and Bansal (1989) reported plantlet formation from callus cultures of cowpea, derived from leaf explants on MS medium containing 100 μM indole-3-butyric acid (IBA) and 10 μM kinetin, but not on medium with indole-3-acetic acid (IAA), 2,4-D and kinetin. The regenerated plants showed among the normal diploid chromosome number with an occurrence of about 1% aneuploid ones.

Muthukumar *et al.* (1995) made use of primary leaf explants including the intact petiole, in order to regenerate plants of *V. unguiculata*. When cultured on B_5 medium supplemented with 0.8 μM 2,4,5-trichlorophenoxyacetic acid (2,4,5-T), 10 mM L-glutamine and 0.1 mM adenine sulphate, callus developed at the petiole ends. Shoot regeneration occurred, when this callus was sub-cultured on B_5 medium without growth regulators. The shoots were rooted and established in soil.

Plants regenerated from callus, cell suspensions are known to exhibit somaclonal variation, which may be genetic or epigenetic in nature. Potentially useful somaclonal variants have been identified in many crop plants including legumes (Jaiwal and Gulati, 1995; Sonia *et al.*, 2000; Singh *et al.*, 2002). It remains to harness the potential of somaclonal variations extending further the gene pool of the cowpea.

2.1.3. Somatic embryogenesis

In *V. unguiculata* leaf explants from seedlings on MS medium containing 2,4-D (2 mg/L) developed callus. Suspension cultures were established from the callus using the same medium and development of somatic embryos up to cotyledonary stage was observed. Plantlets were obtained after transfer of the cotyledonary stage embryos on MS medium supplemented with 0.5 mg/L 2,4-D (Kulothungan *et al.*, 1995).

2.1.4. Protoplast culture

Bharal and Rashid (1980) have used hypocotyl, stem, and stem callus and leaf of *V. unguiculata* seedlings to identify the source that yields maximum viable protoplasts. They have reported that hypocotyl and stem yield 70–80% viable protoplasts, in contrast to 20% in leaf tissue. The enzyme mixture comprised of macerozyme R-10, cellulase R-10, 13% mannitol and the salts (mg/L) – KH_2PO_4 – 27.2, KNO_3 – 101.0, $CaCl_2.2H_2O$ – 148.0, $MgSO_4.7H_2O$ – 240.0, KI – 0.16, $CuSO_4.5H_2O$ – 0.025. For obtaining protoplasts from hypocotyl and stem segments, 8 h incubation with 1.5% macerozyme and 2% cellulase was adequate. However, for leaves and callus tissue the incubation time had to be doubled and the cellulase concentration enhanced. Jha and Roy (1980) have also isolated the protoplasts from leaves of cowpea by treating with the enzyme mixtures of meicelase 4% and macerozyme 0.5%. The protoplasts rapidly developed into vigorously growing protoclones (callus) on reducing the mannitol level in the media after a short interval. Gill *et al.* (1987a) studied callus induction from protoplasts of *V. unguiculata*, *V. sublobata* and *V. mungo*. They did not succeed in obtaining plants from protoclones. Plants of *V. sinensis* were regenerated from protoplasts of immature cotyledons via callus formation and somatic embryogenesis by a series of media changes. On medium containing MS salts, B_5 vitamins, NaCl (500 mg/L), casein hydrolysate (500 mg/L), 2,4-D (2 mg/L) and BAP (0.5 mg/L), about 5% of the calli become embryogenic. When these calli were transferred to medium containing MS salts, B_5 vitamins, IAA (0.1 mg/L), kinetin (0.5 mg/L), 3–5% mannitol and 2% sucrose in light, many somatic embryos originated and some of them developed into shoots or plants. After rooting on $^1/_2$ strength MS medium supplemented with IBA (0.1–0.3 mg/L), the plants grew well in soil and were fertile.

Plant regeneration system from protoplasts still needs to be perfected in *V. unguiculata* so as to obtain interspecific and intergeneric somatic hybrids of wild and related species and transgenic plants by protoplast transformation.

2.1.5. Anther culture

V. unguiculata young anthers containing thin walled microspores at uninucleate stage on semi-solid medium supplemented with 1 mg/L kinetin/BAP and 0.5 mg/L IAA or IBA under weak light (250 lux) or continuous dark conditions for initial 20 days at relatively low temperature (23 ± 1 °C) produce callus in 50–60% of cultures (Arya and Chandra, 1989). However, those cultured in liquid medium under continuous light conditions failed to induce anther callusing. The anther callus sub-cultured either in liquid or semi-solid MS medium with 1 mg/L BAP and 0.5 mg/L IBA or MS with varied concentrations of growth regulators could not regenerate shoots (Arya and Chandra, 1989). The exact nature of anther callus and its origin could not be traced out.

2.1.6. Wide hybridization

Improvement of *V. unguiculata* through the introgression of desirable traits from wild non-cultivated *Vigna* species, such as *V. vexillata* has been hampered due to incompatibility between them. Interspecific crossing has been attempted between *V. unguiculata* and *V. vexillata* and fertilized ovules were achieved, but embryos abort shortly after their formation, suggesting the failure of endosperm to develop following fertilization. A defined embryo rescue medium could substitute for the failed endosperm by providing the interspecific hybrid embryo with the necessary nutrients for the *in vitro* development. All attempts to rescue hybrid embryos *in vitro* failed mainly due to technical difficulty of excising the embryos at an early stage and the physiological needs of these embryos in respect to nutrients and growth regulators to enable development.

Although it was observed that commonly used culture media did not support the growth of very young excised embryos of *Vigna* spp.; media have now been formulated which reliably support the development and sprouting of 4 day-old excised embryos resulting from self-pollination of *V. unguiculata* and *V. vexillata* respectively. Relative to standard tissue culture media, these media contain only small amounts of nitrogen, low concentrations of sucrose (3% or less), a cytokinin (zeatin) as growth hormone, and a balance of mineral nutrients which mimics the composition and concentration of minerals in the cotyledons of cowpea grain (Pellegrineshi *et al.*, 1997). Although successful rearing of excised 4 day-old selfed embryos *in vitro* of both species is now routine, germ lines from interspecific hybridizations have not been obtained, in spite of sustained efforts at IITA; Pellegrineshi *et al.* (1997). Therefore, in addition to embryo rescue, other approaches in gene technology are being perused in the effort to overcome interspecific incompatibility.

2.2. *VIGNA MUNGO*

2.2.1. Direct shoot organogenesis

Cotyledons with intact cotyledonary nodes, excised from 1 day-old seedlings, on culture either on moist filter paper or water agar or MS basal medium developed into single shoots, which subsequently developed roots resulting into plantlets. Addition of various cytokinins at 1 mg/L to MS basal medium induced multiple shoot buds originating from cells adjacent to the existing buds (Gill *et al.*, 1987b). Excised mature cotyledons produced multiple shoot buds via *de novo* meristem formation in sub-epidermal cell layers on MS salts + B_5 vitamins medium supplemented with thidiazuron (TDZ, 0.05 mg/L) (Sen and Guha-Mukharjee, 1998). TDZ induced shoots were incapable of elongation and therefore, once the shoot initiation was observed, the explants were transferred to basal medium with reduced hormones for proper growth of regenerated shoots as well as induction of more shoot initials. The mode of action of TDZ is yet not known but there is a suggestion that TDZ increases the biosynthesis as well as accumulation of endogenous purine, cytokinins leading to plantlet differentiation (Thomas and Katterman, 1986). Seed derived mature cotyledon and embryonic axis explants produced multiple shoot initials on MS medium containing 1 $\times$ macronutrients and 4 $\times$ micronutrients with B_5 vitamins supplemented with 12 mM proline and BAP (13.3 μM) and naphthalene acetic acid (NAA, 0.16 μM) within 15 days of dark incubation. The shoot initials developed into shoots on transfer to MS basal medium (Ignacimuthu and Franklin, 1999). The exposure of the explants to dark period for the initial induction of shoot formation was essential. Cotyledons showed much higher frequency of shoot regeneration as well as the number of shoots per explants than embryonic axes. On cotyledon explants, shoots developed from pre-existing meristem while on embryonal axes, shoot appeared to develop *de novo* from sub-epidermal parenchyma cells (Ignacimuthu and Franklin, 1999).

Cotyledonary node with or without cotyledon and/or shoot tip developed multiple shoot initials/shoots on MS salts, B_5 vitamins and 13.3 μM BAP (Sen and Guha-Mukharjee, 1998; Franklin and Ignacimuthu, 2000). The multiple shoot initials developed into shoots on transfer to basal medium with a low concentration of BAP or kinetin (0.1 mg/L) (Sen and Guha-Mukharjee, 1998). In another study, kinetin was found to be more effective than BAP for shoot induction (Sen and Guha-Mukharjee, 1998). Multiple shoot formation was dependent on the number and size of cotyledons, presence or absence of shoot tip and genotype. The explant with both the cotyledons but lacking shoot tip was found to be more efficient for regeneration than those with one or no cotyledons. Complete removal of both cotyledons caused delayed response of shoot regeneration and multiple shoots were not produced. It appears that cotyledons provide endogenous signals for bud development following decapitation of shoot tip. The morphology of leaves was also dependent on initial explants (presence/absence of cotyledons). In the presence of cotyledons, the axillary shoots produced from node were healthy and possessed trifoliate leaves whereas in the absence of cotyledons the shoots were weak and possessed a simple leaf.

2.2.2. Shoot organogenesis from callus

Geetha *et al.* (1997b) induced calli from different seedling explants (hypocotyl, epicotyl, axillary bud, cotyledonary node and immature leaf) on MS salts + B_5 vitamins medium supplemented with 2.2 μM BAP in combination with 22.8 μM IAA or 16.1 μM NAA. Among the explants, hypocotyl was found to be more efficient in producing callus. Shoots were induced from callus culture of epicotyl, cotyledonary node and immature leaf with varying frequency in the medium containing kinetin (2.3–9.3 μM) or BAP (2.2–8.8 μM) together with IAA (2.8 μM) or NAA (2.6 μM). The highest frequency of shoot bud regeneration (80.8%) and highest number of shoots per callus was observed using cotyledonary node explant with BAP (6.6 μM) and NAA (2.6 μM).

Das *et al.* (1998) reported development of shoot buds via callus from the stem (with or without apex) and petiole explants excised from the axillary shoots developed at nodes of seedlings germinated on MS medium containing TDZ (0.5 mg/L). The explants produced callus along with shoot buds on $^1/_2$ MS supplemented with 0.1 mg/L of NAA. Full strength MS salts in the regeneration medium inhibited the shoot bud formation. The nature of explants and pH of medium influenced the efficacy of the regeneration medium. The stem explants proved superior to petiole explants for shoot induction. Maximum number of shoot buds was obtained on medium with pH adjusted to 5.5. Seed germination in the presence of a cytokinin may have a preconditioning effect on the physiological status of the axillary shoots, which in turn yield stem, and petiole explants amenable for regeneration.

2.2.3. *Somatic embryogenesis*

Cotyledonary nodes without cotyledons from 1 day-old seedlings developed nodular calli on MS + picloram (1 mg/L) + zeatin (1 mg/L) + IAA (0.1 mg/L). Such calli on transfer to liquid medium of the same composition but with reduced concentration of picloram, developed embryoid like structures which could not develop further (Gill *et al.*, 1987b). Geetha *et al.* (1997c) demonstrated the induction of somatic embryos from hypocotyl derived cell suspension cultures of *V. mungo* cv. LBG. Callus induced on MS salts with B_5 vitamins, 2,4-D (3 mg/L), NAA (2.0 mg/L) and kinetin (0.5 mg/L) on transfer to MS liquid medium with BAP (0.5 mg/L) developed somatic embryos after two subcultures. Five to seven percent of the embryos on transfer to semi-solid MS with 0.1 mg/L BAP converted into plantlets.

2.3. *VIGNA RADIATA*

2.3.1. *Direct shoot organogenesis*

Direct regeneration of shoots without an intervening callus phase has been achieved from shoot tips (Bajaj and Dhanju, 1979; Goel *et al.*, 1983a; Singh *et al.*, 1985; Mathews, 1987; Gulati and Jaiwal, 1992), cotyledons (Mathews, 1987; Gulati and Jaiwal, 1990; Patel *et al.*, 1991) and cotyledonary nodes (Mathews, 1987; Gulati and Jaiwal, 1994) and leaf petioles (A. Gulati and P. K. Jaiwal, unpublished data). The efficiency of shoot formation and regeneration depended upon type and size of the explant, age of the donor seedlings, orientation of the explant on medium, hormonal and nutrient regimes, and genotype (Gulati and Jaiwal, 1990, 1992, 1994). Among different cytokinins, BAP alone was found to be the most effective in inducing multiple shoots at high frequency. However, the optimal concentration of BAP varied with the type of explants. Addition of auxins to BAP containing medium did not improve the efficiency of BAP for shoot induction. Cotyledonary node explants with both the cotyledons excised from 4 day-old *in vitro* raised seedlings produced the highest number of shoots on B_5 medium supplemented with 0.5 μM of BAP. The rooted shoots (plantlets) were successfully established in soil with 60–90% success (Gulati and Jaiwal, 1990, 1992, 1994; Avenido and Hattori, 1999).

Histological observations of cotyledons cultured *in vitro* revealed that shoots arise from epidermal and sub-epidermal cells present on the basal adaxial region of the petiole residue (Mendoza *et al.*, 1993). Chandra and Pal (1995) reported the differential regeneration responses of two mungbean cotyledons of mature seed, i.e. cotyledons, that were loosely attached to the embryonal axis, were found more responsive to shoot regeneration.

The regeneration from cotyledon and cotyledonary node explants was highly site specific. Shoots developed from the adaxial side at proximal ends of the cotyledons and from the nodal regions in cotyledonary node (Gulati and Jaiwal, 1990, 1994; Avenido and Hattori, 1999; Chandra and Pal, 1995). Tivarekar and Eapen (2001) reported high frequency plant regeneration from entire surface of immature cotyledons on MS medium supplemented with BA (2 mg/L) and IAA (0.5 mg/L).

The shoots were rooted on either MS (half- to full-strength basal medium) (Gulati and Jaiwal, 1992) or with addition of IAA (Gulati and Jaiwal, 1990, 1994; Avenido and Hattori, 1999) or IBA at 1.0 mg/L (Jaiwal *et al.*, 2001).

Histological observations were conducted to evaluate the origin, initiation and development of callus in mungbean cotyledon culture (see review Jaiwal and Gulati, 1995 and references therein). It has been reported that the endogenous callus formation is due to the rapid cell division of cells in the peripheral zone of the pro-vascular tissue of cotyledons and if this callus type is cultured in an appropriate culture medium and condition, could possess high regeneration efficiency.

In case of cotyledonary nodes, differentiation of shoots occurred directly from epidermal and sub-epidermal cells without callus formation at the node. These explants have been suggested for genetic transformation of *V. radiata* as regenerable cells are mainly surface cells, readily accessible to *Agrobacterium* (Gulati and Jaiwal, 1994).

2.3.2. Organogenesis from callus

Calli derived from hypocotyl and root undergoes rhizogenesis only (Singh *et al.*, 1985; Gulati, 1993). Shoot organogenesis has been reported from callus derived from immature primary leaves (Mendoza *et al.*, 1992) or from leaves of 7 day-old seedlings (Patel *et al.*, 1991; Gulati and Jaiwal, 1993a). Regeneration of shoots from callus derived from cotyledons, cotyledonary nodes, stem and root segments has not been reported (Mathews, 1987; Patel *et al.*, 1991; Gulati and Jaiwal, 1990, 1992, 1994).

2.3.3. Somatic embryogenesis

In *Vigna radiata*, immature cotyledons and leaf explants have been reported to be the most responsive explants in cultures producing somatic embryos (Eapen and George, 1990; Patel *et al.*, 1991). Somatic embryos have developed up to globular (Patel *et al.*, 1991) or cotyledonary stages (Eapen and George, 1990). However, they could not achieve maturation and conversion of somatic embryos into plants (Eapen and George, 1990). Somatic embryogenesis also occurred on cotyledons of interspecific embryos of *V. glabrescens* and *V. radiata* crosses, when cultured on MS medium without growth regulators (Chen *et al.*, 1990). Recently, Girija *et al.* (2001) induced embryogenic callus (yellowish nodular friable) from immature cotyledons on MS medium with NAA 5 mg/L which on transfer to MS liquid medium containing 2,4-D (1.5 mg/L) along with proline (50 mg/L) produced globular, heart and torpedo shaped embryos. The torpedo embryos on transfer to MS liquid medium undergo maturation and germination. The germinated embryos developed further on transfer to MS half strength basal (solid) medium.

2.3.4. *Anther culture*

Anthers of *V. radiata* cultured at uninucleate stage on MS supplemented with IAA, 2,4-D and kinetin (each at 10 μM) produced callus in 28.5% of the cultures, out of which only 1.98% showed pollen embryos (Bajaj and Singh, 1980). However, reduction in concentration of 2,4-D and kinetin to 1 μM resulted callusing in 78% of the anthers (A. Gulati, unpublished data). Regeneration of plants from such calli has not been obtained.

2.3.5. *Wide hybridization*

Mungbean improvements through the introgression of desirable genes from other Asiatic *Vigna* species and wild relatives have been hampered due to incompatibility between them. The incompatibility has been partly due to hybrid inviability and partly due to chromosomal differentiation (Rashid *et al.*, 1988). Wild mungbean, *Vigna radiata* subspecies *sublobata* hybridizes readily with cultivated mungbean (Ahuja and Singh, 1977; Singh *et al.*, 1983; Kitamura *et al.*, 1988) and constitutes the primary gene pool of the cultigen. Bruchid resistance gene from wild mungbean has been incorporated into cultigen by conventional breeding (Tomooka, 1992). *Vigna mungo* constitutes the secondary gene pool, and *V. umbellata* and *V. angularis* make up the tertiary gene pool (Smartt, 1990). However, interspecific hybrids between *V. radiata* and *V. angularis*, *V. mungo* and *V. glabrescens* and *V. trilobata* have been obtained by culturing hybrid embryos excised from immature seeds on artificial medium (embryo rescue technique).

Wild species, *V. radiata* subsp. *sublobata* and *V. vexillata* and other closely related species *V. umbellata* and *V. aconitifolia* have to be added in hybridization programme to incorporate desirable genes to mungbean.

2.3.6. *Protoplast culture and somatic hybridization*

In *V. radiata*, protoplasts have been obtained from roots (Xu *et al.*, 1981), leaves (Goel *et al.*, 1983b; Joshi and Schieder, 1987; Eapen, 1988) and hypocotyls (Eapen, 1988). Protoplasts divide to form callus (Joshi and Schieder, 1987; Eapen, 1988). Protoclones differentiated roots on MS medium supplemented with hormones. However, shoot differentiation was not obtained. Callus induction from protoplast of hypocotyl calli (Gill *et al.*, 1987a) and seedling explants (Bhadra *et al.*, 1984) was also observed in *V. sublobata*. Recently, plants have been regenerated from such calli (Bhadra *et al.*, 1994). On the other hand, protoplasts-to-plant regeneration system still needs to be perfected in *V. radiata* so as to obtain intergeneric and interspecific somatic hybrids of wild and related species and transgenic plants by protoplast transformation.

2.3.7. *Cell suspension culture*

In *V. radiata*, suspension culture was established by transferring actively growing root callus to PC-L2 liquid medium. The cells in suspension continued to grow up to 7 days, thereafter, they turned brown, probably due to the depletion of nutrients. Frequent subculture (between 4–6 days) prevented the browning of cells. There was 6.5 fold increase in number of cells in 197 hours (Chowdhury *et al.*, 1983).

2.4. *VIGNA ACONITIFOLIA*

The work on *in vitro* regeneration of *V. aconitifolia* has been described in Chapter 2 of Volume 10A.

2.5. *VIGNA ANGULARIS*

Azuki bean (*Vigna angularis* Ohwi and Ohashi) is one of the major crops in Japan. It is cultivated almost all over Japan and the main producing district is Hokkaido. This crop is widely used for the production of bean jam for confectionary in Japan. Azuki bean is susceptible to cold weather and diseases. The breeding of high yielding varieties for resistance to cold and disease are required.

Avenido and Hattori (1999) studied the differences in shoot regeneration from cotyledonary node explants of Asiatic *Vigna* species. The cotyledonary node explants of 4 cultivars of *V. angularis* excised from 4 day-old seedlings on medium MS salts, B_5 vitamins, 3% sucrose and 1.0 mg/L of BA developed neither shoot nor callus in the nodal region of the explants. However, multiple adventitious shoots developed at the basipetal cut (hypocotyl) of the explants (35.7 to 66.7%), which is in contrast to all epigeal species (*V. radiata* and *V. mungo*) where multiple shoots developed from the node. In *V. angularis*, cotyledonary–radicle junctions possessed a pair of unique clamp-like connecting tissues termed as cotyledon flap (Endo and Ohashi, 1997), as a result the cotyledons do not form abscises from the node. The shoots were rooted on MS basal medium and 60–100% of the resulting plantlets survived on establishment to soil and grow up to maturity with normal in gross morphology as compared to the seed derived plants. The plantlets attained sexual maturity and produced viable seeds (R_1 generation). Sato (1995) developed adventitious shoots from the epicotyl explants (10 mm) derived from 7 day-old *in vitro* etiolated seedlings on MS basal medium supplemented with BAP 1 mg/L. This direct organogenesis method was used in conjunction with *Agrobacterium tumefaciens* to develop transgenics, Ishimoto *et al.* (1996) introduced α-amylase inhibitor 1 gene in azuki bean to develop transgenics resistant to bruchids. Takaheshi *et al.* (1998) regenerated azuki bean plants from epicotyl and the callus derived from seedling explants.

Ge *et al.* (1989) reported a high frequency (63%) of shoot regeneration from calli derived from protoplasts of mesophyll cells of sterilized seedlings. The protoplasts showed growth and division in liquid medium with 0.5 mg/L each of 2,4-D, 6-BA and zeatin, and produced calli within 2 months which subsequently regenerated into shoots on transfer to MS with 6 mg/L of BA or MS with 5 mg/L of BA, 0.2 mg/L of NAA and 0.1 mg/L of IAA. The shoots developed into plantlets on transfer to half-strength MS with 1 mg/L IAA. The composition and concentration of sugar, microelements, vitamins and phytohormones played an important role in protoplast growth, division, callus formation and plantlet regeneration.

Sato *et al.* (1993) regenerated fertile plants from protoplasts isolated from suspension cell culture raised from the friable and soft calli induced at both cut ends of epicotyl on MS medium containing 2,4-D. The isolated protoplasts produced calli on MS medium containing 1 mg/L of 2,4-D and BAP with plating efficiency 30% after 10 day. Such calli on transfer to 2,4-D (0.1 mg/L) and BAP (1 mg/L) for 2 weeks and subsequently onto solid MS medium supplemented with BAP, Kinetin trans-zeatin and IAA produced shoots. The shoots were rooted on MS basal medium and established in pots where they produced normal seeds.

2.6. OTHER *VIGNA* SPECIES

The protoplasts isolated from mesophyll and hypocotyls of *V. umbellata* developed into protoclones, on MS medium supplemented with 0.5 mg/L each of different auxins and cytokinins, which could not regenerate into plantlets (Eapen, 1988). The calli initiated from *V. umbellata* on MS medium produced single or multiple shoots on transfer to medium supplemented with BA and NAA (Hadiuzzaman and Miah, 1989). Bhadra *et al.* (1990) induced calli from seedling protoplasts of *V. gracilis* and *V. trilobata*, and also succeeded in culturing tissues and protoplasts of *V. umbellata* (Bhadra *et al.*, 1991) and, finally, regenerated plants from protoplasts of *V. sublobata* (Bhadra *et al.*, 1994).

3. Genetic transformation of *Vigna* species

Recent advances in gene transfer techniques have opened new avenues for crop improvement complementing traditional breeding programmes. These techniques demand efficient procedures for the routine introduction of foreign DNA into plant genomes. The potential of these techniques, to introduce defined single genes in specific and controlled ways, without being accompanied by unwanted features, and with no genetic barriers as to the source of gene pool and the promise of transfer of multiple genes for any desired trait, is significant considering the limitations and time required in generating new genotypes by breeding programmes. The most widely used DNA delivery systems, which have potential practical applications, include those based on the natural gene transfer mechanism of *Agrobacterium*, with techniques such as particle bombardment and electroporation and/or chemical treatment of isolated protoplasts providing alternative approaches. Whilst these procedures differ in their way in which DNA is delivered into plant cells (De Block, 1993), the use of compatible cultured cells and tissues *in vitro*, as recipients of foreign DNA, is a prerequisite for all procedures followed.

While most of the crops are more or less amenable to transformation, the legumes, *Vigna* species in particular, remain notoriously difficult to regenerate *in vitro* and transformation of these species a challenging task.

Of the different approaches followed for transformation of *Vigna* species, the most exploited are mediated by *Agrobacterium tumefaciens* and *Agrobacterium rhizogenes* besides considerable progress by particle bombardment and electroporation.

The *Agrobacterium*-mediated genetic transformation is considered so far advantageous over direct DNA delivery techniques in its capacity for high frequency of stable gene integration, the transfer of relatively large segments of DNA and single/low copy numbers of gene(s) insertion (Hiei *et al.*, 1994). The relative ease with which *Agrobacterium* may transform is likely due to the wound response these species exhibit (Potrykus, 1990); however, providing induced *Agrobacterium* with access to cells capable of dedifferentiation followed by regeneration limits its success. Thus, combining transformation competence with totipotency (Birch, 1997) remains imperative in these species. While several leguminous species are notably susceptible to *Agrobacterium* infection, relatively few or no species of *Vigna* have been transformed mediated by *Agrobacterium tumefaciens*. A summary of the key transformation events relating to *Vigna* species is presented (Table 3). Severe host-genotype specificity, lack of wounding response, the choice of suitable explant source with appropriate cells potentially competent for regeneration and/or *Agro* inoculation with selection compatible regeneration mode, has so far, limited the success of the *Agrobacterium*-mediated transformation. The ultimate choice is to target as many competent cells as possible, and moreover to develop ways to maximize the number of cells (Christou, 1995). The compatibility can be improved by subjecting plant tissues to short periods of ultrasound in the presence of bacteria, the treatment called Sonication-Assisted *Agrobacterium*-mediated transformation (SAAT), which induces the formation of channels in the target plant tissues, facilitating bacterial access to internal cells of the explants. Such an approach has been implicated in being especially helpful in transforming meristems which normally are physically impermeable to *Agrobacteria*. Indeed, stable transformation has been reported in soybean using this technique (Meurer *et al.*, 1998).

There has been considerable interest in developing *Agrobacterium rhizogenes* as an alternative system to *A. tumefaciens* for inducing foreign DNA into plant cells. The ability to recover plants from transformed roots is an essential feature of this system (Porter, 1991; Riazuddin and Hussain, 1993; Siefkes-Boer *et al.*, 1995; Rech *et al.*, 1989; Kumar *et al.*, 1991). However, plant regeneration from transformed, cultured roots of grain legumes has, so far, remained difficult and erratic.

The ability to engineer organized tissue, comprising different cell populations that are developmentally and positionally distinct, considerably limits the time duration of introduction of foreign genes into elite germplasm. Regeneration systems based on pre-existing meristems, precisely shoot apex, are promising as they develop directly into various tissues and organs of a mature plant including flowers (Schoneberg *et al.*, 1994; Ritala *et al.*, 1994; Aragao and Rech, 1997) without an intervening callus phase (Babaoglu, 2000). The avoidance of a callus phase, followed by regeneration to shoots or somatic embryos, ensures a low incidence of somaclonal variation.

Biolistic, particle bombardment or gene gunning technique, which relies upon the acceleration of DNA coated particles into target cells, is much preferred option for gene introduction into grain legumes, circumventing the host specificity as well as genotype dependence by *Agrobacterium* (Christou, 1994, 1995, 1997).

Shoot apex with its very high regeneration potential, offers an ideal target for bombardment, overcoming many of the regeneration problems (Potrykus, 1990; Kim and Minamikawa, 1996; Aragao and Rech, 1997). However, the main limitation to this approach is the limited access of researchers to particle bombardment instruments and the required skills.

Although transformation based on such tissue carries the risk of producing chimeric plants containing both transformed and untransformed sectors, regardless of whether regeneration is achieved through axillary shoot growth of organogenesis (Pigeaire *et al.*, 1997), production of chimeric transformants is not always undesirable as there is evidence to indicate that wholly transformed plants can be recovered from chimeric plants (McHughen and Jordan, 1989) in the next generation.

Electroporation, which facilitates the entry of DNA into protoplasts/cells by the use of high-voltage electric pulses, has potential application for enhancing transient as well

Table 3. Current status of genetic transformation of Vigna *species*

Species and variety	Explant used	Mode of transformation	Vector/ construct	Selection/ remark	Reference
Vigna unguiculata	Leaf disc	*A. tumefaciens*-mediated	—	—	Garcia *et al.*, 1986
	Leaf disc	*A. tumefaciens*-mediated	—	—	Garcia *et al.*, 1987
	Mature embryo	*A. tumefaciens*-mediated	—	—	Penza *et al.*, 1991
	Shoot apices, embryonic axes, cotyledonary segments	*A. tumefaciens*-mediated and particle bombardment	—	—	Kononowicz 1993 *et al.*,
	Protoplast	PEG	—	—	Nagl and Ehemann, 1994
	Mature de-embryonated cotyledon	*A. tumefaciens*-mediated	pUCD2614, pUCD2340	Hygromycin	Muthukumar *et al.*, 1996
V. sesquipedalis	Cotyledonary nodes	*A. tumefaciens* (oncogenic strain)	—	—	Ignacimuthu *et al.*, 1997
V. unguiculata	Shoot apices	*A. tumefaciens*-mediated	pBin9Gus-Int, pIG121-Hm, pTOK-233	Kanamycin	Sahoo *et al.*, 2000
V. unguiculata	Embryos	Particle bombardment	pBI221, pBc1GUS	—	Matsuoka *et al.*, 1997
V. mungo cv. T-9 cv. RV-19	Germinating mature embryo	Particle bombardment	pBI221 (GUS and *nptII*)	Kanamycin for rooting	Bhargava and Smigocki, 1994
V. aconitifolia cv. IPC MO-560, cv. IPCMO-909	Protoplasts	*A. tumefaciens*-mediated	—	Kanamycin	Eapen *et al.*, 1987
V. aconitifolia Genotypes Line 560 and Line 909	Protoplasts, protoplasts and cells	HS/PEG Electroporation	Plasmid DNA containing (*nptII*), Plasmid pCAP 212 CAT	Kanamycin/ Chloramphenicol	Kohler *et al.*, 1987
V. radiata cv. BI cv. T44	Cotyledon	*A. tumefaciens*-mediated	LBA 4404, pBI121 binary (*npt II* and GUS) GV2260, pGV2260; pGSGluc1 cointegrative	Kanamycin higher concentration 300 mg/L	Pal *et al.*, 1991

Table 3. (Continued)

Species and variety	Explant used	Mode of transformation	Vector/ construct	Selection/ remark	Reference
V. aconitifolia		*A. rhizogenes*-mediated	(sbPRP1, GUS) Up-stream flanking DNA sequences from a soybean proline rich cell wall protein	—	Suzuki *et al.*, 1993
V. aconitifolia cv. RMO-40 cv. TCVA-1	Germinating mature embryos	Particle bombardment	pBI221 (GUS and *nptII*)	Rooting in the presence of kanamycin	Bhargava and Smigocki, 1994
V. aconitifolia	—	—	nodulin-35 metabolic pathway gene	Transgenic root nodules expressing lower uricase activity	Lee *et al.*, 1993
V. radiata ML-5, K-851	Mature germinating embryos	Particle bombardment helium-driven gene gun (Biolistics PDS-1000, Dupont)	pBI121 (GUS and *nptII*)	Putative transformed shoots were rooted on kanamycin selective medium	Bhargava and Smigocki, 1994
V. radiata	Seedling without one cotyledon	*A. tumefaciens*-mediated	LBA4404, pBI121, pWCI-2 GUS fusion (*nptII* and GUS)	Kanamycin	Peyachoknagul *et al.*, 1996
V. angularis	Epicotyl	*A. tumefaciens*-mediated	*nptII*, GUS and α*ai-1*	Transgenic seeds resistant to bruchids	Ishimoto *et al.*, 1996
V. radiata	Hypocotyl and primary leaves	*A. rhizogenes*	Ri-plasmid and pBin9 GUSInt	Transgenic roots and calli	Jaiwal *et al.*, 1998
V. radiata	Hypocotyl and primary leaves	*A. tumefaciens*-mediated	LBA4404 pTOK233, EHA105 pBin9Gus Int), C58C1 (pG121Hm)	Kanamycin, transgenic calli and plants obtained	Jaiwal *et al.*, 2001
V. angularis	Epicotyl	*A. tumefaciens*-mediated	*nptII*, GUS or GFP	Transgenic plants	Yamada *et al.*, 2001

as stable expression of genes. An electroporation-mediated DNA transfer system described by Chowira *et al.* (1995, 1996), unlike others, uses intact nodal meristems and relies on the plant to develop its reproductive structures in a more or less normal fashion and is successfully applied to diverse legume crops like pea, soybean, cowpea and lentil.

Although these methods (polyethylene glycol or electroporation, and particle bombardment) are used routinely in a number of laboratories worldwide, these direct DNA up-take procedures generally result in insertion of multiple copies of the desired genes, with a high-frequency of sterile plants (Birch, 1997).

3.1. *AGROBACTERIUM*-MEDIATED TRANSFORMATION

An attempt was made to introduce foreign genes into *V. unguiculata* tissues using *Agrobacterium* (Garcia *et al.*, 1986, 1987) in a program towards the study of interaction between cowpea mosaic virus and its natural host. Kanamycin-resistant calli were obtained from cultured leaf discs on selective medium. Whole plants were not recovered from the transformed calli, but Northern blot analysis revealed the presence of more than 3500 nucleotides long full length transcripts in the integrated copy.

Penza *et al.* (1991) unsuccessfully attempted to regenerate plants from chimerical transgenic calli. The procedure involved co-cultivation of longitudinal slices of embryos, with *Agrobacterium* A281, wounding both apical and root meristems in the process. In some cases the presence of transgenes, in chimerical transgenic calli obtained, confirmed by dot blots and Southern blots. Although GUS activity was detected in some of the three week old shoots developed from the co-cultivated mature embryos, the shoots did not survive. Hence stable integration of foreign genes could not be conclusively proven. However, Kononowicz *et al.* (1993) reported the recovery of chimeric plants of *V. unguiculata* under kanamycin selection. Embryonic axes and cotyledonary segments of immature seeds were employed for *Agrobacterium* infection.

Muthukumar *et al.* (1996) successfully recovered transgenic shoots, of *Vigna unguiculata* on hygromycin selection medium, from mature de-embryonated cotyledons infected with *A. tumefaciens* harbouring a binary vector pUCD2340 with high copy stable virulence helper *A. tumefaciens* plasmid pUCD2694. Stable integration of transgenes was confirmed by Southern analysis; however, T_1 seeds were found not to be viable.

Sahoo *et al.* (2000) recovered chimeric plants of *Vigna unguiculata* on kanamycin selection medium showing GUS expression in various parts of plant including stem, employing wounded shoot apices in conjunction with *A. tumefaciens* strain EHA105 (pBin9GUSInt). Out of the three different *Agrobacterium* strains, EHA105 (pBin9GUSInt), C58C1 (pIG121Hm) and LBA4404 (pTOK233) and two procedures of co-cultivation (co-cultivation with bacterial suspension and application of *Agrobacterium* suspension at the tip of wounded short apices cultured) compared, strain EHA105 (pBin9GUSInt) applied at the tip of pre-wounded shoot apices cultured was found to be optimal.

Karthikeyan *et al.* (1996) obtained transformed calli of *Vigna mungo* by co-cultivating segments of primary leaves with *Agrobacterium* strains LBA4404 and EHA105 carrying binary vector pGA472. Transformation frequency with LBA4404 and EHA105 was 23% and 10% respectively. Expression of *nptII* gene in transformed calli was demonstrated by neomycin phosphotransferase assay and stable integration confirmed by Southern blot analysis.

Eapen *et al.* (1987) investigated the effect of two genotypes (cultivars) on transformation rates in *Vigna aconitifolia*. A large number of kanamycin-resistant colonies were recovered by co-cultivation of protoplasts with *Agrobacterium tumefaciens*, with 23% of the transformed lines showing expression of non-selected, co-transferred nopaline synthase gene. The transformation rate displayed was found to be cultivar-dependent. Out of some 12,000 colonies, cv. IPCMO-560 showed a 85-fold higher transformation rate than cv. IPCMO-909. Regeneration of some plants from antibiotic-resistant lines was reported, but no molecular or genetic analysis was presented to demonstrate their transgenic nature.

Expression of an antisense nodulin-35 cDNA encoding nodulin-specific uricase in *Vigna aconitifolia* was detected (Lee *et al.*, 1993) in transgenic root nodules, resulting in lower uricase activity and a reduction in the nodulin size. Peroxisome development was retarded and nitrogen deficiency symptoms appeared in the plants, suggesting that the availability of symbiotically reduced nitrogen in legumes is limited by a reduction in ureide biosynthesis.

Pal *et al.* (1991) in a preliminary attempt recovered primary transformants of *Vigna radiata* from cotyledons of cv. B1 and cv. T44 upon co-cultivation with *A. tumefaciens* strain LBA4404 with binary vector pBI121 and GV2260 with cointegrate vector pGV2260 pGSGluc1. The experimental evidence was supported by dot blot assay and MUG assay. However, the absence of support for stabilization of genes in primary transformants and flowering caused speculation as to their true transgenic nature as well as sexually viable character and thus ruled out the stringent criteria for true transformation.

Generation of transgenic calli from primary leaf explants of *V. radiata* was reported by Phogat *et al.* (1999) employing *A. tumefaciens* strain LBA4404 pKIWI105. The transformed calli, selected on 100 mg/L kanamycin, exhibited β-glucuronidase activity and were found to be resistant up to 750 mg/L kanamycin. The T-DNA transfer and integration were proved by Southern hybridization analysis of junction fragments. However, the calli failed to regenerate under different culture conditions tested.

An attempt to transform a Thai variety of *Vigna radiata* by Peyachoknagul *et al.* (1996) resulted in recovery of transgenic T_0 plantlets. The seedlings with one intact cotyledon were co-cultivated with *A. tumefaciens* strain LBA4404 pWC12-GUS. The putative transformed plants selected on kanamycin containing mediums were found GUS positive and the presence of transgene was confirmed by PCR and Southern hybridization.

Transgenic plants of *Vigna radiata* were reported recently (Jaiwal *et al.*, 2001) from cotyledonary nodes co-cultivated with *A. tumefaciens* strain LBA4404 (pTOK233). The plants were selected on kanamycin containing medium, subsequently developed flowers and pods with viable seeds showing GUS activity in stamens, pollen grains and T_0 seeds. Molecular analysis revealed the integration and expression of transgenes in T_0 plants. An effort to transform hypocotyl and primary leaves, employing *Agrobacterium* strains LBA4404 (pTOK233), EHA105 (pBin9GUSInt) and C58C1 (pIG121Hm), resulted in kanamycin resistant calli. Although NPTII assay and GUS histochemical assay demonstrated the expression and stable integration of T-DNA confirmed by Southern blot hybridization, regeneration of plants from the transformed calli could not be accomplished after several subcultures on different media compositions.

Ishimoto *et al.* (1996) transformed azukibean (*Vigna angularis*) with bean α-amalyse inhibitor gene driven by the promoter of phytohaemagglutinin leading to high levels of α-amalyse inhibitor proteins in the transgenic seeds and subsequently complete block of bruchid development. The system relied on *Agrobacterium* infection of epicotyl explants and their subsequent selection on kanamycin containing medium. Using *Agrobacterium* binary vectors carrying either *gus* gene or modified *gfp* gene [sGFP(S65T)], transgenic plants of *Vigna angularis* were recovered with viable seeds from epicotyl explants, showing stable integration and expression of foreign gene (Yamada *et al.*, 2001).

Suzuki *et al.* (1993) introduced an upstream flanking DNA sequences from a soybean proline-rich cell wall protein, SbPRP1, fused to the *gus* gene into mothbean (*Vigna aconitifolia*) by *Agrobacterium rhizogenes*. Histochemical GUS staining in transformed hairy roots indicated that SbPRP1 was expressed in the apical and elongating region of both primary and lateral roots, with the strongest activity observed in the epidermis. Following experiments demonstrated that this expression was temporally regulated during very early stages of seedling growth.

Transformed roots and calli of *Vigna radiata* expressing *gus* gene were generated (Jaiwal *et al.* 1998), by using a wild type *Agrobacterium rhizogenes* strain harbouring a binary vector pBin9GUSInt, from hypocotyl explants. Transfer of *gus* gene into calli was demonstrated by Southern blot analysis. Further analysis ruled out the possible *Agrobacterium* contamination. Both the root as well as the calli failed to regenerate shoots even after prolonged culture on shoot regeneration medium.

3.2. CRITICAL ASSESSMENT OF *AGROBACTERIUM*-MEDIATED GENE DELIVERY

The more directed and stable method for introduction of a gene is probably transformation using the Ti-plasmid of *A. tumefaciens*, as it is known to have a preferential mechanism of insertion (Zupan and Zambryski, 1997). The consideration in choosing a combination of plant regeneration system and *Agrobacterium* mediated transformation method (Mukhopadhyay *et al.*, 1992), tuned to its maximum compatibility, decides success or failure owing to the natural recalcitrance of *Vigna* species exhibits to *Agrobacterium* infection.

Most of the transformation methods developed so far for these species (Table 3) use tissue from developing, mature or germinating embryonic axes. This tissue permits adventitious regeneration in some cases, while in others regeneration depends on axillary shoot growth from the axis. These embryonic explants, although requiring initial propagation in gel media under sterile conditions, do not involve callus induction or the culture of cells or protoplasts. They are more promisingly used for both *Agrobacterium* and direct DNA transfer methods with a high degree of success and in a more or less variety independent manner. The short time duration required for development to complete fertile plant, from this tissue, considerably limits the time required to generate transgenic plants.

However, the tissue, from the developmentally young or mature embryonic axes, best suitable for transformation, needs to be carefully screened. Although the transformation resulting in transgenic calli, with foreign gene stably integrated into calli, has been reported in most of the *Vigna* species discussed (Table 3), further regeneration from these calli could not be achieved in any of the cases. Refinement of these protocols with more emphasis on successful regeneration of plants from these calli can possibly lead to recovery of true transgenics at the first instance.

Whilst use of dedifferentiated tissue, with a *de novo* regeneration mode, in transformation has optimal chance of recovery of true transgenics, such a system neither has been developed nor adequately applied to *Agrobacterium*-mediated transformation except in azukibean (Ishimoto *et al.*, 1996; Yamada *et al.*, 2001).

Owing to the natural resistance of many plants, legumes in particular, to selective agents (Dekeyser *et al.*, 1990), an effective antibiotic/herbicide for a particular species can only be determined empirically (Babaoglu *et al.*, 2000). Further, the concentration of selective agent, choosen so as to allow a lesser number of escapants to grow, is vital in generating transformants. Moreover the dose of selective agent applied (lethal or sublethal), at an early stage or more precisely immediately after co-culture, or at a much delayed stage whether with gradual increasing or decreasing increments, is to be carefully assessed for a particular species and a variety of explants. Judicious choice of selective levels may be an important criterion for the recovery of transformed cells at initial stage of screening (Sharma and Anjaiah, 2000). The various analogs of antibiotic available for selective marker genes and comparing their performance for selecting transformants without interference in regeneration is vital and can provide a distinct advantage. The use of aminoglycoside antibiotics, kanamycin sulphate in particular, for selecting transformed plants has been extensively exploited in the species discussed (Table 3), except in few cases, viz. (Muthukumar *et al.*, 1996) where hygromycin is used to select plants. In the case of herbicide-based selection, the *bar* gene for bialaphos resistance has given tight selection in several related crops (Schroeder *et al.*, 1993; Pigeaire *et al.*, 1997; Mohapatra *et al.*, 1999; Zhang *et al.*, 1999) due to the rapid translocation mechanism of glufosinate through xylem and phloem (Shelp *et al.*, 1992). Use of *bar* gene for selection of transformed plants in *Vigna* species can improve the efficiency.

Although *gus* or *uidA* gene (Jefferson *et al.*, 1987) is most widely used as a reporter for rapid assessment of the susceptible tissue as well as transient transformation, use of *gfp* gene (Ponappa *et al.*, 1999) can provide a non-destructive approach, and more precisely, monitoring gene expression, as well as screening shoots at the preliminary stage, is beneficial particularly in case of high degree of chimeras. Light stimulated GFP fluorescence does not require any co-factors, substrates, or additional gene products, and moreover, GFP accumulation in transgenic crops produced, so far, does not appear to have a toxic effect (Maximova *et al.*, 1998). None of the methods developed so far for these species, except in *Vigna angularis* (Yamada *et al.*, 2001), explored this possibility.

In the case of *Agrobacterium*-mediated transformation, some strains are more effective with a particular species than others (Christou, 1994). Evaluation of strain-cultivar compatibility is an important step in the establishment of a transformation protocol. Whilst the natural virulence of *Agrobacterium* varies and hence the ability to infect plants, the virulence of some strains can be increased by the introduction of supervirulent plasmid, such as pTOK47 carrying extra copies of some of the virulence genes, into the *Agrobacterium* cell, alongside the binary vector. Such supervirulent strains of *Agrobacterium* have been useful in transforming certain dicots, such as lettuce (Curtis *et al.*, 1994). Superbinary vectors, in which extra copies of virulence genes are present on the binary vector itself, have also proven useful in the transformation of cereals, as in the case of rice (Hiei *et al.*, 1994,

1997) and also in *Vigna radiata* (L. Sahoo and coworkers, unpublished data). This is an important consideration in the transformation of least amenable *Vigna* species using *Agrobacterium* with super-binary vectors or binary vector with additional copies of *vir* genes. The problem encountered with co-integrative vectors is their size, which makes their manipulation difficult, so the binary vector has been the choice. Certain disarmed strains of *Agrobacterium tumefaciens* have been used extensively for several years to carry the binary and super-binary vectors, an excellent example being LBA4404 (Hoekema *et al.*, 1983).

Inclusion of plant phenolic compounds, such as acetosyringone, in the culture medium prior to or during co-cultivation (Godwin *et al.*, 1991; Orczyk and Orczyk, 2000) apparently influences the efficiency of transformation with varying degrees, by inducing the virulence genes (Stachel *et al.*, 1985). Addition of such phenolic compounds in inducing *vir* genes in hypervirulent or moderately virulent strains can improve the susceptibility to the tissue of these recalcitrant species described.

Access of bacteria to susceptible cells has been accomplished by means of extra wounding treatment (Park *et al.*, 1996; Bean *et al.*, 1997; Sahoo *et al.*, 2000) in explant like shoot apex, which are normally impermeable to *Agrobacteria*. Lack of competence of meristematic cells may be due to absence of a key surface component (receptor) on totipotent cells in the apical meristem for recognition of *Agrobacterium* (Escudero *et al.*, 1996) and thus can be overcome by wounding several times with fine needle or bisecting the embryonic axes longitudinally up to cotyledonary nodes or completely in cowpea (Sahoo *et al.*, 2000), pea (Schroder *et al.*, 1993) and sunflower (Schrammeijer *et al.*, 1990; Knittel *et al.*, 1994; Schoneberg *et al.*, 1994; Burrus *et al.*, 1996). Prewounding or bisecting may circumvent the attachment step or make available compounds activating bacterial virulence *in situ* and thus may result in T-DNA transfer (Sahoo *et al.*, 2000).

vir gene activity and transformation have also been shown to be affected by diverse conditions during the co-cultivation stage such as concentration of bacteria, pH of the co-cultivation medium (Stachel *et al.*, 1986; Godwin *et al.*, 1991), light conditions and temperature maintained during infection and co-culture (Stachel *et al.*, 1985), are vital although these effects may not be consistent across species.

Several groups have demonstrated the optimal induction of *vir* genes attained when the pH is lower than 5.8–6.0 (Holford *et al.*, 1992). Alt-Moerbe *et al.* (1988) demonstrated that *vir* induction is strictly dependent on acidic pH with an optimum between 5.0 and 5.7 and in the presence of acetosyringone. Thus, maximal transformation with individual genotype may be dependent on the correct choice of inducer compound and its use at the appropriate pH. In majority of the cases maximum transformation has been achieved with the combination of low pH and the presence of acetosyringone.

For *in vitro* co-culture experiments, 25°C is a commonly employed temperature and literature reports mentioning higher temperatures abound (Dillen *et al.*, 1997). In many systems, the actual temperature inside the culture vessel is well above the temperature of the growth chamber due to radiation heating from lamps (a difference of two degrees is not exceptional). Hence, the species for which regeneration is often performed at relatively high temperatures, lowering the temperature during *Agrobacterium* co-culture, may prove beneficial. A probable explanation for the observed thermo-sensitivity of gene transfer would thus be that the regulation of the *vir* regulon is temperature dependent in a similar fashion and the temperature range (20 to 25°C) is in agreement with the temperature optimum established for transient expression in most of the crops (Dillen *et al.*, 1997).

In the majority of the transformation processes, *Agrobacterium* concentration (10^7–10^9 cells per ml) was found to be optimal for infection.

The beneficial effect of preculture/preconditioning on transformation has been reported in several plant species although it is not clear why regeneration of transgenic shoots requires a preculture period, especially in light of the fact that no preculture period is required for cellular transformation (McHughen *et al.*, 1989). The insertion of T-DNA does not interfere with regeneration because regeneration from transformed cells does occur (when associated with a preculture period). During the preculture time, the tissues undergo a physiological and developmental shift to become competent to regenerate; when the T-DNA is inserted after such a period, the recipient cells are already on the developmental pathway for regeneration. In such cases, if preculture not allowed, the presence of the selection agent perhaps interferes with the expected physiological shift in the cultured cells from non-regenerable to competent for regeneration and leads to failure of transformation.

Explant size, explant orientation, gelling agent and plate sealant were found to affect transformation efficiency (Frary and Earle, 1996).

Since with *Agrobacterium*-mediated transformation, the introduction of foreign DNA into the plant tissues is dependent on the infection by *Agrobacterium*, increasing the concentration of *Agrobacterium* cells (10^7–10^9 cells) in the co-cultivation process should increase the number of plant cells infected. However, *Agrobacterium* cells aggregate when more than 10^9 cells are added during co-cultivation and substantially reduce the transformation efficiency (L. Sahoo and coworkers, unpublished data).

The report of bacteria applied aerially to the tissue during co-culture has been demonstrated to influence the rate of transformation in cowpea (Sahoo *et al.*, 2000) and lead to successful recovery of transgenics in cotton (Zapata *et al.*, 2000).

The fact that not all plant cells are uniformly transformed in crown gall tissues, but all tissues are uniformly transformed in hairy roots (Petit *et al.*, 1986), leads to development of transformation protocols using (Ri) plasmids of *Agrobacterium rhizogenes* with core system of regenerating shoots from transformed hairy roots in legumes (Kumar and Davey, 1991), and indeed, no success in *Vigna* species (Table 3).

3.3. DIRECT DNA DELIVERY

Kononowicz *et al.* (1993) reported the production of chimeric plants of *Vigna unguiculata* using the microprojectile bombardment technique under kanamycin selection.

Matsuoka *et al.* (1997) evaluated the efficiency of gene transfer by particle bombardment of embryos in *V. unguiculata*, using transient GUS expression coating the gold microprojectiles with *gus* reporter gene constructs pBI221 and pBC1. Different factors for bombardment influencing transient GUS expression were optimized. Preculture of embryos for one day prior to bombardment significantly enhanced the transient GUS expression.

Bhargava and Smigocki (1994) observed transient expression of *gus* gene in the germinating embryos of *Vigna aconitifolia V. mungo* and *V. radiata*, 18 to 24 h after particle bombardment. The raised plantlets were rooted on a kanamycin selection medium. Bombarding embryos two to three times with DNA-coated particles increased the level of expression upto 4-fold.

V. unguiculata embryos isolated mechanically from mature seeds and incubated in the presence of plasmid-DNA harboring chimerical *gus* genes germinated into seedlings which exhibited apparent GUS activity (Akella and Lurquin, 1993). Embryo electroporation in the presence of DNA and protectants such as spermidine and Lipofectin, increased the proportion of embryo-derived seedlings expressing the chimeric gene. The authors, however, did not attempt to grow the seedlings into transformed plants. Whether the observed activity in the seedlings was a result of some transient activity of episomal DNA, as the assays were performed three days after DNA introduction or partly due to prolonged incubation period (72 h), is not clear.

Nagl and Ehemann (1994) have reported the transient and stable expression of the *uidA* gene in the calli obtained from transformed protoplasts of *V. unguiculata* sp. *Sesquipedalis*. The transformation was carried out using the plasmid pRT66GUSR and PEG. However, the calli, failed to regenerate to shoots.

Dillen *et al.* (1995) recovered chimeric plants of *V. unguiculata* showing GUS expression from intact embryonic axes by electroporation. The method was demonstrated to be applicable to a variety of other grain legumes.

In an innovative approach, Chowira *et al.* (1995, 1996) transformed shoot meristems of young seedlings of *V. unguiculata*, using *gus* as reporter gene, which relied on both *Agrobacterium* and electroporation. The method not only resulted in a high frequency of transformation but also proved to be applicable to a wide range of cultivars and species, raising the possibility of their wide application. As the transformation was carried out *in planta*, thus the need for tissue-culture and special conditions for propagation are essentially avoided.

Kohler *et al.* (1987a) generated kanamycin resistant callus of *V. aconitifolia* by treating heat shocked protoplasts with PEG and plasmid DNA carrying *nptII* gene. Putative transgenic plants were recovered from the callus. As with earlier studies, the plant cultivar used was an important factor in attaining higher transformation frequencies (cv. IPMCO-560 showed about 8-fold higher transformation frequency than cv. IPMCO-909). Transformation was confirmed by Southern blot. Attempts to transform mesophyll and suspension culture cells by this method were found to be unsuccessful. Protoplasts electroporated with a plasmid encoding chloramphenicol acetyltransferase (Kohler *et al.*, 1987b), exhibited transient expression of the gene, while electroporated cells did not show any enzyme activity. The use of different expression signals derived from gene V1 of the cauliflower mosaic virus or from the nopaline synthase gene of *Agrobacterium* also resulted in different frequencies of stable transformed colonies. However, molecular or genetic data demonstrating transformed nature and regeneration of plants from protoplasts was not presented.

3.4. CRITICAL ASSESSMENT OF DIRECT DNA DELIVERY SYSTEMS

Direct DNA delivery by particle bombardment is more promising as no specific DNA regions are required for the DNA integration, as in the case of *Agrobacterium*, thus allowing the introduction of only the target genes into the plant genome without any extra redundant plasmid DNA (Fu *et al*., 2000). The high degree of specificity (severe species specific and frequently cultivar specific) reflected in *Agrobacterium*-mediated transformation has precluded the use of this method. Thus particle bombardment may be the preferred option for gene introduction into grain legumes, *Vigna* species in particular, circumventing the host specificity of these species to *Agrobacterium* infection.

Shoot apical meristems of mature seeds or whole embryos have been used extensively as target tissues by particle bombardment in legumes (Kim and Minamikawa, 1996, 1997; Aragao *et al*., 2000) and, with more limited success in *Vigna* species (Bhargava and Smigocki, 1994) (Table 3). In majority of the cases, explants from near the shoot apex or the apex itself, have been the targets of choice (Christou, 1997) as apical meristems permit rapid regeneration of complete fertile plant with minimum culture manipulation in a variety-independent manner.

The major problem encountered with organized tissue, shoot apex in particular, is that transformed cells permit proliferation of non-transformed tissue in their vicinity by effectively detoxifying the selective agent. This results in the creation of chimeric tissue, which subsequently gives rise to both transformed and non-transformed plants, the latter being the overwhelming majority. Although regeneration of intact plants from transformed tissue is not always an easy task; the problem can be overcome by using a highly specific antibiotic or herbicide that effectively mobilized to apical tissues, making selection of these remote tissues more efficient. Using imidazilinone classes of herbicides for selection of transformed shoots, for mutant acetolactate synthase *ALS* genes, from bombarded shoot apices has successfully led to recovery of transgenics in common bean (Aragão *et al*., 2000).

Although the use of particle bombardment for transformation of *Vigna* species is still at an infancy stage (Table 2), mostly restricted to transient GUS expression in bombarded tissue, success in generating transgenics using shoot apex-particle bombardment with subsequent highly specific selection system seems to be practical.

Electroporation-mediated gene transfer system described by Chowira *et al*. (1995, 1996), unlike using protoplasts or cells, uses intact nodal meristems and relies on the plant to develop its reproductive structures from the treated apices in a more or less normal fashion. Transformation of intact embryonic explants by electroporation (Dillen *et al*., 1995) obtained high frequency of transformation with wider applicability to diverse legume species. *In planta* techniques which do not rely on shoot production from tissues cultured *in vitro* are potentially more rapid, avoiding the complexities of cell and tissue culture and reduce the likelihood of somaclonal variation. *In planta* techniques in conjunction with electroporation are perhaps most promising and can circumvent the high recalcitrance these species exhibit to *in vitro* regeneration and their transformation using routine methods.

4. Targets for biotechnological improvement

The extreme importance of grain legumes, *Vigna* species in particular, as a protein source, is an important consideration in developing countries, which derive most of their nutritional calories from grain legumes and suffer from severe malnutrition. Although these crops occupy unique position with regard to their total area of cultivation as well as production, the yield is severely limited due to diseases (viral, fungal, insects and storage pests) besides considerable yield loss due to drought and salinity.

In view of the severity of insect-pest related grain yield losses and the expectation of only a slow rate of progress by routine conventional breeding, transfer of ideal genes coding for the insect-specific toxins of *Bacillus thuringiensis* (Bt), α-amylase inhibitor for storage pests resistance (Ishimoto *et al.*, 1996), chitinase, glucanase, RIP for resistance to fungal pathogens can check the damage.

Quality improvement of these crops with enhanced protein production as well as correct nutritional balance is to be addressed and necessitates transfer of foreign genes encoding methionine-rich proteins such as 2S-albumin of brazil nut/sunflower to these crops.

Following the spectacular advances in agricultural biotechnology with the advantage of cutting-edge tools of recombinant technology, improvement of these crops for various biotic and abiotic stresses is inevitable. However, lack of coordination between academic institutions pioneering technologies and major agribio organizations, slackness in government policies in permitting significant research allocation and a virtual establishment of multi-disciplinary research teams, has so far dogged progress, and needs to be addressed without much delay.

5. Conclusions and future prospects

Different species of *Vigna* are cultivated throughout the world for their protein rich seeds. The production of these crops has remained stagnant over time and strategies for the development of improved varieties are under way. Breeding programs, limited by the available gene pool and restricted potential for hybridization between species, need to be supplemented by biotechnological means through *in vitro* regeneration and transfer of the desirable genes from sources across the sexual boundaries. The recalcitrant nature of these crops to *in vitro* regeneration has so far hindered the successful application of routine transformation strategies. Success in plant regeneration from various seedling explants via direct organogenesis is recent, and regeneration from callus, suspension and protoplast cultures, except in *Vigna aconitifolia*, still remain as a breakthrough. Most of the *Vigna* species have shown their susceptibility to *Agrobacterium* and subsequent recovery of transgenics, with integrated marker and reporter genes, reported in *V. unguiculata, V. angularis* and *V. radiata*, but with a very low frequency. The stability and expression of transgenes in progenies have not been investigated in either case. The refinement in regeneration and *Agrobacterium*-mediated transformation protocols with emphasis on their compatibility is vital, and exploring the different analogs of antibiotic for the currently used marker genes for the efficient selection of transformants will be an advantage. Transformation through bombardment of DNA coated particles to intact regenerable tissues such as seedling meristems is still in its infancy and needs thorough investigation. Electroporation of intact plants and recent approaches like Agrolistics can be explored. A high frequency of transformation resulting from either of the techniques in *Vigna* species is essential before incorporating an array of desirable genes for better yield and nutritional quality.

Acknowledgement

PKJ is grateful to Department of Biotechnology, New Delhi for financial assistance to his laboratory for gene transfer in legumes.

References

Ahuja M R and Singh B V (1977) Induced genetic variability in mungbean through interspecific hybridization. *Indian J. Genet. Plant Breed.*, **37**: 133–137.

Akella V and Lurquin (1993) Expression of cowpea seedlings of chimeric transgenes after electroporation into seed-derived embryos. *Plant Cell Rep.*, **12**: 110–117.

Aragão F J L and Rech E L (1997) Morphological factors influencing recovery of transgenic bean plants (*Phaseolus vulgaris* L.) of a carioca cultivar. *Intl. J. Plant Sci.*, **158**: 157–163.

Aragão F J L, Sarokin L, Vianna G R and Rech E L (2000) Selection of transgenic meristematic cells utilizing a herbicidal molecule results in the recovery of fertile transgenic soybean (*Glycine max* (L.) Merrill) plants at high frequency. *Theor. Appl. Genet.*, **101**: 1–6.

Arya I D and Chandra N (1989) Organogenesis in anther derived callus cultures of cowpea (*Vigna unguiculata* L. Walp.). *Curr. Sci.*, **58**: 257–259.

Att-Moerbe J, Neddermann P, von Lintig J, Weiler E W and Schroeder J (1988) Temperature-sensitive step in Ti plasmid *vir*-region induction and correlation with cytokinin secretion by *Agrobacterium. Mol. Gen. Genet.*, **213**: 1–8.

Avenido R A and Hattori K (1999) Differences in shoot regeneration response from cotyledonary node explants in Asiatic *Vigna* species support genomic grouping within subgenus *Ceratotropis* (Piper) Verdc. *Plant Cell Tiss. Org. Cult.*, **58**: 99–110.

Babaoglu M, Davey M R and Power J B (2000) Genetic engineering of grain legumes: key transformation events. *Ag. Biotech Net*, **2**: 4BN050.

Bajaj Y P S and Dhanju M S (1979) Regeneration of plants from apical meristem tips of some legumes. *Curr Sci.*, **48**: 906–907.

Bajaj Y P S and Singh H (1980) *In vitro* induction of androgenesis in mungbean (*Phaseolus aureus* L.). *Indian J. Exp. Biol.*, **18**: 1316–1318.

Bean S J, Gooding P S, Mullineaux P M and Davis D R (1997) A simple system for pea transformation. *Plant Cell Rep.*, **16**: 513–519.

Bhadra S K, Hammatt N and Davey M R (1990) Callus induction from seedling protoplasts of *Vigna gracilis* and *V. trilobata. SABRAO J.*, **22**: 25–33.

Bhadra S K, Hammatt N and Davey M R (1991) Tissue and protoplast culture of rice bean (*Vigna umbellata*). *Trop Agric.*, **68**: 344–348.

Bhadra S K, Hammatt N, Power J B and Davey M R (1994) A reproducible procedure for plant regeneration from seedling protoplasts of *Vigna sublobata. Plant Cell Rep.*, **14**: 175–189.

Bharal S and Rashid (1980) Isolation of protoplasts from stem and hypocotyl of the legume *Vigna sinensis* and some factors affecting their regeneration. *Protoplasma*, **102**: 307–313.

Bhargava S C and Smigocki A C (1994) Transformation of tropical grain legumes using particle bombardment. *Curr Sci.*, **66**: 439–442.

Birch R G (1997) Plant transformation: problems and strategies for practical application. *Annu. Rev. Plant. Physiol. Mol. Biol.*, **48**: 297–326.

Burrus M, Molinier J, Himber C, Hunfold R, Bronner R, Rousselin P and Hahne G (1996) *Agrobacterium*-mediated transformation of sunflower (*Helianthus annuus* L.) shoot apices transformation patterns. *Mol. Breed.*, **2**: 329–338.

Chandra M and Pal A (1995) Differential response of the two cotyledons of *Vigna radiata in vitro. Plant Cell Rep.*, **15**: 248–253.

Chen H K, Mok M C and Mok D W S (1990) Somatic embryogenesis and organogenesis from interspecific hybrid embryos of *Vigna glabrescens* and *V. radiata. Plant Cell Rep.*, **9**: 77–79.

Chowdhury V K, Sareen P K, Sharma D R, Chowdhury J B and Gupta V K (1983) Establishment of callus and cell suspensions and isolation of mutant cell lines in mungbean (*Vigna radiata* var. *aureus*). In: *Plant Cell Culture in Crop Improvement* (Eds. Sen S K and Giles K L), Vol. 22. Plenum Press, New York and London, 405–409.

Chowira G M, Akella V, Fuerst P E and Lurquin P F (1996) Transgenic grain legumes obtained by *in planta* electroporation-mediated gene transfer. *Mol. Biotechnol.*, **5**: 85–96.

Chowira G M, Akella V and Lurquin P F (1995) Electroporation-mediated gene transfer into intact nodal meristems *in planta*. *Mol. Biotechnol.*, **3**: 17–23.

Christou P (1994) Biotechnology of crop legumes. *Euphytica*, **74**: 165–185.

Christou P (1995) Strategies for variety-independent genetic transformation of important cereals, legumes and woody species utilizing particle bombardment. *Euphytica*, **85**: 13–27.

Christou P (1997) Biotechnology applied to grain legumes. *Field Crops Res.*, **53**: 187–204.

Curtis I S, Power J B, Blackhall N W, De Laat A M M and Davey M R (1994) Genotype-independent transformation of lettuce using *Agrobacterium tumefaciens*. *J. Exp. Bot.*, **45**: 1441–1449.

Das D K, Shiva Prakash and Bhalla-Sarin N (1998) An efficient regeneration system of blackgram (*Vigna mungo* L.) through organogenesis. *Plant Sci.*, **143**: 199–206.

De Block M (1993) The cell biology of plant transformation: current state, problems, prospects and implications for plant breeding. *Euphytica*, **71**: 1–14.

Dekeyser R A, Claes B, De Rycke RMU, Habets M E, Montagu M V and Caplan A B (1990) Transient gene expression in intact and organized rice tissues. *Plant Cell*, **2**: 591–602.

Dillen W, Clercq J De, Kapila J, Zambre M, Montagu M V and Angenon G (1997) The effect of temperature on *Agrobacterium tumefaciens*-mediated gene transfer to plants. *Plant J.*, **12**: 1459–1463.

Dillen W, Engler G, Montagu M V and Angenon G (1995) Electroporation-mediated DNA delivery to seedling tissues of *Phaseolus vulgaris* L. (common bean). *Plant Cell Rep.*, **15**: 119–124.

Eapen S (1988) Callus induction from mesophyll and hypocotyl protoplasts of mungbean (*Vigna radiata* L.). *Ann. Bot.*, **62**: 441–443.

Eapen S and George L (1990) Ontogeny of somatic embryos of *Vigna aconitifolia*, *Vigna mungo* and *Vigna radiata*. *Ann. Bot.*, **66**: 219–226.

Eapen S, Kohler F, Gerdemann M and Schieder O (1987) Cultivar dependence of transformation rates in moth bean after co-cultivation of protoplasts with *Agrobacterium tumefaciens*. *Theor. Appl. Genet.*, **75**: 207–210.

Escudero J, Neuhaus G, Schlappi M and Hohn B (1996) T-DNA transfer in meristematic cells of maize provided with intracellular *Agrobacterium*. *Plant J.*, **5**: 355–360.

Evans D A (1989) Somaclonal variation-genetic basis and breeding applications. *Trends in Genetics*, **5**: 46–50.

Fillipone E (1993) To improve resistance against diseases and pests. *Grain Legumes*, **2**: 20–21.

Franklin G and Ignacimuthu S (2000) Differential morphogenetic response of cotyledonary explants of *Vigna mungo*. *Biol. Plant.*, **43**: 1–4.

Frary A and Earle E D (1996) An examination of factors affecting the efficiency of *Agrobacterium*-mediated transformation of tomato. *Plant Cell Rep.*, **16**: 235–240.

Freytag A H, Rao-Arelli A P, Anard S C, Wrather J A and Owens L D (1989) Somaclonal variation in soybean plants regenerated from tissue culture. *Plant Cell Rep.*, **8**: 199–202.

Fu X, Due L T, Fontana S, Bong B B, Tinjuangjun P, Sudhakar D, Twyman R M, Christou P and Kohli A (2000) Linear transgene constructs lacking vector backbone sequences generate low copy-number transgenic plants with simple integration patterns. *Transgenic Res.*, **9**: 11–19.

Garcia T A, Hille J and Goldbach R (1986) Transformation of cowpea, *Vigna unguiculata* cells with an antibiotic resistance gene using a Ti-plasmid derived vector. *Plant Sci.*, **44**: 37–46.

Garcia J A, Hille J, Vos P and Goldbach R (1987) Transformation of cowpea (*Vigna unguiculata* Walp.) cells with a full length DNA copy of cowpea mosaic virus mRNA. *Plant Sci.*, **48**: 89–98.

Ge K, Wang Y, Yuang P, Yang J, Nie Z, Testa D and Lee N (1989) Plantlet regeneration from protoplasts isolated from mesophyll cells of adzuki bean (*Phaseolus angularis* Wright). *Plant Sci.*, **63**: 209–246.

Geetha N, Venkatachalam P and Rao G R (1997a) Plant regeneration and propagation of blackgram (*V. mungo* L. Hepper) through tissue culture. *Trop. Agric.*, **74**: 73–76.

Geetha N, Venkatachalam P and Rao G R (1997b) *In vitro* plant regeneration from different seedling explants of blackgram (*V. mungo* L. Hepper) via organogenesis. *Breed. Sci.*, **47**: 311–315.

Geetha N, Venkatachalam P and Rao G R (1997c) Somatic embryogenesis and plant regeneration from cell suspension cultures of blackgram (*Vigna mungo* L. Hepper). *Physiol. Mol. Biol. Plants*, **3**: 25–30.

Gill R, Eapen S and Rao P S (1987a) Callus induction from protoplasts of *V. unguiculata*, *V. sublobata* and *V. mungo*. *Theor. Appl. Genet.*, **74**: 100–103.

Gill R, Eapen S and Rao P S (1987b) Morphogenetic studies of cultured cotyledons of urd bean (*V. mungo* L. Hepper). *J. Plant. Physiol.*, **130**: 1–5.

Girija S, Ganapathi A and Ananthakrishanan G (2001) Somatic embryogenesis in *Vigna radiata* (L.) Wilczek. *Indian J. Exp. Biol.*, **38**: 1241–1244.

Godwin I, Gordon T, Ford-Lloyd B and Newbury H J (1991) The effects of acetosyringone and pH on *Agrobacterium*-mediated transformation vary according to plant species. *Plant Cell Rep.*, **9**: 671–675.

Goel S, Mudgal A K and Gupta S C (1983a) Development of plants from *in vitro* cultured shoot tips of *Vigna mungo* and *Vigna radiata*. *Trop. Plant Sci. Res.*, **1**: 31–33.

Goel S, Mudgal A K and Gupta S C (1983b) Isolation of protoplasts from leaves of *Vigna radiata*-legume. *Trop. Plant Sci. Res.*, **1**: 339–341.

Gosal S S and Bajaj Y P S (1984) Isolation of sodium-chloride resistant cell lines in some grain-legumes. *Indian J. Exp. Biol.*, **22**: 209–214.

Gulati A (1993) Isolation and characterization of salt tolerant cell lines of *Vigna radiata* (L.) Wilczek. Ph.D Thesis submitted to Dept. of Bio-sciences, M. D. University, Rohtak (India).

Gulati A and Jaiwal P K (1990) Culture conditions affecting plant regeneration from cotyledon of *Vigna radiata* (L.) Wilczek. *Plant Cell Tiss. Org. Cult.*, **23**: 1–7.

Gulati A and Jaiwal P K (1992) *In vitro* induction of multiple shoots and plant regeneration from shoot tips of mungbean (*Vigna radiata* (L.) Wilczek). *Plant Cell Tiss. Org. Cult.*, **29**: 199–205.

Gulati A and Jaiwal P K (1994) Plant regeneration from cotyledonary node explants of mungbean (*Vigna radiata* L. Wilczek). *Plant Cell Rep.*, **15**: 500–505.

Hadiuzzaman S and Miah M A K (1989) *In vitro* organogenesis in rice bean (*Vigna umbellata*). *Bangladesh J. Bot.*, **18**: 157–162.

Hiei Y, Komari T and Kubo T (1997) Transformation of rice mediated by *Agrobacterium tumefaciens*. *Plant Mol. Biol.*, **35**: 205–218.

Hiei Y, Ohta S, Komari T and Lumashire T (1994) Efficient transformation of rice (*Oryza sativa* L.) mediated by *Agrobacterium* and sequence analysis of the boundaries of the T-DNA. *Plant J.*, **6**: 271–282.

Hoekema A, Hirsch P R, Hooykaas P J J and Schilperoort R A (1983) A binary plant vector strategy based on separation of *vir* and T-region of the *Agrobacterium tumefaciens* Ti Plasmid. *Nature*, **303**: 179–180.

Holford P, Hernandez N and Newburg H T (1992) Factors influencing the efficiency of T-DNA transfer during co-cultivation of *Antirrhinum majus* with *Agrobacterium tumefaciens*. *Plant Cell Rep.*, **11**: 196–199.

Ignacimuthu S and Arockiaswamy (1999) Plant regeneration from callus and somatic embryogenesis in *Vigna mungo* L. Hepper. In: *Proc. Natl. Sym. on Role of Plant Tissue Culture in Biodiversity and Economic Development*, held at Pantnagar, India, 7–8.

Ignacimuthu S and Franklin G (1999) Regeneration of plantlets from cotyledons and embryonic axis explants of *Vigna mungo* L. Hepper. *Plant Cell Tiss. Org. Cult.*, **55**: 75–80.

Ignacimuthu S, Franklin G and Melchias G (1997) Multiple shoot formation and *in vitro* fruiting from cotyledonary nodes of *Vigna mungo* L. Hepper. *Curr. Sci.*, **73**: 733–735.

Ishimoto M, Sato T, Chrispeels M J and Kitamura K (1996) Bruchid resistance of transgenic azuki bean expressing seed alpha-amylase inhibitor of common bean. *Entomologia Experimentalis et Applicata*, **79** (3): 309–315.

Jaiwal P K and Gulati A (1995) Current status and future strategies of *in vitro* culture techniques for genetic improvement of mungbean (*Vigna radiata* (L.) Wilczek). *Euphytica*, **86**: 167–181.

Jaiwal P K, Kumari R, Ignacimuthu S, Potrykus I and Sautter C (2001) *Agrobacterium*-mediated transformation of mungbean (*Vigna radiata*) – a recalcitrant grain legume. *Plant Sci.*, **161**: 239–247.

Jaiwal P K, Sautter C and Potrykus I (1998) *Agrobacterium rhizogenes*-mediated gene transfer in mungbean (*Vigna radiata* L. (Wilczek)). *Curr. Sci.*, **75**: 41–45.

Jefferson R A, Kavanagh T A and Bevan M W (1987) GUS fusions: β-glucuronidase as a sensitive and versatile gene fusion marker in higher plants. *EMBO J.*, **6**: 3901–3907.

Jeswani L M and Baldev B (1990) Advances in pulse production technology, Publication and Information Division, ICAR, New Delhi.

Jha T B and Roy S C (1980) Rapid callus formation at low mannitol level from protoplasts of *Vigna sinensis*. *Indian J. Exp. Biol.*, **18**: 87–89.

Joshi C P and Schieder O (1987) Isolation, culture and regeneration of legume protoplasts. In: *Proc. Sym. Plant Cell and Tissue Culture of Economically Important Plants*. (Ed Reddy G M), Hyderabad, India, 33–36.

Kartha K K, Paul K, Neung N L and Mroginski L A (1981) Plant regeneration from meristems of grain legumes: Soybean, Cowpea, Peanut, Chickpea and Bean. *Can. J. Bot.*, **59**: 1671–1679.

Karthikeyan A, Sarma K S and Veluthambi K (1996) *Agrobacterium tumefaciens*-mediated transformation of *Vigna mungo* (L.) Hepper. *Plant Cell Rep.*, **15**: 328–331.

Kim D H (1994) Highlights of mungbean at AVRDC in the 1990s. In: *Mungbean Session. Intl. Sym. on Pulses Res.*, New Delhi, April 6.

Kim J W and Minamikawa T (1996) Transformation and regeneration of French bean plants by particle bombardment process. *Plant Sci.*, **117**: 131–138.

Kim J W and Minamikawa T (1997) Stable delivery of a concanavalin promoter-β-glucuronidase gene fusion into French bean by particle bombardment. *Plant Cell Physiol.*, **38**: 70–75.

Kitamura K, Ishimoto M and Sawa M (1988) Inheritance of resistance to infestation with adzuki bean weevil in *Vigna radiata*. *Japan J. Breed.*, **38**: 459–464.

Knittel N, Grubber V, Hahne G and Lenee P (1994) Transformation of sunflower (*Helianthus annuus* L.): A reliable protocol. *Plant Cell Rep.*, **14**: 81–86.

Kohler F, Golz C, Eapen S and Schieder O (1987b) Influence of plant cultivar and plasmid DNA on transformation rates in tobacco and moth bean. *Plant Sci.*, **53**: 87–91.

Kohler F, Golz C, Eapen S, Kohn H and Schieder O (1987a) Stable transformation of moth bean (*Vigna aconitifolia*) via direct gene transfer. *Plant Cell Rep.*, **6**: 313–317.

Koluthangan S, Ganapathi A, Shajahan A and Kathiravan K (1995) Somatic embryogenesis in cell suspension culture of cowpea (*Vigna unguiculata* L. Walp). *Israel J. Plant Sci.*, **43**: 385–390.

Kononowicz A K, Narasimhan M L, Reuveni M, Moclatchey G, Bressan P H, Zhang Y, Larosa P C, Murdock L L, Chrispeels M J, Bressan R A and Hasegawa P M (1993) Genetic transformation of cowpea (*Vigna unguiculata*) using microprojectile bombardment and *Agrobacterium tumefaciens* infection. *Plant Physiol.*, **102**, suppl., Abstract no. 945.

Kumar V, Jones, B and Davey M R (1991) Transformation by *Agrobacterium rhizogenes* and regeneration of transgenic shoots of the wild soybean *Glycine argyrea. Plant Cell Rep.*, **10**: 135–138.

Kumar V and Davey M R (1991) Genetic improvement of legumes using somatic cell and molecular techniques. *Euphytica*, **55**: 159–169.

Lee N G, Stein B, Suzuki H and Verma D P S (1993) Expression of antisense nodulin-35 RNA in *Vigna aconitifolia* transgenic root nodules retards peroxisome development and affects nitrogen availability to the plant. *Plant J.*, **3**: 599–606.

Malik K A and Saxena P K (1992) Somatic embryogenesis and shoot regeneration from intact seedlings of *Phaseolus acutifolius* (L.), *P. aureus* (L.) Wilczek, *P. coccineus* (L.). and *P. whightii* (L.). *Plant Cell Rep.*, **11**: 163–168.

Marty D (1988) La tomate dans tous ses états. *Biofutur*, **72**: 43–48.

Mathews H (1987) Morphogenetic responses from *in vitro* cultured seedling explants in mungbean (*Vigna radiata* (L.) Wilczek). *Plant Cell Tiss. Org. Cult.*, **11**: 233–240.

Matsuoka M, Kai Y and Yoshida T (1997) Induction of adventitious buds and gene transfer by particle gun in cowpea (*Vigna unguiculata*). *Bulle of the Chugoku-Nat-Agri. Expt. Stat.*, **18**: 31–39.

Maximova S N, Dandekar A M and Guiltinan M J (1998) Investigation of *Agrobacterium*-mediated transformation of apple using green fluorescent protein: high expression and low stable transformation suggest that factors other than T-DNA transfer are rate-limiting. *Plant Mol. Biol.*, **37**: 549–559.

McHughen A and Jordan M C (1989) Recovery of transgenic plants from escape shoots. *Plant Cell Rep.*, **7**: 611–614.

McHughen A, Jordan M and Feist G (1989) A preculture period prior to *Agrobacterium* inoculation increases production of transgenic plants. *Plant Physiol.*, **135**: 245–248.

Mendoza A B, Hattori K and Futsuhara Y (1992) Shoot regeneration from callus of immature primary leaves in mungbean *Vigna radiata* (L.) Wilczek. *Japan J. Breed.*, **42**: 145–149.

Mendoza A B and Futsuhara Y (1990) Varietal differences on plant regeneration by tissue culture in mungbean (*Vigna radiata*). *Japan J. Breed.*, **40**: 457–467.

Mendoza A B, Hattori K, Nishimura T and Futsuhara Y (1993) Histological and scanning electron microscopic observations on plant regeneration in mungbean cotyledon (*Vigna radiata* (L.) Wilczek) cultured *in vitro. Plant Cell Tiss. Org. Cult.*, **32**: 137–143.

Meurer C A, Dinkins R D and Collins G B (1998) Factors affecting soybean cotyledonary node transformation. *Plant Cell Rep.*, **18**: 180–186.

Mohapatra U, McCabe M S, Power J B, Schepers F, Arend A V D and Davey M R (1999) Expression of *bar* gene confers herbicide resistance in transgenic lettuce. *Transgenic Res.*, **8**: 33–44.

Mukhopadhyay A, Arumugam N, Nandakumar P B A, Pradhan A K, Gupta V and Pentel D (1992) *Agrobacterium*-mediated genetic transformation of oil seed *Brassica campestris*: Transformation frequency is strongly influenced by the mode of shoot regeneration. *Plant Cell Rep.*, **11**: 506–513.

Muthukumar B, Mariamma M and Gnanam A (1995) Regeneration of plants from primary leaves of cowpea. *Plant Cell Tiss. Org. Cult.*, **42**: 153–155.

Muthukumar B, Mariamma M, Veluthambi K and Gnanam A (1996) Genetic transformation of cotyledon explants of cowpea (*Vigna unguiculata* L. Walp.) using *Agrobacterium tumefaciens. Plant Cell Rep.*, **15**: 980–985.

Nagl W and Ehemann A (1994) Transformation of protoplasts from asparagus bean, *Vigna unguiculata* ssp. *Sesquiedalis*. L., and attempts of regeneration. *Life Sci. Adv. Plant Physiol.*, **13**: 177–182.

Nagl W, Ignacimuthu S and Becker J (1997) Genetic engineering and regeneration of *Phaseolus* and *Vigna*. State of the art and new attempts. *J. Plant Physiol.*, **150**: 625–644.

Ohwi J and Ohashi H (1969) Adzuki beans of Asia. *J. Jap Bot.*, **44**: 29–33.

Orczyk A N and Orczyk W (2000) Study of the factors influencing *Agrobacterium*-mediated transformation of pea (*Pisum sativum* L.). *Mol. Breed.*, **6**: 185–194.

Pal M, Ghosh U, Chandra M, Pal K and Biswas B B (1991) Transformation and regeneration of mungbean (*Vigna radiata*). *Indian J. Biochem. Biophys.*, **28**: 428–435.

Pandey P and Bansal Y K (1989) Plantlet formation from callus cultures of cowpea (*Vigna sinensis* L.). *Curr Sci.*, **58**: 394–395.

Park S H, Pinson S R M and Smith R H (1996) T-DNA integration into genomic DNA of rice following *Agrobacterium* inoculation of isolated shoot apices. *Plant Mol. Biol.*, **32**: 1135–1198.

Patel M B, Bhardwaj R and Joshi A (1991) Organogenesis in *Vigna radiata* (L.) Wilczek *Indian J. Exp. Biol.*, **29**: 619–622.

Pellegrineshi A (1997) *In vitro* plant regeneration via organogenesis of cowpea (*V. unguiculata* L. Walp.). *Plant Cell Rep.*, **17**: 89–95.

Penza R, Lurquin P F and Fillipone E (1991) Gene transfer by co-cultivation of mature embryos with *Agrobacterium tumefaciens*: Application to cowpea (*Vigna unguiculata* Walp.). *J. Plant Physiol.*, **138**: 39–42.

Petit A, Berkaloff A and Tempe J (1986) Multiple transformation of plant cells by *Agrobacterium* may be responsible for complex organization of T-DNA in crown gall and hairy root. *J. Mol. Genet.*, **202**: 388–398.

Peyachoknagul S, Phonjun C, Pongtongkam P, Suputtitad S and Ngernsiri L (1996) Gene transformation of mungbean (*Vigna radiata*). *Kasetsart J. (Nat. Sci.)*, **30**: 303–311.

Phogat S K, Karthikeyan A S and Veluthambi K (1999) Generation of transformed calli of *Vigna radiata* (L.) Wilczek by *Agrobacterium tumefaciens*-mediated transformation. *J. Plant Biol.*, **26** (1): 77–82.

Pigeaire A, Abernethy D, Smith P M, Simpson K, Fletcher N, Lu Chin Yi, Atkins C A and Cornish E (1997) Transformation of a grain legume (*Lupinus angustifolius* L.) via *Agrobacterium tumefaciens* mediated gene transfer to shoot apices. *Mol. Breed.*, **3**: 341–349.

Ponappa T, Brzozowski A E and Finer J J (1999) Transient expression and stable transformation of soybean using jellyfish green fluorescent protein. *Plant Cell Rep.*, **19**: 6–12.

Porter J R (1991) Host range and implications of plant infection by *Agrobacterium rhizogenes*. *Crit. Rev. Plant Sci.*, **10**: 387–421.

Potrykus I (1990) Gene transfer to plants: assessment and perspectives. *Physiol Plant*, **79**: 125–134.

Prem Anand R, Ganapathi A, Vengadesan G, Selvaraj N, Anbazhagan V R and Kulothungan S (2001) Plant regeneration from immature cotyledon derived callus of *Vigna unguiculata* (L.) Walp. *Curr. Sci.*, **80**: 671–674.

Rashid K A, Smartt J and Haq N (1988) Hybridisation in the genus *Vigna*. In: *Proc. Second International Symposium on Mungbean*. AVRDC, Bangkok, Thailand, 205–214.

Rech E L, Golds T J, Husnain T, Vainstein M H, Jones B, Hammatt N, Mulligan B J and Davey M R (1989) Expression of a chimeric kanamycin resistance gene introduced into the wild soybean *Glycine canescens* using a cointegrate Ri plasmid vector. *Plant Cell Rep.*, **8**: 33–36.

Riazuddin S and Husnain T (1993) Transformation in chickpea (*Cicer arietinum* L.). In: *Biotechnology in Agriculture and Forestry, Vol. 23. Plant Protoplasts and Genetic Engineering. IV*. (Ed Bajaj Y P S). Springer-Verlag, Berlin, 183–193.

Ritala A, Aspergren K, Kurten U, Marttila M S, Mannonen L, Hannus R, Kauppinen V, Teeri T H and Enari T M (1994) Fertile transgenic barley by particle bombardment of immature embryos. *Plant Mol. Biol.*, **24**: 317–325.

Sahoo L, Sushma, Sugla T, Singh N D and Jaiwal P K (2000) *In vitro* plant regeneration and recovery of cowpea (*Vigna unguiculata*) transformants via *Agrobacterium*-mediated transformation. *Plant Cell Biotech. Mol. Biol.*, **1**: 47–54.

Sato T (1995) Basic study of biotechnology in adzuki bean (*Vigna angularis* Ohwi & Ohashi). *Report of Hokkaido Pref Agri. Exp. Station*, **87**: 1–68.

Sato T, Asaka D, Harada T and Matsukawa I (1993) Plant regeneration from protoplasts of Adzuki bean (*Vigna angularis* Ohwi & Ohashi). *Japan J. Breed.*, **43**: 183–191.

Schoneberg J M, Scelonge C J, Burrus M and Bidney D L (1994) Stable transformation of sunflower using split embryonic axis explants. *Plant Sci.*, **103**: 199–207.

Schrammeijer B, Sijmons P C, Elzen P J M and Hoekema A (1990) Meristem transformation of sunflower by *Agrobacterium*. *Plant Cell Rep.*, **9**: 55–60.

Schroeder H, Schotz A, Richardson W T, Spencer D and Higgins T (1993) Transformation and regeneration of two cultivars of pea (*Pisum sativum* L.) *Plant Physiol.*, **101**: 751–757.

Sen J and Guha-Mukherjee S (1998) *In vitro* induction of multiple shoots and plant regeneration in *Vigna. In Vitro Cell. Dev. Biol. Plant.*, **34**: 276–280.

Sharma K K and Anjaiah V (2000) An efficient method for the production of transgenic plants of peanut (*Arachis hypogaea*. L) through *Agrobacterium tumefaciens*-mediated genetic transformation. *Plant Sci.*, **159**: 7–19.

Shelp B J, Swanton C J and Hall J C (1992) Glufosinate (Phosphinothricin) mobility in young soybean shoots. *J. Plant Physiol.*, **139**: 626–628.

Siefkes-Boer H J, Noonam M J, Bullock D W and Connor A J (1995) Hairy root transformation system in large seeded grain legumes. *Israel J. Plant Sci.*, **43**: 1–5.

Singh B D, Rao G S R L and Singh R P (1982) Polyphenol accumulation in callus cultures of cowpea (*Vigna sinensis*). *Indian J. Exp. Biol.*, **20**: 387–389.

Singh D P, Sharma B L and Dwivedi S (1983) Inheritance of hard seeds in interspecific crosses in mungbean. *Indian J. Genet.*, **43**: 378–379.

Singh N D, Kumar P A and Jaiwal P K (2002) *In vitro* regeneration and genetic transformation of pigeonpea. In: *Applied Genetics of Leguminoseae Biotechnology* (Eds Jaiwal P K and Singh R P). Kluwer Acad. Publ., Dordrecht, Netherlands (in press).

Singh R P, Singh B D, Singh R M and Jaiswal H (1985) Genotypic differences in callus growth and organogenesis in green gram. *Indian J. Agri. Sci.*, **55**: 612–615.

Smartt J (1990) *Grain legumes: Evolution and Genetic Resources*. Cambridge University Press, Cambridge.

Sonia, Preeti, Sharma P, Ragini and Jaiwal P K (2000) Application of biotechnology and molecular biology for improvement of chickpea (*Cicer arietinum* L.). In: *Recent Advances in Biotechnology* (Ed Trivedi P C). Panima Publ., New Delhi, 135–153.

Stachel S E, Messens E, Van Montzgu M and Zambryski P (1995) Identification of the signal molecules produced by wounded plant cells that activate T-DNA transfer in *Agrobacterium tumefaciens. Nature*, **318**: 624–629.

Stachel S E, Nester E W and Zambryski P C (1986) A plant cell factor induces *Agrobacterium tumefaciens vir* gene expression. *Proc. Natl. Acad. Sci. USA*, **83**: 379–383.

Suzuki H, Fowler T and Tierney M (1993) Deletion analysis and localization of sbPRP1, a soybean cell wall protein gene, in roots of transgenic tobacco and cowpea. *Plant Mol. Biol.*, **21**: 109–119.

Takahashi W, Matsushita J, Kobayashi T, Tanaka O and Beppu T (1998) Plant regeneration from epicotyl segment and callus of *Vigna angularis* cv Tanbadainagon. *Japan J. Crop. Sci.*, **67**: 561–567.

Thomas J C and Khatterman F R (1986) Cytokinin activity induced by thidiazuron. *Plant Physiol.*, **81**: 681–683.

Tomooka N, Lariungreang C, Nakeeraks P, Egawa Y and Thavarasook C (1992) Development of bruchid-resistant mungbean line using wild mungbean germplasm in Thailand. *Plant Breed.*, **109**: 60–66.

Verdcourt B (1969) New combinations of *Vigna* savi. (Leguminoseae-Papilionoideae). *Kew Bull.*, **23**: 464.

Xu Z H, Davey M R and Cocking E C (1981) Isolation and sustained division of *Phaseolus aureus* root protoplasts. *Z. Pflanzenphysiol.*, **104**: 289–298.

Yamada T, Teraishi M, Hattori K and Ishimoto M (2001) Transformation of azuki bean by *Agrobacterium tumefaciens. Plant Cell Tiss. Org. Cult.*, **64**: 47–54.

Zapata C, Park S H, El-Zik K M and Swith R H (1999) Transformation of Texas cotton cultivar by using *Agrobacterium* and the shoot apex. *Theor. Appl. Genet.*, 98: 252–256.

Zhang Z, Xing A, Staswick P and Clemente T E (1999) The use of glufosinate as a selective agent in *Agrobacterium*-mediated transformation of soybean. *Plant Cell Tiss. Org. Cult.*, **56**: 37–46.

Zupan J and Zambryski P (1997) The *Agrobacterium* DNA transfer complex. *Crit. Rev. Plant Sci.*, **16**: 279–295.

IN VITRO REGENERATION AND GENETIC TRANSFORMATION OF *VICIA* SPECIES

THOMAS PICKARDT
Institute for Applied Genetics, Free University of Berlin,
Albrecht-Thaer-Weg 6, 14195 Berlin, Germany
e-mail: pickardt@zedat.fu-berlin.de

Abstract

Vicia species have been subjected to cell and tissue culture experiments for approximately 40 years. The economically most important species of this genus, the faba-bean, continuously exhibited (like many other grain legumes) a rather low amenability to *in vitro* conditions. Not only the low capability to regenerate shoots or somatic embryos from dedifferentiated tissue, but also the fact that explants and callus cells of *Vicia faba* tend to produce high amounts of phenolic compounds resulting in subsequent mortifying of the tissue, has strongly limited the application of genetic engineering techniques in this species. As a consequence, among the major grain legume crops *Vicia faba* was the last species where the production of transgenic plants has (only recently) been reported. In contrast, the closely related species *Vicia narbonensis* could relatively easily be manipulated *in vitro*. In this chapter, an overview on tissue culture and genetic manipulation of *Vicia* species has been described.

1. Introduction

The genus *Vicia* comprises approximately 150 species of annual and perennial herbaceous legumes distributed in temperate and subtropical areas of the world (Maxted *et al.*, 1991; Maxted, 1993). Some *Vicia* species have been utilized in agriculture since ancient times. Today *Vicia faba* (faba-, field-, or broad bean) is by far the economically most important species of the genus. With a world production of 3.5 million tons in 1999 (FAO), *Vicia faba* ranks among the most important grain legume crops. The largest producer in 1999 was China (2 million tons), followed by Egypt (0.3), Ethiopia (0.28), Australia (0.15) and Morocco (0.1). Other *Vicia* species that have been used as grain legumes for animal feeding are *Vicia sativa* L., *V. narbonensis* L., *V. villosa* Roth, *V. monantha*, and *V. ervilia* (Becker-Dillingen, 1929). Certain *Vicia* species are also used as cover crops, green manure, or livestock forage (Allen and Allen, 1981).

P. K. Jaiwal and R. P. Singh (eds.), Applied Genetics of Leguminosae Biotechnology, 121–131.

Today breeding programs of *Vicia faba* could be supplemented by recombinant DNA-technlogy, which requires, however, the development of reproducible protocols for *in vitro*-manipulation of single cells or explant tissue and the subsequent regeneration of plants. *Vicia faba* exhibits, like most other grain legumes, a rather low amenability to tissue culture conditions, mainly owing to difficulties in the regeneration of shoots or somatic embryos from dedifferentiated tissues. The evaluation of other species of the genus revealed that at least one of the closest relatives of *Vicia faba, Vicia narbonensis*, shows a considerably higher amenability to *in vitro* manipulation. The aim of this review is to provide an overview on the present state in tissue culture and genetic manipulation of *Vicia faba* and related species.

2. Regeneration from explant tissue

The first attempts to cultivate *Vicia faba in vitro* were focused on the optimal growth of callus tissue or suspension cultures rather than the induction of shoot morphogenesis and plant regeneration (Venketesvaran, 1962; Grant and Fuller, 1968; Mitchell and Gildow, 1975). The influence of media composition and explant source on the initiation and maintenance of cultures were tested and the conditions were optimized for a maximum increase in callus fresh weight. Poor growth rates and an increase of necrotic tissue during cultivation were described. Suspension cultures could also be established in *Vicia hajastana* (Singh *et al.*, 1972) which served as source for protoplast isolation in later studies (Kao *et al.*, 1974; see below). Since *Vicia faba* is an excellent karyological object, studies on callus growth were also performed by Cionini *et al.* (1978) and Jalaska *et al.* (1981) to investige the pattern of chromosomal instability during callus development. The low morphogenic potential of *Vicia faba* cells cultured *in vitro* was first mentioned by Röper (1979). With the aim to develop a system for plant regeneration from single cells, he established callus and cell suspension cultures over a long period of time, but all attempts to initiate shoot regeneration were unsuccessful.

During the following decade a number of reports were published describing the cultivation of tissues containing shoot apical meristems and the subsequent recovery of shoots (Martin *et al.*, 1979; Cheyne and Dale, 1980; Galzy and Hamoui, 1981; Schulze *et al.*, 1985; Busse, 1986; Fakhrai *et al.*, 1989; Selva *et al.*, 1989; Taha and Francis, 1990). Excised apical meristems, nodal buds and cotyledonary nodes were exposed to media containing cytokinins (in most cases BA) alone or in combination with low amounts of an auxin. Under these conditions shoots preferentially develop from preexisting meristems. Since cytokinins are effective in removing apical dominance (Skoog and Schmitz, 1972) this treatment continuously promotes the development of young meristems of the apical dome(s) (Steves and Sussex, 1989) to lateral buds and shoots, which are again themselves suppressed in their further growth. This cycle of simultaneous induction and suppression finally gives rise to multiple bud/shoot proliferation. Similar protocols have repeatedly been described for many other grain legumes (e.g. Cheng *et al.*, 1980; Kartha *et al.*, 1981; Martins, 1983; Griga *et al.*, 1986; Jackson and Hobbs, 1990; Malik and Saxena, 1992a, b; Brandt and Hess, 1994). The origin of shoots (axillary vs. adventitious) in these regeneration systems has frequently been discussed and investigated in histological studies. The development of shoot initials from superficial layers and the absence of a vascular connection to axillary buds are generally taken as evidence for a *de novo* formation of shoots (e.g. Fakhrai *et al.*, 1989). Since both attributes apply also for the young shoot meristems of the apical dome, these observations do not provide proof for an adventitious origin of shoots. However, even if the distinction between *de novo* organogenesis and shoot proliferation from preexisting meristems is not trivial, this question may finally be less relevant, considering the development of transformation systems as the main purpose of these studies. Today several transformation protocols in grain legumes are based on the 6-benzyladenine (BA) induced shoot development from embryo axes and cotyledonary nodes, using *Agrobacterium tumefaciens* (e.g. Bean *et al.*, 1997/pea; Sato *et al.*, 1993/soybean) or the biolistic approach (McCabe *et al.*, 1988/soybean; Brar *et al.*, 1994/peanut; Russel *et al.*, 1993; Aragão *et al.*, 1996/common bean). The recovery of both, clonal as well as chimeric primary transformants were repeatedly described in these studies (e.g. Christou and McCabe, 1992; Sato *et al.*, 1993), indicating the occurrence of single- and multiple-cell origin of shoots. Chimeric transformants do not necessarily limit the value of these systems. A number of chimeric individuals are germline (L2)-transformants giving rise to clonal transformants in the progeny. A serious constraint in *Vicia faba* tissue culture is the deterioration of explant material and cultivated tissue as a result of the action of phenolic compounds. Bieri *et al.* (1984) and Selva *et al.* (1989) examined the effect of various chemical and physical parameters in axillary shoot cultures. In their studies low temperatures were found to limit the formation of phenolics.

Plantlet regeneration from explants lacking preexisting shoot meristems was claimed by Thynn and Werner (1987). Callus was initiated from epicotyl segments on B_5 basal medium (Gamborg *et al.*, 1968) supplemented with 0.2 mg/L, naphthaleneacetic acid (NAA), and shoot development was achieved on a subsequent transfer to B_5 medium containing 0.05 mg/L NAA and 0.5 mg/L kinetin. With the exception of the study of Tegeder *et al.* (1995) on protoplast regeneration (see below), this is the only report of shoot regeneration from an explant without apical or axillary shoot meristems in *Vicia faba*. However, further informations didn't follow and attempts to reproduce these results in our laboratory were unsuccessful.

Somatic embryogenesis in callus and suspension cultures derived from immature cotyledons of *Vicia faba* was reported by Griga *et al.* (1987). They followed the traditional pathway of somatic embryogenesis induction: 2,4-dichlorophenoxyacetic acid (2,4-D) initiation of callus and subsequent lowering or removal of 2,4-D (Ammirato, 1983). The formation of bipolar structures were observed. These structures did obviously not contain shoot meristems, as only root development occurred on further cultivation.

Somatic embryogenesis was also described for *Vicia narbonensis* in two independent reports. In a study of Albrecht and Kohlenbach (1989) leaf-derived callus was cultured for 5–6 months in a series of different Murashige and Skoog's (MS) media supplemented with combinations of picloram/BA, 2,4-D/BA and 2,4-D/Kinetin. Somatic embryo development resulted from a last step on hormone free-MS medium. Further development to plants was not reported. Pickardt *et al.* (1989) described a protocol in which shoot tips from young seedlings cultivated on MS medium containing 0.1–10 mg/L 2,4-D gave rise to callus which formed somatic embryos if 2,4-D was removed and replaced by 1 mg NAA in a subsequent cultivation step (Fig. 1a and b). Plantlets derived from somatic

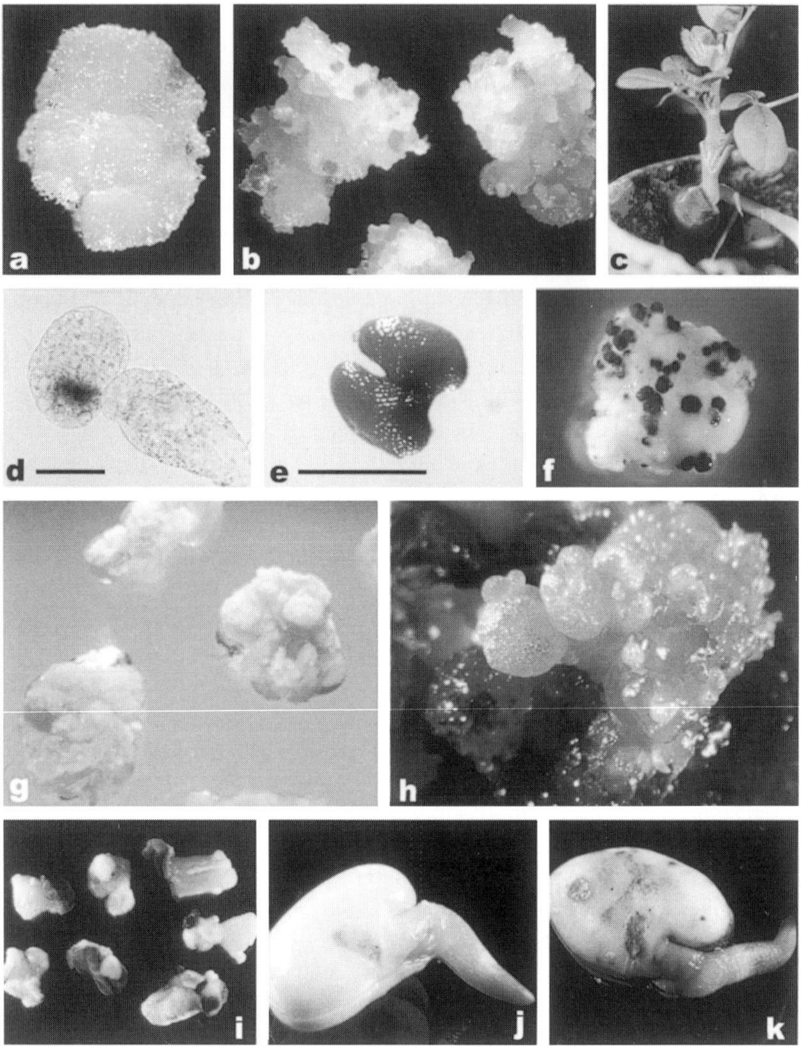

Figure 1. Regeneration and transformation in Vicia narbonensis *and* Vicia faba. *a–c: somatic embryogenesis in* V. narbonensis*, (a) dedifferentiated callus on picloram-containing medium, (b) somatic embryo development on medium with reduced levels of auxin, (c) recovery of mature plants by grafting somatic embryo-derived shoots on young seedlings; d–f: expression of GUS under control of the embryo-specific usp-promoter in zygotic and somatic embryos of transgenic* V. narbonensis, *(d) globular stage zygotic embryo (suspensor cells still attached, bar represents 0.1 mm), (e) early cotyledonary stage zygotic embryo (bar 0.5 mm), (f) somatic embryos of the same transgenic line; g–h: thidiazuron-induced shoot morphogenesis in* Vicia faba, *(g) callus induced from stem segments, (h) regeneration of shoot initials on thidiazuron-containing medium; i–k: TR-promoter-GUS expression in* Vicia faba, *(i) kanamycin-selected calli subjected to histochemical GUS analysis, X-Gluc-stained seed from WT (j) and from a transgenic line of* Vicia faba *(k).*

embryos could be grown to maturity. In this regeneration system, the frequency of explants producing embryogenic callus as well as the mean number of embryos per embryogenic callus increased with the 2,4-D concentration initially used for callus induction. Later studies of T. Pickardt and coworkers demonstrated that shoot tips could be replaced by epicotyl- and stem segments (unpublished data, 1990). An improved protocol was subsequently combined with the *Agrobacterium tumefaciens*-mediated gene transfer (Pickardt *et al.*, 1991; see below).

The results of Albrecht and Kohlenbach (1989) and of Pickardt (1991) revealed that, at least in *Vicia narbonensis*, cells of a differentiated tissue (lacking apical or axillary meristems) like leaf sections or stem segments can be triggered back to an embryogenic state. Compared to *Vicia faba*, where by that time regeneration of plants had occurred most probably only from preexisting shoot meristems, the *in vitro* cultured cells of the less domesticated *Vicia narbonensis* obviously possess a higher morphogenic potential.

3. Regeneration from protoplasts

Vicia hajastana protoplasts were isolated from suspension cultures in studies on protoplast fusion (Kao and Michayluk, 1974; Kao *et al.*, 1974) and in order to evaluate the nutritional requirements for growth of protoplasts at a low population density (Kao and Michayluk (KM), 1975). In these studies division of protoplast and callus formation was achieved. Donn (1978) described the isolation of protoplasts from leaves in *Vicia narbonensis*. A 7-day preculture of leaves on MS-medium containing 0.5 mg/L of each BA and *p*-chlorophenoxyacetic acid appeared to be crucial to obtain cell division and callus formation. The addition of asparagine, glutamine, and serine enhanced the rate of division. Attempts to initiate shoot regeneration in protoplast derived calli were unsuccessful. Only root formation occurred in a low frequency on a medium supplemented with 0.1 mg/L BA and 0.1–0.5 mg/L *p*-chlorophenoxyacetic.

Binding and Nehls (1978a, b) isolated protoplast from leaves and shoot apices of *Vicia faba*. Division of protoplasts could be initiated and maintained in KM medium or a combination of V-47 medium (Binding, 1974) and KM medium containing 0.5 mg/L BA, 1 mg/L NAA and 0.1 mg/L 2,4-D. Highest plating efficiencies were achieved if less than 3×10^3 protoplasts per ml of *Vicia faba* were cocultured with $0.5–10 \times 10^4$ protoplasts of *Petunia hybrida*. The authors suggest that the coculture effect is probably congruent to a feeding layer sytem (Binding and Nehls, 1978a). In further experiments on somatic cell hybridization between *Vicia faba* and *Petunia hybrida* the same authors obtained three hybrid clones, one of them could be propagated at least for 9 months. The fusion hybrids contained predominantly nuclei or chromosomes of one or the other species and a few chromosomes of the second parent (Binding and Nehls, 1978b). In both studies initiation of shoot or root morphogenesis was not reported.

Protoplasts were also isolated from suspension cells of *Vicia faba* (Röper, 1981). In KM medium supplemented with 0.5 mg/L BA, 0.2 mg/L 2,4-D and 0.5 mg/L IAA protoplasts divided and formed cell colonies that gave rise to proliferating calli. Regeneration of shoots was not described. During the next 14 years no reports on protoplast regeneration in *Vicia* species were published. Tegeder *et al.* (1995) demonstrated for the first time the recovery of mature plants from protoplasts of *Vicia faba*. Protoplasts were isolated from shoot tips of etiolated seedlings and embedded in alginate discs. In a screening of 10 cultivars of *Vicia faba* plating efficiencies, survival rates, and the regenerative competence were evaluated. Depending on cultivar, division rates of up to 40% were obtained in a KM medium containing 0.5 mg/L of each BA, 2,4-D and NAA, and protocallus development occurred at high frequency. The apparent key step in this study was the application of the phenyl-urea herbicide thidiazuron (Mok *et al.*, 1982) in a subsequent culture phase on solidified medium. In the german cultivar 'Mythos' shoot initials appeared with a frequency of 5–8% after 3–8 months (varying between individual calli), from which fertile plants could be recovered. A comparison of 30 different phytohormone compositions revealed that shoot morphogenesis occurred only in calli exposed to thidiazuron.

The extension of these studies to *Vicia narbonensis* (Tegeder *et al.*, 1996) showed that also in this species plant regeneration from protoplast-derived calli via shoot morphogenesis can be achieved using the thidiazuron pathway. In addition, the protocol for the induction of somatic embryogenesis (originally developed for shoot tip-derived calli of *Vicia narbonensis*, Pickardt *et al.*, 1989) was also successful. It is important to mention, that in *Vicia narbonensis* regeneration via somatic embryogenesis occurs with a considerable higher efficiency compared to shoot morphogenesis induced by thidiazuron. Unfortunately all attempts to induce somatic embryogenesis in *Vicia faba* were unsuccessful so far (T. Pickardt, 1988 unpublished; Tegeder *et al.*, 1995).

One of the problems in breeding of *Vicia faba* is its sexual incompatibility with other *Vicia* species, limiting broadening of germplasm and the creation of new variability. Desirable agronomic traits like drought tolerance, insect and fungal resistance are available, e.g. in *Vicia narbonensis* (Lawes *et al.*, 1983). Hybridization between *Vicia faba* and *Vicia narbonensis* by conventional techniques, including embryo rescue, has not been successful so far (Cubero, 1982; Ramsay and Pickersgill, 1986; Lazaridou and Roupakias, 1993). The development of effective protocols for plant regeneration from protoplasts in both species establishes the prerequisites for attempting somatic hybridization between *Vicia faba* and *Vicia narbonensis*. Moreover, the availability of transgenic lines containing different selection markers in both species (Pickardt *et al.*, 1991; Böttinger *et al.*, 1997; see below) will allow an efficient selection of hybrid cells.

4. Transformation

The first report on the transfer of chimeric genes into *Vicia faba* was published by Schiemann and Eisenreich (1989). Seedlings were inoculated with *Agrobacterium tumefaciens*/*rhizogenes* strains containing the binary vector pGSGluc1 transferring neomycin phosphotransferase-II (NPTII) and β-glucuronidase (GUS) under control of the bidirectional TR1/2-promoter. GUS positive roots developing at the inoculation sites could be propagated on hormone-free medium. Callus established from these roots maintained GUS activity. Regeneration of shoots was not reported. A similar study was performed by Ramsay and Kumar (1990), using *A. rhizogenes*/pBin19 (Bevan *et al.*, 1984) transferring *nptII* to cotyledons and stem tissues of *Vicia faba*. The transgenic nature of established root clones were confirmed by hormone autotrophy and NPTII dot blot assays. Attempts to initiate shoot morphogenesis in root cultures were not described. Quandt *et al.* (1993) reported *A. rhizogenes*-mediated transformation and hairy root induction on wounded epicotyls of *Vicia hirsuta*. The aim of these investigations was to study gene expression in transgenic root nodules. Regeneration of shoots was not attempted.

Pickardt *et al.* (1991) reported the recovery of five transgenic plantlets (T_0) of *Vicia narbonensis*, transformed with *Agrobacterium tumefaciens* strain C58C1/3850*hpt* carrying the gene for hygromycin phosphotransferase and the nopaline synthase gene. The integration of the T-DNA into the plant genome of the primary transformants was confirmed by Southern analysis. In this study the regeneration protocol for somatic embryogenesis (Pickardt *et al.*, 1989) was successfully combined with the *Agrobacterium tumefaciens*-mediated gene transfer. Initially, failure in rooting of transformed shoots prevented the recovery of mature plants. This bottleneck was overcome by grafting shoots on wildtype seedlings under *in vitro* conditions (see Pickardt *et al.*, 1995 for details, Fig. 1c). Three out of the five plantlets (each containing a single T-DNA integration) could be grown to maturity and produced seeds. The progeny analysis revealed a 3 : 1 segregation of the foreign genes, confirming the stable integration of the T-DNA at a single locus (Meixner *et al.*, 1994, 1996). The transformation system developed for *Vicia narbonensis* has been further optimized and is now a routine technique. Up to now more than 200 independent transgenic lines containing various selection markers, NPTII, hygromycin phosphotransferase (HPT), phosphinothricin acetyltransferase (PAT), methionine rich proteins (2S albumin from brazil nut) (Saalbach *et al.*, 1994; Pickardt *et al.*, 1995; Saalbach *et al.*, 1995) or the *uidA* gene under control of different seed specific promoters (Pickardt *et al.* 1998, Fig. 1d–f) have been developed. The protocol could successfully be reproduced by other research groups for the production of transgenic *Vicia narbonensis* lines (Weber *et al.*, 1998; Czihal *et al.*, 1999).

In order to develop a transformation system for *Vicia faba* we followed a procedure based on a protocol for *de novo* regeneration of shoot initials from dedifferentiated cells that was initially developed for plant regeneration from protoplasts (Tegeder *et al.*, 1995, Fig. 1g and h), combined with the *Agrobacterium*-mediated gene transfer. Shoot morphogenesis is achieved from internodal stem segments after a callus phase and a shoot initiation phase, using thidiazuron as growth regulator. A subsequent transfer of shoot-buds to BA-containing medium is necessary for stem-elongation and leaf-development. As described for *Vicia narbonensis*, plants are finally recovered from regenerated shoots by grafting.

The first transgenic line of *Vicia faba* transmitting *nptII* and *uidA* to the progeny in a Mendelian fashion was presented by Böttinger *et al.* (1997). Stem segments of cultivar 'Mythos' were inoculated with the *Agrobacterium tumefaciens*-strain EHA101/pGSGluc1 (transferring *uidA*/*nptII* under control of the bidirectional TR1/2 promoter) and cultivated on MS-medium containing 0.5 mg/L of each thidiazuron, 2,4-D and NAA for callus induction and 100 mg/L kanamycin as selection agent (Fig. 1i). In order to obtain shoot regeneration resistant calli were subsequently exposed to MS medium supplemented with 7.5 mg/L thidiazuron and 0.75 mg/L NAA (Tegeder *et al.*, 1995). Four out of 200 selected calli exposed to regeneration medium developed shoots. Shoots from one resistant callus line could be grown up to a fertile plant while the shoots recovered from the other lines showed morphological aberrations. Stable transmission of the T-DNA to the progeny could be demonstrated by GUS analysis (Fig. 1j and k) and Southern hybridization. The current transformation protocol for *Vicia faba* is still time consuming and of relative low efficiency, but could repeatedly be applied for the stable introduction of various foreign genes into faba-bean (Böttinger *et al.*, 2001).

5. Conclusions and future prospects

While in several grain legumes like soybean, peanut and pea, a considerable progress towards the development of transformation protocols to a routine level could be achieved during the recent years, *in vitro* and transformation techniques are only poorly developed in faba bean. This can be attributed not only to the particular low amenability of *Vicia faba* to *in vitro* conditions, but to a lack of sufficient funding as well. Nevertheless, the first steps in establishing a transformation protocol for faba bean were recently reported (Böttinger *et al.*, 1997). Possible applications of recombinant DNA-technology in breeding of *Vicia faba* are obvious: e.g. fungal resistance (Hain *et al.*, 1993; Jach *et al.*, 1995), nuclear encoded, artificial male sterility (Denis *et al.*, 1993), or the generation of glyphosate resistant cultivars (Padgette *et al.*, 1995) to control *Orobanche* infestation.

Vicia narbonensis has initially been chosen as a 'model' grain legume, due to its relatively high amenability for tissue culture (Pickardt *et al.*, 1989) and genetic manipulation (Pickardt *et al.*, 1995). Today the *Agrobacterium tumefaciens*-based transformation protocol for *Vicia narbonensis* belongs to the most efficient gene transfer methods in grain legumes. Moreover, even if presently only of low economical importance, *Vicia narbonensis* has become attractive as grain and forage crop for dryland farming (Siddique *et al.*, 1996). Breeding programs, e.g. aiming the reduction of antinutritive factors like glutamyl-ethenyl-cystein (Enneking, 1995), could already benefit from the advanced state of the methodology.

As a main source of protein in human consumption, the importance of grain legumes is likely to increase during the next century. Biotechnology offers the opportunity to supplement conventional breeding practices in broadening the natural gene pool and in reducing time and costs of breeding programs. In the most important species of the genus *Vicia*, the faba bean, techniques for *in vitro* culture and genetic manipulation are still lagging behind those for many other crop plants. However, the recently obtained results encourage us to dedicate further research to a plant species which is known to be very difficult to manipulate *in vitro*.

References

Albrecht C and Kohlenbach H W (1989) Induction of somatic embryogenesis in leaf-derived callus of *Vicia narbonensis* L. *Plant Cell Rep.*, **8**: 267–269.

Allen O N and Allen E K (1981) *The Leguminosae*. Macmillan Publishers, London.

Ammirato P V (1983) Embryogenesis. In: *Handbook of Plant Cell Culture* Vol 1 (Eds Evans D A, Sharp W R, Ammirato P V and Yamada Y), Macmillan, New York, 82–123.

Aragão F J L, Barros L M G, Brasileiro A C M, Ribeiro S G, Smith F D, Sanford J C, Faria J C and Rech E L (1996) Inheritance of foreign genes in transgenic bean (*Phaseolus vulgaris* L.) co-transformed via particle bombardment. *Theor. Appl. Genet.*, **93**: 142–150.

Bean S J, Gooding P S, Mullineaux P M and Davies D R (1997) A simple system for pea transformation. *Plant Cell Rep.*, **16**: 513–519.

Becker-Dillingen J (1929) Handbuch des Hülsenfrüchteanbaues und Futterbaus. Paul Parey, Berlin, Germany.

Bevan M W (1984) Binary *Agrobacterium tumefaciens* vectors for plant transformation. *Nucleic Acids Res.*, **12**: 8711–8721.

Bieri V, Schmid J and Keller E R (1984) Shoot tip culture in *Vicia faba* L. In: *Efficiency in Plant breeding* (Eds Lange W, Zeven A C and Hogenboom N F), Proceedings of the 10th Congress of the European Association for Research on Plant Breeing, EUCARPIA, Wageningen, The Netherlands, 295.

Binding H (1974) Cell cluster formation by leaf protoplasts from axenic cultures of haploid *Petunia hybrida* L. *Plant Sci. Lett.*, **2**: 185–188.

Binding H and Nehls R (1978a) Regeneration of isolated protoplasts of *Vicia faba* L. *Z. Pflanzenphysiol.*, **88**: 327–332.

Binding H and Nehls R (1978b) Somatic cell hybridization of *Vicia faba* and *Petunia hybrida. Mol. Gen. Genet.*, **164**: 137–143.

Böttinger P, Gebhardt D, Steinmetz A, Schieder O and Pickardt T (1997) *Agrobacterium*-mediated transformation of *Vicia faba* cv. 'Mythos'. III. International Food Legume Research Conference, Adelaide, Australia, Abstracts, 84.

Böttinger P, Steinmetz A, Schieder O and Pickardt T (2001) *Agrobacterium*-mediated transformation of *Vicia faba. Mol. Breed.*, **8**: 243–254.

Brandt E B and Hess D (1994) *In vitro* regeneration and propagation of chickpea (*Cicer arietinum* L.) from meristem tips and cotyledonary nodes. *In Vitro Cell Dev. Biol. Plant.*, **30**: 75–80.

Brar G S, Cohen B A, Vick C L and Johnson G W (1994) Recovery of transgenic peanut (*Arachis hypogaea* L.) plants from elite cultivars utilizing ACCELL(R) technology. *Plant J.*, **5**: 745–753.

Busse G (1986) *In vitro* cultivation of *Vicia faba* and induction of morphogenesis. *Biol. Zentralbl.*, **105**: 97–104.

Cheng T, Saka H and Voqui-Dinh T H (1980) Plant regeneration from soybean cotyledonary node segments in culture. *Plant Sci. Lett.*, **19**: 91–99.

Cheyne V and Dale P J (1980) Shoot tip culture in forage legumes. *Plant Sci. Lett.*, **19**: 303–309.

Christou P and McCabe D E (1992) Prediction of germ-line transformation events in chimeric R_O transgenic soybean plantlets using tissue-specific expression patterns. *Plant J.*, **2**: 283–290.

Cionini P, Bennici A and D'Amato F (1978) Nuclear cytology of callus induction and development *in vitro. Protoplasma*, **96**: 101–112.

Cubero J I (1982). Interspecific hybridization in *Vicia*. In: *Faba Bean Improvement* (Eds Hawtin G and Webb C), ICARDA, The Netherlands, 91–108.

Czihal A, Conrad B, Buchner P, Brevis R, Farouk A A, Manteuffel R, Adler K, Wobus U, Hofemeister J and Baeumlein H (1999) Gene farming in plants: Expression of a heat-stable *Bacillus* amylase in transgenic legume seeds. *J. Plant Physiol.*, **155**: 183–189.

Denis M, Delourme R, Gourret J P, Mariani C and Renard M (1993) Expression of engineered nuclear male sterility in *Brassica napus*. Genetics, morphology, cytology and sensitivity to temperature. *Plant Physiol.*, **101**: 1295–1304.

Donn G (1978) Cell division and callus regeneration from leaf protoplasts of *Vicia narbonensis. Z. Pflanzenphysiol*, **86**: 65–75.

Enneking D (1995) The toxicity of *Vicia* species and their utilisation as grain legumes. Ph.D. Thesis, Clima, Adelaide, Australia.

Fakhrai H, Fakhrai F and Evans P K (1989) *In vitro* culture and plant regeneration in *Vicia faba* ssp. equina cultivar spring blaze. *J. Exp. Bot.*, **40**: 813–818.

Galzy R and Hamoui M (1981) Induction de l'organogénèse sur des cals de *Vicia faba* minor provenant d'apex. *Can. J. Bot.*, **59**: 203–207.

Gamborg O L, Miller R and Ojima K (1968) Nutrient requirements of suspension cultures of soybean root cells. *Exp. Cell. Res.*, **50**: 148–151.

Grant M and Fuller K W (1968) Tissue culture of root cells of *Vicia faba. J. Exp. Bot.*, **19**: 667–680.

Griga M, Kubalakova M and Tejklova E (1987) Somatic embryogenesis in *Vicia faba* L. *Plant Cell Tiss. Org. Cult.*, **9**: 167–171.

Griga M, Tejklova E, Novak F and Kubalakova M (1986) *In vitro* clonal propagation of *Pisum sativum* L. *Plant Cell Tiss. Org. Cult.*, **6**: 95–104.

Hain R, Reif H J, Krause E, Langebartels R, Kindl H, Vornam B, Wiese W, Schmelzer E, Schreier P H and Stenzel K (1993) Disease resistance results from foreign phytoalexin expression in a novel plant. *Nature*, **361**: 153–156.

Jach G, Gornhardt B, Mundy J, Logemann J, Pinsdorf E, Leah R, Schell J and Maas C (1995) Enhanced quantitative resistance against fungal disease by combinatorial expression of different barley antifungal proteins in transgenic tobacco. *Plant J.*, **8**: 97–109.

Jackson J A and Hobbs S L A (1990) Rapid multiple shoot production from cotyledonary node explants of pea (*Pisum sativum* L). *In Vitro Cell Dev. Biol.*, **26**: 835–838.

Jelaska S, Pevalek B, Papes D and Devide Z (1981) Developmental aspects of long-term callus culture of *Vicia faba* L. *Protoplasma*, **105**: 285–292.

Kao K, Constabel F, Michayluk M R and Gamborg O L (1974) Plant protoplast fusion and growth of intergeneric hybrid cells. *Planta*, **120**: 215–227.

Kao K and Michayluk M R (1974) A method for high-frequency intergeneric fusion of plant protoplasts. *Planta*, **115**: 355–367.

Kao K and Michayluk M R (1975) Nutritional requirements for growth of *Vicia hajastana* cells and protoplasts at a very low population density in liquid media. *Planta*, **126**: 105–110.

Kartha K, Pahl K, Leung N and Mroginski L A (1981) Plant regeneration from meristems of grain legumes: soybean, cowpea, peanut, chickpea, and bean. *Can. J. Bot.*, **59**: 1671–1679.

Lawes D A, Bond D A and Poulsen M H (1983) Classification, origin, breeding methods and objectives. In: *The Faba Bean* (*Vicia faba* L.) (Ed Hebblethwaite P D), Butterworth, London, 24.

Lazaridou T B, Roupakias D G (1993) Intraspecific variation in mean endosperm cell cycle time in *Vicia faba* L. and interspecific hybridisation with *Vicia narbonensis* L. *Plant Breed.*, **110**: 9–15.

Malik K A and Saxena P K (1992a) Thidiazuron induces high-frequency shoot regeneration in intact seedlings of pea (*Pisum sativum*), chickpea (*Cicer arietinum*) and lentil (*Lens culinaris*). *Austr. J. Plant Physiol.*, **19**: 731–740.

Malik K A, Saxena P K (1992b) Regeneration in *Phaseolus vulgaris* L – High-frequency induction of direct shoot formation in intact seedlings by N-6-benzylaminopurine and thidiazuron. *Planta*, **186**: 384–389.

Martin C, Carré M and Duc G (1979) Note sur les cultures de tissus de féverole (*Vicia faba* L.). Bouturage, culture de cals, culture de méristèmes. *Ann. Amelior. Plant*, **29**: 277–287.

Martins I (1983) Multiple shoot formation from shoot apex cultures of *Phaseolus vulgaris* L. *J. Plant Physiol.*, **115**: 205–208.

Maxted N (1993) A phenotypic investigation of *Vicia* L. subgenus *Vicia* (Leguminosae, Vicieae). *Bot. J. Linn. Soc.*, **111**: 155–182.

Maxted N, Callimassia M A and Bennett M D (1991) Cytotaxonomic studies of eastern Mediterranean *Vicia* L. (Leguminosae). *Plant Systematics and Evolution*, **177**: 221–234.

McCabe D, Swain W, Martinell B and Christou P (1988) Stable transformation of soybean (*Glycine max*) by particle acceleration. *Biotechnology*, **6**: 923–926.

Meixner M, Brinkmann S, Schieder O and Pickardt T (1996) Genetic stability and expression of foreign genes in transgenic lines of the legume *Vicia narbonensis*. In: *Transgenic Organisms and Biosafety* (Eds Schmidt E R and Hankeln T), Springer, Berlin, Heidelberg, 249–260.

Meixner M, Schneider U, Schieder O and Pickardt T (1995) Studies on the stability of foreign genes in the progeny of transgenic lines of *Vicia narbonensis*. In: *Current Issues in Plant Molecular and Cellular Biology* (Ed Terzi M), Kluwer Academic Publ., The Netherlands, 285–290.

Mitchell J P and Gildow F E (1975) The initiation and maintenance of *Vicia faba* tissue cultures. *Physiol. Plant.*, **34**: 250–253.

Mok M C, Mok D W S, Armstrong D J, Shudo K, Isogai Y and Okamoto T (1982) Cytokinin activity of N-phenyl-N′-1,2,3-thiadiazol-5-ylurea (thidiazuron). *Phytochemistry*, **21**: 1509–1511.

Murashige T and Skoog F (1962) A revised medium for rapid growth and bioassays with tobacco tissue cultures. *Physiol. Plant.*, **15**: 473–497.

Padgette S R, Kolacz K H, Delannay X, Re D B, La Vallee B J, Tinius C N, Rhodes W K, Otero Y I, Barry G F, Eichholtz D A, Peschke V M, Nida D L, Taylor N B and Kishore G M (1995) Development, identification, and characterization of a glyphosphate-tolerant soybean line. *Crop Sci.*, **35**: 1451–1461.

Pickardt T, Huancaruna Perales E and Schieder O (1989) Plant regeneration via somatic embryogenesis in *Vicia narbonensis*. *Protoplasma*, **149**: 5–10.

Pickardt T, Meixner M, Schade V and Schieder O (1991) Transformation of *Vicia narbonensis* via *Agrobacterium tumefaciens*-mediated gene transfer. *Plant Cell Rep.*, **9**: 535–538.

Pickardt T, Saalbach I, Waddell D, Meixner M, Müntz K and Schieder O (1995) Seed specific expression of the 2S albumin gene from Brazil nut (*Bertholletia excelsa*) in transgenic *Vicia narbonensis*. *Mol. Breed.*, **1**: 295–301.

Pickardt T, Ziervogel B, Schade V, Ohl L, Bäumlein H and Meixner M (1998) Developmental-regulation and tissue-specific expression of two different seed promoter GUS-fusions in transgenic lines of *Vicia narbonensis*. *J. Plant Physiol.*, **152**: 621–629.

Quandt H J, Pühler A and Broer I (1993) Transgenic root nodules of *Vicia hirsuta*: A fast and efficient system for the study of gene expression in indeterminate-type nodules. *Mol. Plant-Micr. Interact.*, **6**: 699–706.

Ramsay G and Kumar A (1990) Transformation of *Vicia faba* cotyledon and stem tissues by *Agrobacterium rhizogenes* – infectivity and cytological studies. *J. Exp. Bot.*, **41**: 841–847.

Ramsay G, Pickersgill B (1986) Interspecific hybridisation between *Vicia faba* and other species of *Vicia*: Approaches to delaying embryo abortion. *Biol. Zentralbl.*, **105**: 171–179.

Röper W (1979) Growth and cytology of callus and cell suspension cultures of *Vicia faba* L. *Z. Pflanzenphysiol.*, **93**: 245–257.

Röper W (1981) Callus formation from protoplasts derived from cell suspension cultures of *Vicia faba* L. *Z. Pflanzenphysiol.*, **101**: 75–78.

Russell D R, Wallace K M, Bathe J H, Martinell B J and McCabe D E (1993) Stable transformation of *Phaseolus vulgaris* via electric-discharge mediated particle acceleration. *Plant Cell Rep.*, **12**: 165–169.

Saalbach I, Pickardt T, Machemehl F, Saalbach G, Schieder O and Müntz K (1994) A chimeric gene encoding the methionine-rich 2S albumin of Brazil nut (*Bertholletia excelsa* H.B.K.) is stably expressed and inherited in transgenic grain legumes. *Mol. Gen. Genet.*, **242**: 226–236.

Saalbach I, Waddell D, Pickardt T, Schieder O and Müntz K. (1995) Stable expression of the sulphur-rich 2s albumin gene in transgenic *Vicia narbonensis* increases the methionine content of seeds. *J. Plant Physiol.*, **145**: 674–681.

Sato S, Newell C, Kolacz K, Tredo L, Finer J, and Hinchee M (1993) Stable transformation via particle bombardment in two different soybean regeneration systems. *Plant Cell Rep.*, **12**: 408–413.

Schiemann J and Eisenreich G (1989) Transformation of field bean *Vicia*-faba l. cells expression of a chimeric gene in cultured hairy roots and root-derived callus. *Biochem. Physiol. Pflanzen*, **185**: 135–140.

Schulze S, Grunewaldt J and Schmidt H (1985) Zur *in vitro* Regeneration von *Vicia faba* L. *Z. Pflanzenzüchtung*, **94**: 244–250.

Selva E, Stouffs M and Briquet M (1989) *In vitro* propagation of *Vicia faba* L. by micro-cutting and multiple shoot induction. *Plant Cell Tiss. Org. Cult.*, **18**: 167–179.

Siddique K H M, Loss S P and Enneking D (1996) Narbon bean (*Vicia narbonensis* L.) – a promising grain legume for low rainfall areas of south-western Australia. *Austr. J. Exp. Agricult.*, **36**: 53–62.

Singh B D, Harvey B L, Kao K N and Miller R A (1972) Selection pressure in cell populations of *Vicia hajastana* cultured *in vitro*. *Can. J. Genet. Cytol.*, **14**: 65–70.

Skoog F, Schmitz R Y (1972) Cytokinins. In: *Plant Physiology: A Treatise*, Vol VIB (Ed Steward F C). Academic Press, New York, London, 181–212.

Steeves T A and Sussex I M (1989) *Patterns in Plant Development*. Cambridge University Press, Cambridge.

Taha R M and Francis D (1990) The relationship between polyploidy and organogenetic potential in embryo and root-derived tissue cultures of *Vicia faba* L. *Plant Cell Tiss. Org. Cult.*, **22**: 229–236.

Tegeder M, Gebhardt D, Schieder O and Pickardt T (1995) Thidiazuron-induced plant regeneration from protoplasts of *Vicia faba* cv. Mythos. *Plant Cell Rep.*, **15**: 164–169.

Tegeder M, Kohn H, Nibbe M, Schieder O and Pickardt T (1996) Plant regeneration from protoplasts of *Vicia narbonensis* via somatic embryogenesis and shoot organogenesis. *Plant Cell Rep.*, **16**: 22–25.

Thynn M and Werner D (1987) Plantlet regeneration and somatic differentiation in faba bean (*Vicia faba* L.) from callus culture of various explants. *Angew. Bot.*, **61**: 483–492.

Venketeswaran S (1962) Tissue culture studies on *Vicia faba* L. Establishment of culture. *Phytomorphology*, **12**: 300–306.

Weber H, Heim U, Golombek S, Borisjuk L, Manteuffel R and Wobus U (1998) Expression of a yeast-derived invertase in developing cotyledons of *Vicia narbonensis* alters the carbohydrate state and affects storage functions. *Plant J.*, **16**: 163–172.

IN VITRO REGENERATION AND GENETIC TRANSFORMATION OF LENTIL

ANJU GULATI* AND ALAN McHUGHEN
Crop Development Centre, University of Saskatchewan, 51 Campus Drive, Saskatoon SK S7N 5A8, Canada
**e-mail: gulati25@hotmail.com*

Abstract

Lentil (*Lens culinaris* Medik.) is an important pulse crop in many diverse regions around the world. As such, farmers face diverse agronomic problems that have not been adequately addressed using conventional plant breeding methods. Biotechnology holds the promise of overcoming many problems with the lentil crop, including insect resistance and disease tolerance, weed control, nutritional enhancements and reduction of antinutritional components. In order to apply biotechnology to lentil crop improvement, however, the underlying technologies, including tissue culture, *in vitro* regeneration, and gene transfer technologies, need to be elucidated and formalized. In this chapter we review the history and the current state of biotechnology relating to lentil. Like many grain legumes, lentil has a long but often frustrating history of tissue culture and regeneration *in vitro*. Lentil plant regeneration has been reported using both direct and indirect organogenic processes, as well as via somatic embryogenesis. We analyze the important factors in the tissue culture and regeneration of lentil *in vitro*. Gene transfer into lentil has been attempted using *Agrobacterium* based techniques and also by particle gun techniques. We also discuss the factors influencing gene transfer technology for lentil and prospects for future successes.

1. Introduction

Lentil (*Lens culinaris* Medik.) is a widely grown crop in the Middle East, West Asia, North Africa and Indian sub-continent. India leads the world in lentil production and most of the lentils are consumed by India and its neighbors in the form of split lentil (dal). Currently, Canada and Turkey are the world's largest exporters and Saskatchewan alone produces about 97% of the Canadian lentil crop. The nutritional value of lentil is higher (20–36% protein) than cereals, meat and fish. Among the cool season legume crops, lentil is the richest in the important amino acids (lysine, arginine, leucine, and sulphur

P. K. Jaiwal and R. P. Singh (eds.), Applied Genetics of Leguminosae Biotechnology, 133–147.

containing amino acids) (Williams *et al.*, 1994). In addition, it is high in carbohydrate, calories, fibre, vitamin A, calcium, starch, iron, phosphorus, copper and manganese. Lentils are also grown to improve soil status (as green manure). It has the capability of producing abundance of foliage and *Rhizobium* associated di-nitrogen fixation, and therefore, contributes to the fertilizer budget of farmers. All *Lens* species are diploid herbaceous annuals with 2n = 14 chromosomes. Plant height ranges from 25 to 75 cm with soft hairy branches with pinnately compound leaves and numerous oval leaflets. Roots are taproots with fibrous laterals. Germination is hypogeal, flowers are white, lilac or pale blue and naturally self-pollinated. Pods typically have one or two seeds and occasionally three seeds. Seeds are lens shaped hence the name 'lentil'. Depending on cultivar, cotyledons can be red, yellow or green.

The average lentil yield range from 450 to 675 kg/ha in dry areas may increase to 2000 kg/ha with irrigation and yields over 3000 kg/ha have been recorded. Among the agronomic problems, pod shedding, pod shattering, lodging and the spreading growth habit are important constraints in production (Robertson *et al.*, 1996).

Lentil is attacked by variety of insects during crop development and storage. Seed corn maggots (*Delia platura* meigen), wire worm (*Lemonius* spp and *Cteniera* spp) and cut worms (*Agriotis* spp) are major pests during crop development and storage. Bean weevil (*Sitonia lineatus*) attack the root nodule and affect nitrogen fixation; aphids (*Aphid craccivora*), thrips (*Frankliniella* spp) and grasshoppers feed on leaves, stem and flowers; lygus bugs (*Lygus* spp), and bruchid beetles (*Bruchus* spp, *Callosobruchus* spp) attack pods and seeds. Among the viruses, pea seedborne mosaic virus (PSbMV) is a potential threat to lentil as it is seed borne and is transmissible by aphids. Ascochyta blight and anthracnose are two important fungal diseases recorded on lentil.

An alternative and novel approach for the improvement of this crop is to complement the traditional breeding with techniques to regenerate plants from single cells and organized tissues and to transfer desirable genes from other sources. Until recently, gene transfer in grain legumes has been difficult and challenging because of their recalcitrant nature to *in vitro* regeneration and competency of regenerating cells for transformation is not great. However, success has been achieved in obtaining transgenic plants in many other grain legumes (Christou, 1994, 1997), soybean (Hinchee *et al.*, 1988; Meurer *et al.*, 1998), bean (Dillen *et al.*, 1997), pea (Bean *et al.*, 1997; Puonti-Kaerlas *et al.*, 1990), peanut (McKently *et al.*, 1995) and chickpea (Fontana *et al.*, 1993). Transgenic plants for agronomically important traits like insect resistance and improved nutritional quality have been produced in some legume species, e.g. soybean (Stewart *et al.*, 1996), pea (Schroeder *et al.*, 1995; Shade *et al.*, 1994), chickpea (Kar *et al.*, 1996), adzuki bean (Ishimoto *et al.*, 1996), common bean (Aragão *et al.*, 1996) and *Vicia* (Pickardt *et al.*, 1995, 1998). Lentil is considered to be an underexploited crop since very little research has been done so far to improve it in comparison with other pulse crops. Genetic transformation in lentil may provide resistance towards biotic and abiotic stresses and also qualitatively better yield by metabolic engineering may prove to be a boon for lentil production. Therefore, *in vitro* culture may provide an opportunity to improve plant types through selection of cell lines and/or cells with superior traits and by introducing desired genetic information.

Tissue culture technology is a prerequisite for the application of cellular and molecular genetic manipulation to crop improvement programs. The legume family includes many important crop species, but regeneration *in vitro* has, until recently, proven especially difficult in grain legumes. The available literature on tissue culture of lentil suggests that all approaches of *in vitro* propagation, i.e., propagation from pre-existing meristematic tissues, differentiation of adventitious shoot buds, indirect organogenesis and somatic embryogenesis have been tried. More advanced tissue culture techniques such as protoplast fusion, haploid production and genetic transformation are still in their infancy.

2. *In vitro* studies

In vitro culture studies of lentil are summarized in Tables 1 and 2. The emphases of these studies are: (1) establishment of callus, (2) regeneration of plants (via direct or indirect organogenesis), (3) somatic embryogenesis, (4) embryo culture, and (5) protoplast culture.

2.1. ESTABLISHMENT OF CALLUS CULTURES

Almost every part of the lentil is callogenic. Shoot tip, epicotyl, first node and first leaves show good cell proliferation. By using proper explants, it is possible to induce organogenesis or embryogenesis from callus.

2.2. REGENERATION OF PLANTS

2.2.1. Direct shoot organogenesis

Direct regeneration of shoots without intervening callus phase has been achieved from shoot tip (Bajaj and Dhanju, 1979; Polanco and Ruiz, 1997; Singh and Raghuvanshi, 1989), immature seeds, first node, bractlets and first pair of leaves (Polanco *et al.*, 1988; Polanco and Ruiz, 1997), intact seedling cultures (Malik and Saxena, 1992), nodal segments (Ahmad *et al.*, 1997; Singh and Raghuvanshi, 1989) and cotyledonary nodes (Warkentin and McHughen, 1993). Singh and Raghuvanshi (1989) also reported direct plantlet regeneration from nodal segments and shoot tips of lentil. Addition of cytokinin to the regeneration medium seems important for lentil shoot regeneration. Direct germination of excised nodal segments and shoot tips was observed on MS (Murashige and Skoog, 1962) basal media and addition of kinetin resulted in the formation of multiple shoots from both explants (Singh and Raghuvanshi, 1989). Bajaj and Dhanju (1979) and Williams and McHughen (1986) also observed the best shoot regeneration response when lentil explants were cultured on media containing kinetin. Polanco *et al.* (1988) observed lentil shoots regeneration on media containing N^6-benzylaminopurine (BAP) alone or in combination with other growth regulators. However, Malik and Saxena (1992) obtained a high frequency of shoot regeneration in intact seedlings on medium containing thidiazuron (TDZ). Ahmad *et al.* (1997) reported plant regeneration in lentil and two of its wild relatives from nodal segments via *in vitro* clonal multiplication and used the same protocol for F_1 interspecific hybrid multiplication.

Table 1. In vitro *regeneration studies in lentil*

Variety	Basal medium	Growth regulators	Explant	Morphogenetic response	Reference
—	MS	IAA (2.0 mg/L) + KIN (0.5 mg/L)	Apical meristem	Plantlets	Bajaj and Dhanju, 1979
Laird	−/+ activated charcoal	—	Leaf protoplast	Microcalli	Stiff *et al.*, 1986
Eston, Laird, NEL-48	MS	KIN (10.0 mg/L) + GA_3 (1.0 or 0.1 mg/L)	Shoot meristem and epicotyl	Callus, shoots	Williams and McHughen, 1986
	MS	-Do-	Cotyledon	Callus	
Laird	MS or B_5A (B_5 medium + ammonium nitrate (500 mg/L))	2,4-D (1 mg/L)	Embryo	Callus	Saxena and King, 1987
	B_5A	BAP (1.0 mg/L) + IAA (0.25 mg/L)	Embryo-derived callus	Embryoids	
	B_5A	Glutamine (70 mg/L)	Embryoids	Bipolar embryos with cotyledons, shoots and roots	
Verdina, Pardina, Castellana	MS or MS salts + B_5 vitamins	2,4-D (2.0 mg/L) or BAP (2.25 mg/L) + NAA (0.186 mg/L)	Shoot tip, first node and first pair of leaves	Callus	Polanco *et al.*, 1988
—	-Do-	BAP (2.25 mg/L) + NAA (0.186 mg/L)	Callus	Shoots	—
—	-Do-	BAP (2.25 mg/L)	Shoot tip and node	Shoots	—
L. culinaris 2n = 4X, 2n = 2X	MS	KIN (1.0 mg/L) + 2,4-D (10.0 mg/L)	Nodal segment and shoot tip	Callus	Singh and Raghuvanshi, 1989
L. culinaris 2n = 4X, 2n = 2X	MS	KIN (10.0 mg/L)	Sh oot tip and nodal calli	Shoots	Singh and Raghuvanshi, 1989
Eston	Liquid KM (lacking coconut water, ammonium nitrate, and amino acids) + glucose (0.35 M) + sucrose (1.0 g/L) or $CaCl_2 . 2H_2O$ (1.0 g/L)	2,4-D (2.2 μM) + NAA (2.7 μM) + KIN (2.3 μM) + BAP (2.2 μM) + GA_3 (1.4 μM) or 2,4-D (2.2 μM) + NAA (5.4 μM) + BAP (2.2 μM)	Epicotyl protoplast	Microcalli	Rozwadowski *et al.*, 1990

Table 1. (Continued)

Variety	Basal medium	Growth regulators	Explant	Morphogenetic response	Reference
	B_5 solid + glutamine (5 mM)	-Do-	Microcalli	Calli	
Eston	MS or K_3 liquid layered on solid media (0.2% gelrite) + mannitol (0.5 M)	NAA (2.0 mg/L) + BAP (0.5 mg/L)	Leaf protoplast	Microcalli	Stiff *et al.*, 1991
	B_5 + ammonium nitrate (500 mg/L)	2,4-D (1.0 mg/L) + IAA (0.25 mg/L) or BAP (1.0 mg/L) + IAA (0.25 mg/L)	Microcalli	Calli	
		B_5 liquid or solid (0.2% gelrite) + ammonium nitrate (500 mg/L) + glutamine (70 mg/L)		Callus, no organo-genesis	
Eston	MS salts + B_5 vitamins	TDZ (1–50 μM)	Intact seedlings	Shoots	Malik and Saxena, 1992
Laird	MS	BAP (1.0 mg/L)	Cotyledo-nary node	Shoots	Warkentin and McHughen, 1993
L. culinaris ssp. *culinaris*, *L. ervoides*, *L. culinaris* ssp. *orientalis*	MS (without sucrose)	BAP (1.11 μM) + GA_3 (2.89 μM)	Nodal segment	Shoots	Ahmad *et al.*, 1997
Verdina	MS	BAP (2.25 or 0.225 mg/L)	Shoot tip, first node and bractelet	Shoots	Polanco and Ruiz, 1997

KM: Kao and Michayluk (1975) medium; K_3: Nagy and Maliga (1976) medium.

2.2.2. *Indirect organogenesis*

Regeneration of plants from the callus may result in genetic variability. The periodic subculture of callus results in genetic mutation and change in ploidy number, etc. Though undesirable for maintaining clones, it is a novel and rich source of genetic variability. The adventitious regeneration directly from organs generally gives a limited number of propagules. This may be increased several fold by an intervening callus phase.

Shoot organogenesis in lentil has been reported from calli derived from shoot meristem and the epicotyls (Williams and McHughen, 1986), and shoot tips and nodal explants (Polanco *et al.*, 1988; Singh and Raghuvanshi, 1989). Calli derived from the cotyledon

among auxins, 2,4-D, NAA and indole-3-acetic acid (IAA) have been used to initiate callus and shoot and root regeneration. The regeneration frequency was found higher with TDZ and BAP than with KIN (Malik and Saxena, 1992; Saxena and King, 1987; Singh and Raghuvanshi, 1989; Williams and McHughen, 1986). Polanco *et al.* (1988) reported that BAP in combination with NAA supported shoot regeneration but inhibited root formation in regenerated shoots; media supplemented with 2,4-D formed callus irrespective of cultivar and explant type, but were ineffective in inducing morphogenesis and media containing IAA generally produced least callus formation and growth. The callus induced on medium containing 2,4-D and kinetin could subsequently regenerate shoot buds on kinetin containing media (Singh and Raghuvanshi, 1989). BAP and kinetin in combination with gibberellic acid (GA_3) have been used for shoot regeneration with or without a callogenic stage, respectively (Ahmad *et al.*, 1997; Williams and McHughen, 1986). Among all the growth regulators, BAP was the most effective in inducing shoot formation directly from the explants or from the callus (Polanco *et al.*, 1988).

2.6.3. Explants

Explant type influence the callus and shoot formation in lentil. Shoot meristem, epicotyl, shoot tip and nodal explants had regenerated shoots from callus more frequently than leaves. Cotyledon calli did not regenerate under tested conditions (Polanco *et al.*, 1988; Williams and McHughen, 1986).

2.6.4. Genotypes

The efficiency of callus formation and shoot regeneration from diverse explants varied with genotype. Among the three lentil cultivars tested, Verdina generally formed callus at a lower frequency than the other two cultivars Castellana and Pardina on media with no 2,4-D, while the percentages between Castellana and Pardina varied from medium to medium (Polanco *et al.*, 1988). However, Ahmad *et al.* (1997) noticed no significant differences in direct shoot regeneration from nodal explants of cultivated lentil species, *L. culinaris* ssp. *culinaris* (Medikus) Williams and two wild species, *L. ervoides* and *L. culinaris* ssp. *orientalis*.

3. Genetic transformation

The grain legumes have been shown to be less amenable to genetic transformation than most of the other dicotyledonous crop species. However, transformation reports especially for soybean and pea via *Agrobacterium* or particle bombardment revealed that it is possible to produce transgenic plants in this group of plant species (Christou, 1994, 1997). Genetic transformation studies in lentil are summarized in Table 3. Transformation has been done either via *Agrobacterium*-mediated gene transfer or as direct gene transfer.

3.1. *AGROBACTERIUM*-MEDIATED GENE TRANSFER

This technique relies on the natural ability of *Agrobacterium* to transfer DNA to dicotyledonous plants through a mechanism akin to conjugation. Lentil tissues are susceptible to

Table 3. Genetic transformation in lentil

Species/ cultivar	Vector/method used	Plasmid	Explant	Observation/remarks	Reference
Laird	*Agrobacterium tumefaciens* strains C58, Ach5, GV3111 and A281	pTiC58, pTiBo542, pTiB653, pTiAch5	Stem	Tumor development on inoculated stems *in vivo* and on excised shoot apices *in vitro*	Warkentin and McHughen, 1991
Laird	*Agrobacterium tumefaciens* strain GV2260	p35SGUS INT	Shoot apex, epicotyl, root	Low GUS expression	Warkentin and McHughen, 1992
L. culinaris	Electroporation	pAS6 + anti LOX inhibitory monoclonal antibodies	Leaf or root protoplast	Antibody transfer intact protoplast	Maccarrone *et al.*, 1992a
Laird	*Agrobacterium tumefaciens* strain	p35SGUS INT GV2260	Cotyledonary node	No recovery of transgenic shoots	Warkentin and McHughen, 1993
L. culinaris	Lipofection	pBI221	Leaf or root protoplast	Low expression of foreign gene	Maccarrone *et al.*, 1992b
L. culinaris	Electroporation + Lipofection	pBI221	Leaf or root protoplast	Moderate expression of foreign gene	Maccarrone *et al.*, 1993
L. culinaris	Electroporation	pAS6 + fragment of LOX gene (antisence construct)	Leaf or root protoplast	Enzyme inhibition by antisense mRNA	Maccarrone *et al.*, 1995a
L. culinaris	PEG treatment	pBI121, pBI221, pAS6	Leaf or root protoplast	Moderate expression of foreign gene	Maccarrone *et al.*, 1995b
L. culinaris	Electroporation	pBI121, pBI221 and pAS6 with immuno-globulins	Leaf or root protoplast	High expression of foreign gene. Coelectrotransfer of antibodies and plasmids	Maccarrone *et al.*, 1995b
Laird	Electroporation	pGPT1.0 (the *EcoR Hind III* fragment of the plasmid p35SGUS INT cloned into pUC8)	Nodal axillary bud	*In planta* gene transfer	Chowrira *et al.*, 1996

Table 3. (Continued)

Species/ cultivar	Vector/method used	Plasmid	Explant	Observation/remarks	Reference
L. culinaris	*Agrobacterium tumefaciens*	Plasmid harbouring herbicide (basta) resistance gene	—	Herbicide resistant shoots in culture, their rooting and recovery of T_1 seeds	Barton *et al.*, 1997
ILC 5883, ILC 5582, ILC 5588	*Agrobacterium tumefaciens* strain EHA101	pIBGUS	Epicotyl	Transgenic callus	Halbach *et al.*, 1998
Eston, Palouse, Brewer, Chilean	*Agrobacterium tumefaciens* strain C58 and EHA105	p35SGUS INT	Longitudinal slices of embryonic axes	Expression of foreign gene	Lurquin *et al.*, 1998
Laird	Particle bombardment	pCAMBIA-1201	Cotyledonary node	Expression of *hpt* and *uidA* genes confirmed by PCR	Gulati and McHughen, 2000
Laird	Particle bombardment	Plasmid harbouring herbicide (sulfonylurea) resistance gene	Cotyledonary node	Herbicide resistant T_0 plants; testing of T_1 plants confirmed transformation stability	Gulati and McHughen (unpublished data)

transformation by *Agrobacterium* (Warkentin and McHughen, 1991). Four diverse strains of *A. tumefaciens* (C58, Ach5, GV 3111 and A281) were capable of inducing tumors at a high frequency on inoculated stem of lentil *in vivo* and on excised shoot apices *in vitro*. Tumor formation and opine production were indicative of plant cell transformation. Later, several explants (shoot apex, epicotyl, root and cotyledonary node) derived from lentil seedlings were evaluated for their ability to express a foreign gene (GUSINT) after *A. tumefaciens* inoculation (Warkentin and McHughen, 1992, 1993). Low, but reproducible, levels of GUS expression were obtained in shoot apex, epicotyl and root. However, cotyledonary nodes were not amenable to *A. tumefaciens* mediated transformation, as measured by GUS assay. Different lentil cultivars also vary in their response to *A. tumefaciens* (Lurquin *et al.*, 1998). This has been demonstrated by co-cultivating half embryos of four different lentil cultivars with *A. tumefaciens* strain C58 carrying a *uidA* reporter gene and scoring GUS positive plants as well as the average number of sectors present on the positive plants up to 4 weeks of growth. Both parameters were shown to vary widely with cultivar. However, co-cultivation with hypervirulant *A. tumefaciens* strain EHA 105 had no effect on the frequencies of these same parameters (Lurquin *et al.*, 1998). No transgenic plants were reported. Halbach *et al.* (1998) obtained transgenic callus lines by inoculating epicotyl segments of three lentil cultivars with *Agrobacterium* strain EHA101, harbouring *pat*- and *gusA* gene. In a short note, however, Barton *et al.* (1997)

claimed the regeneration of 'Basta' resistant transgenic lentil via *Agrobacterium*-mediated gene transfer. The rooted shoots were transferred to liquid culture in the greenhouse and T_1 seeds were recovered.

3.2. DIRECT GENE TRANSFER

This method achieves DNA uptake by plant cells and protoplasts through a variety of processes that include the use of facilitator molecules (Shillito *et al.*, 1985), micro-injection (Neuhaus and Spangenberg, 1990), biolistic delivery (Sanford, 1988) or electro-poration (Fromm *et al.*, 1987). In lentil, electroporation-mediated gene transfer of nodal meristems *in planta* was used by Chowrira *et al.* (1996). Transient expression of a *gus* reporter gene was used to study the uptake and expression of the introduced DNA in treated buds. Transgenic R_1 plants were recovered from seeds originating on R_0 chimeric branches. Transgenic R_2 plants were also obtained. Protoplasts were also used for gene transfer by electroporation, lipofection or PEG treatment (Maccarronne *et al.*, 1992a, b, 1993, 1995a, b). Transient GUS and CAT activity was detected in lentil protoplasts following delivery of the genes via liposomes (Maccarronne *et al.*, 1992b). Electro-poration was found more effective than PEG treatment as a transfection procedure and PEG was, in turn, more effective than lipofection (Maccarronne *et al.*, 1992a, b, 1995b). Also, monoclonal antibodies directed against lipoxygenase when introduced into protoplasts, reduced the activity of the target enzyme (Maccarronne *et al.*, 1992a). Protoplasts were also a means for coelectroporation of immunoglobulins, plasmid DNA and antisence RNA delivery (Maccarronne *et al.*, 1995a, b). No transgenic plant was obtained.

There is no published report on transfer of gene in lentil through particle bombard-ment-mediated gene transfer. Therefore, McHughen's group at the University of Saskatchewan recognized the need to utilize this genetic modification technology to trans-fer genes to lentils. They have evaluated particle bombardment mediated gene transfer to lentil var. Laird cotyledonary node explants using plasmid pCAMBIA1201 which carries the β-glucuronidase (GUS) gene *uidA* as a reporter and hygromycin phosphotransferase gene (*hpt*) as a selectable marker (Fig. 1A and B). Transient GUS expression assay, selec-tion of bombarded explants on hygromycin B and polymerase chain reaction (PCR) analy-sis of transformants with *hpt* and *uidA* primers confirmed the putative transformants. Unfortunately no transgenic plant was regenerated from shoot buds (Gulati and McHughen, 2000). Cotyledonary node explants excised from 2–3 day-old seedlings were also bombarded with a gene, which confers sulfonylurea resistance. Initial results from screening of explants on medium supplemented with chlorsulfuron and the regeneration of herbicide resistant putative transformants (Fig. 1C) are encouraging. The T_0 plantlets were transferred to soil and grown in growth cabinets. Three weeks after transfer, one leaflet per plant was painted with 2X field rate metsulfuron to differentiate between putative trans-formants and non-transformants (escapes) (Fig. 1D). The plants, which survived the paint-ing test, were grown to maturity and seeds were collected. The T_1 seeds were germinated and tested by leaf painting to confirm the stability of the engineered trait. The resistant plants survived the test and the sensitive died (Fig. 1E and F). The collection and testing of seeds from other putative transformants is under way.

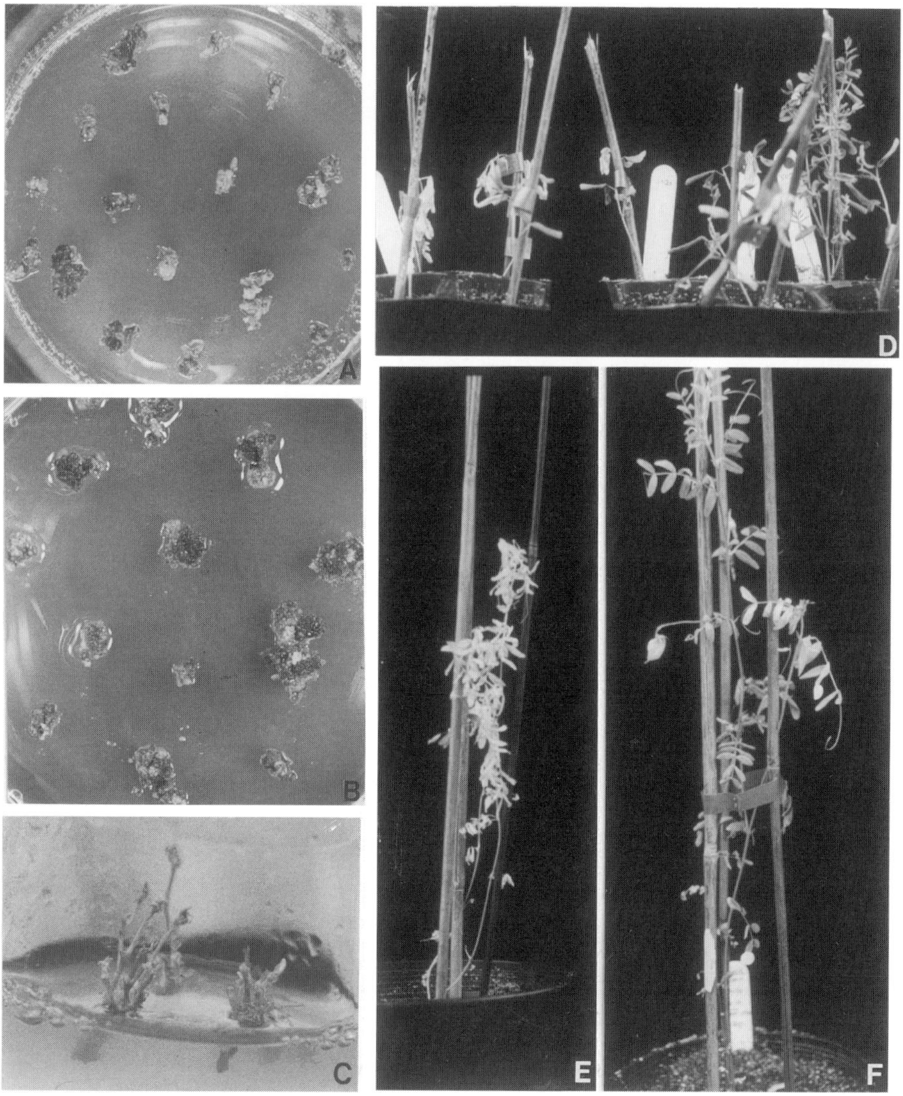

Figure 1. (A–F) Particle bombardment-mediated gene transfer in lentil. A. Putatively transformed explants growing on hygromycin medium produced green callus. Non-transformed explants turned brown; B. Two-month-old cotyledonary node explants with green callus on hygromycin initiate shoot bud; C. Regenerated shoots from bombarded cotyledonary node explants on chlorsulfuron selection medium; D. Herbicide treated (T_0 generation) putative transformants 14 days after treatment. The transformed plant survived and non-transformed died. E and F. Leaf painting of T_1 generation seedlings to identify stable transformed herbicide resistant plants. Sensitive plant died (E) and resistant survived (F).

4. Conclusions and future prospects

Shoot regeneration has been obtained directly or indirectly via callus phase, but root induction in regenerated shoots and transplantation of plantlets to pots is restricted to a few cases. Mature fertile plants of lentil have been regenerated using shoot tips, cotyledonary nodes and nodal segments. Work needs to be done towards regeneration of plants through somatic embryogenesis. Progress in the development of a protocol for transformation of lentil is encouraging. The presence of transgene activity in different tissues after *Agrobacterium*-mediated gene transfer has been noted. Unfortunately no transgenic plant has been recovered. *In planta* electroporation has led to the production of putative transgenic R_2 individuals. Transfer of herbicide resistance to lentil using *Agrobacterium* and particle bombardment is still at its preliminary stage.

References

Ahmad M, Fautrier A G, McNeil D L, Hill G D and Burritt D J (1997) *In vitro* propagation of *Lens* species and their F_1 interspecific hybrids. *Plant Cell Tiss. Org. Cult.*, **47**: 169–176.

Aragão F J L, Barros L M G, Brasileiro A C M, Ribeiro S G, Smith F D, Sanford J C, Faria J C and Rech E L (1996) Inheritance of foreign genes in transgenic bean (*Phaseolus vulgaris* L.) co-transformed via particle bombardment. *Theor. Appl. Genet.*, **93**: 142–150.

Bajaj Y P S and Dhanju M S (1979) Regeneration of plants from apical meristem tips of some legumes. *Curr. Sci.*, **48**: 906–907.

Barton J E, Klyne A, Tennakoon D, Francis C and Hamblin J (1997) Development of a system for gene transfer to lentils. In: *International Food Legume Research Conference III*, Adelaide, 85.

Bean S J, Gooding P S, Mullineaux P M and Davies D R (1997) A simple system for pea transformation. *Plant Cell Rep.*, **16**: 513–519.

Chowrira G M, Akella V, Fuerst P E and Lurquin P F (1996) Transgenic grain legumes obtained by *in planta* electroporation-mediated gene transfer. *Mol. Biotech.*, **5**: 85–96.

Christou P (1994) The biotechnology of crop legumes. *Euphytica*, **74**: 165–185.

Christou P (1997) Biotechnology applied to grain legumes. *Field Crops Res.*, **53**: 83–97.

Cohen D, Ladizinsky G, Ziv M and Muehlbauer F J (1984) Rescue of interspecific *Lens* hybrids by means of embryo culture. *Plant Cell Tiss. Org. Cult.*, **3**: 343–347.

Dillen W, Clercq J De, Goossens A, Montagu M Van and Angenon G (1997) *Agrobacterium*-mediated transformation of *Phaseolus acutifolius* A. Gray. *Theor. Appl. Genet.*, **94**: 151–158.

Fontana G L, Santini L, Frugis G and Mariotti D (1993) Genetic transformation in the grain legume *Cicer arietinum* L. (chickpea). *Plant Cell Rep.*, **12**: 194–198.

Fromm M, Callis J, Taylor L P and Walbot V (1987) Electroporation of DNA and RNA into plant protoplasts. *Methods Enzymol.*, **153**: 351–366.

Gamborg O L, Miller R A and Ojima K (1968) Nutrient requirement of suspension cultures of soybean root cells. *Exp. Cell Res.*, **50**: 151–158.

Gulati A and McHughen A (2000) Regeneration and particle bombardment mediated transformation of lentil (*Lens culinaris* Medik.). In: *Proceedings of 6th International Congress of Plant Molecular Biology*, Quebec, Canada, June 18–24, 2000. P S 03-45.

Halbach T, Kiesecker H, Jacobsen H J and DeKathen A (1998) Tissue culture and genetic engineering of lentil (*Lens culinaris* Medik.). In: *3rd European Conference on Grain Legumes*, Valladolid, 376.

Hinchee M A W, Connor-Ward D V, Newell C A, McDonnell R E, Sato S J, Gaser C, Fischhoff D A, Re D B, Fraley R T and Horsch R B (1988) Production of transgenic soybean plants using *Agrobacterium*-mediated DNA transfer. *Bio/Technology*, **6**: 915–922.

Ishimoto M, Sato T, Chrispeels M J and Kitamura K (1996) Bruchid resistance of transgenic azuki bean expressing seed α-amylase inhibitor of common bean. *Entomologia Experimentalis et Applicata*, **79**: 309–315.

Kao K N and Michayluk M R (1975) Nutritional requirements for growth of *Vicia hajastana* cells at a very low population density in liquid media. *Planta*, **126**: 105–110.

Kar S, Johnson T M, Nayak P and Sen S K (1996) Efficient transgenic plant regeneration through *Agrobacterium*-mediated transformation of chickpea (*Cicer arietinum* L.). *Plant Cell Rep.*, **16**: 32–37.

Ladizinsky G, Cohen D and Muehlbauer F J (1985) Hybridization in the genus *Lens* by means of embryo culture. *Theor. Appl. Genet.*, **70**: 97–101.

Lurquin P F, Cai Z, Stiff C M and Fuerst E P (1998) Half embryo cocultivation technique for estimating the susceptibility of pea (*Pisum sativum* L.) and lentil (*Lens culinaris* Medik.) cultivars to *Agrobacterium tumefaciens*. *Mol. Biotech.*, **9**: 175–179.

Maccarrone M, Veldink G A and Vilegenthart J F G (1992a) Inhibition of lipoxygenase activity in lentil protoplasts by monoclonal antibodies introduced into the cells via electroporation. *Eur. J. Biochem.*, **205**: 995–1001.

Maccarrone M, Dini L, Di Marzio L, Di Giulio A, Rossi A and Finazzi Agrò A (1992b) Interaction of DNA with cationic liposomes: ability of transfecting lentil protoplasts. *Biochem. Biophys. Res. Commun.*, **186**: 1417–1422.

Maccarrone M, Dini L, Rossi A and Finazzi Agrò A (1993) Gene transfer to lentil protoplasts by lipofection and electroporation. *J. Lipid Res.*, **3**: 707–716.

Maccarrone M, Hilbers M P, Veldink G A and Vilegenthart J F G and Finazzi Agrò A (1995a) Inhibition of lipoxygenase in lentil protoplasts by expression of antisense RNA. *Biochem. Biophys. Acta*, **1259**: 1–3.

Maccarrone M, Veldink G A, Finazzi Agrò A and Vilegenthart J F G (1995b) Lentil root protoplasts: a transient expression system suitable for coelectroporation of monoclonal antibodies and plasmid molecules. *Biochim. Biophys. Acta*, **1243**: 136–142.

Malik K A and Saxena P K (1992) Thidiazuron induces high-frequency shoot regeneration in intact seedlings of pea (*Pisum sativum*), chickpea (*Cicer arietinum*) and lentil (*Lens culinaris*). *Aust. J. Plant Physiol.*, **19**: 731–740.

McKently A H, Moore G A, Doostdar H and Niedz R P (1995) *Agrobacterium*-mediated transformation of peanut (*Arachis hypogea* L.) embryo axes and the development of transgenic plants. *Plant Cell Rep.*, **14**: 699–703.

Meurer C A, Dinkins R D and Collins G B (1998) Factors affecting soybean cotyledonary node transformation. *Plant Cell Rep.*, **18**: 180–186.

Murashige T and Skoog F (1962) A revised medium for rapid growth and bioassays with tobacco tissue cultures. *Physiol. Plant.*, **15**: 473–497.

Nagy J I and Maliga P (1976) Callus induction and plant regeneration from mesophyll of *Nicotiana sylvestris*. *Z. PflanzennPhysiol.*, **78**: 453–455.

Neuhaus G and Spangenberg G (1990) Plant transformation by microinjection techniques. *Physiol. Plant.*, **79**: 213–217.

Pickardt T, Saalbach I, Waddell D, Meixner M, Muentz K and Schieder O (1995) Seed specific expression of the 2S albumin gene form Brazil nut (*Bertholletia excelsa*) in transgenic *Vicia narbonensis*. *Mol. Breed.*, **1**: 295–301.

Pickardt T, Ziervogel B, Schade V, Ohl L, Baeumlein H and Meixner M (1998) Developmental-regulation and tissue-specific expression of two different seed promoter GUS-fusions in transgenic lines of *Vicia narbonensis*. *J. Plant Physiol.*, **152**: 621–629.

Polanco M C, Peláez M I and Ruiz M L (1988) Factors affecting callus and root formation in *in vitro* cultures of *Lens culinaris* Medik. *Plant Cell Tiss. Org. Cult.*, **15**: 175–182.

Polanco M C and Ruiz M L (1997) Effect of benzylaminopurine on *in vitro* and *in vivo* root development in lentil, *Lens culinaris* Medik. *Plant Cell Rep.*, **17**: 22–26.

Puonti-Kaerlas J T, Erikson T and Engstrom P (1990) Production of transgenic pea (*Pisum sativum* L.) plants by *Agrobacterium tumefaciens*-mediated gene transfer. *Theor. Appl. Genet.*, **80**: 246–252.

Robertson L D, Singh K B, Erskine W and Abd El Moneim Ali M (1996) Useful genetic diversity in germplasm collections of food and forage legumes from West Asia and North Africa. *Germplasm Resources and Crop Evolution*, **43**: 447–460.

Rozwadowski K L, Saxena P K and King J (1990) Isolation and culture of *Lens culinaris* Medik cv. Eston epicotyl protoplasts to calli. *Plant Cell Tiss. Org. Cult.*, **20**: 75–79.

Sanford J C (1988) The biolistic process. *Trends Biotechnol.*, **6**: 299–302.

Saxena P K and King J (1987) Morphogenesis in lentil: plant regeneration from callus cultures of *Lens culinaris* Medik. via somatic embryogenesis. *Plant Sci.*, **52**: 223–227.

Schroeder H E, Gollasch S, Moore A, Tabe L M, Craig S, Hardie D C, Chrispeels M J, Spencer D and Higgins T J V (1995) Bean alpha-amylase inhibitor confers resistance to the pea weevil (*Bruchus pisorum*) in transgenic peas (*Pisum sativum* L.). *Plant Physiol.*, **107**: 1233–1239.

Shade R E, Schroeder H E, Pueyo J J, Tabe L M, Murdock L L, Higgins T J V and Chrispeels M J (1994) Transgenic pea seeds expressing the alpha-amylase inhibitor of the common bean are resistant to bruchid beetles. *Bio/Technology*, **12**: 793–796.

Shillito R D, Saul M W, Paszkowski J, Müller M and Potrykus I (1985) High frequency direct gene transfer to plants. *Biotechnology*, **3**: 1099–1103.

Singh R K and Raghuvanshi S S (1989) Plantlet regeneration from nodal segment and shoot tip derived explants of lentil. *Lens Newsletter*, **16**: 33–35.

Stewart C N Jr, Adang M J, All J N, Boerma H R, Cardineau G, Tucker D and Parrott W A (1996) Genetic transformation, recovery, and characterization of fertile soybean (*Glycine max* L. Merrill) transgenic for a synthetic *Bacillus thuringiensis cry1Ac* gene. *Plant Physiol.*, **112**: 121–129.

Stiff C M, Kleinhofs A, Lurquin P F and LeTourneau (1986) Isolation and culture of lentil (*Lens culinaris* var. Laird) protoplasts. In: *The Sixth International Congress of Plant Tissue and Cell Culture*, Abstracts–1986, Minnesota, 232.

Stiff C M, Leuba V, Sun Q and LeTourneau (1991) Isolation, culture, and callus regeneration of lentil protoplasts from leaf tissue. *Lens*, 30–33.

Warkentin T D and McHughen A (1991) Crown gall transformation of lentil (*Lens culinaris* Medik.) with virulent strains of *Agrobacterium tumefaciens*. *Plant Cell Rep.*, **10**: 489–493.

Warkentin T D and McHughen A (1992) *Agrobacterium tumefaciens*-mediated beta-glucuronidase (GUS) gene expression in lentil (*Lens culinaris* Medik.) tissues. *Plant Cell Rep.*, **11**: 274–278.

Warkentin T D and McHughen A (1993) Regeneration from lentil cotyledonary nodes and potential of this explant for transformation by *Agrobacterium tumefaciens*. *Lens Newsletter*, **20**: 26–28.

Williams D J and McHughen A (1986) Plant regeneration of the legume *Lens culinaris* Medik (lentil) *in vitro*. *Plant Cell Tiss. Org. Cult.*, **7**: 149–153.

Williams P C, Bhatty R S, Deshpande S S, Hussein L A and Savage G P (1994) Improving nutritional quality of cool season food legumes. In: *Expanding the Production and Use of Cool Season Food Legumes* (Eds Muehlbauer F J and Kaiser W J), Kluwer Academic Publishers, Dordrecht, The Netherlands, 113–129.

TRANSFORMATION AND REGENERATION OF GUAR

MORTEN JOERSBO
Danisco Biotechnology, Langebrogade 1, DK-1001 Copenhagen K, Denmark
e-mail: shmj@danisco.com
Present address: Danisco Seed, Hojbygardvej 14, DK-4960 Holeby, Denmark

Abstract

Production of transgenic plants of the legume guar (*Cyamopsis tetragonoloba* L.) has recently been reported. Transformation was accomplished by *Agrobacterium tumefaciens*-mediated gene transfer of a neomycin phosphotransferase gene and a β-glucuronidase gene to cotyledonary explants of different varieties. The transformation frequencies which were enhanced by thidiazuron and silver thiosulfate were up to about 1%. Carbenicillin and cefotaxime, used for the elimination of *Agrobacterium* after co-culture were toxic to the explants at high concentrations but by reducing the concentrations of the antibiotics and potentiating their effect by addition of the β-lactamase inhibitor sulbactam, complete bactericidal control and no phytotoxic effects were observed. Selection by kanamycin gave highest transformation frequencies at 145 mg/L where about 35% of the selected shoots were transgenic. These could only be rooted by grafting onto non-transgenic seedlings. Transgenes were transmitted to offspring generations in a stable Mendelian manner for about half of the transformants but aberrant inheritance was also observed.

1. Introduction

Guar (*Cyamopsis tetragonoloba* L.) is a high yielding subtropical legume producing seeds with large endosperms which consist mainly of galactomannan. This polysaccharide is primarily used as a viscosity enhancer by the food industry and it is composed of a mannan backbone with about 50% galactose substitutions that affect the functional properties of the galactomannan (Whistler and Hymowitz, 1979). Galactomannans with lower degrees of galactose substitutions such as locust bean gum (22% galactose substitutions) are employed due to their valuable gel forming properties useful for improving the texture of various foods (Morris, 1990) but the supply of this galactomannan is quite limited.

P. K. Jaiwal and R. P. Singh (eds.), Applied Genetics of Leguminosae Biotechnology, 149–152.

Therefore, galactomannans with reduced galactose content produced in transgenic guar is of commercial interest (Joersbo *et al.*, 2001).

The production of transgenic guar plants has recently been published (Joersbo *et al.*, 1999a). Like many other species, and notably legumes, this process required the development and thorough optimization of a new transformation protocol. In this review, the most important features concerning guar transformation are described.

2. Transformation

Transgenic guar shoots have been produced by *Agrobacterium tumefaciens*-mediated gene transfer of a neomycin phosphotransferase gene and a β-glucuronidase (GUS) gene, both driven by 35S-promoters, to cotyledonary explants (Joersbo *et al.*, 1999a). The transformation frequencies, calculated as the number of transgenic shoots obtained from 100 inoculated cotyledons, were up to about 1%. The authors found that this method was applicable to a variety of *Agrobacterium tumefaciens*-strains as well as different cultivars of guar, enabling transformation of the cultivars with the best agricultural performance. Such versatility is not common for legume transformation systems as they have been reported to depend considerably on the optimal combination of cultivar and *Agrobacterium tumefaciens*-strain (Hinchee *et al.*, 1988).

Elimination of *Agrobacterium* after co-culture was achieved by potentiating the effect of low dose carbenicillin and cefotaxim by addition of the β-lactamase inhibitor sulbactam, as high concentrations of these antibiotics proved to be toxic to the guar explants (Joersbo *et al.*, 1999a). In other studies on legume transformation, alternatives to the frequently used carbenicillin and cefotaxim have also been employed, e.g. timentin for pea transformation (Schroeder *et al.*, 1993), suggesting that the used antibiotics should be chosen carefully to avoid phytotoxic effects in legumes.

2.1. SELECTION

Kanamycin selection was employed for the isolation of transgenic guar shoots (Joersbo *et al.*, 1999a). Transformation frequencies reached a maximum at 145 mg/L kanamycin and they declined at either higher or lower concentrations. The selection efficiency, calculated as the number of transgenic shoots among 100 regenerated shoots during selection, increased progressively with the kanamycin concentration, being about 35% at 145 mg/L kanamycin.

It is possible that higher transformation frequencies and/or selection efficiencies might be achieved with other selection systems. Thus, for pea transformation, kanamycin selection has been reported to be inappropriate (Puonti-Kaerlas *et al.*, 1990) while the phosphinothricin/phosphinothricin acetyltransferase system worked efficiently (Schroeder *et al.*, 1993). On the other hand, in these studies an MS medium (Murashige and Skoog, 1962) was employed containing 3.0 mM Ca^{2+}. In the study of Joersbo *et al.* (1999a), a Gamborg B_5 medium (Gamborg *et al.*, 1968) containing only 1.0 mM Ca^{2+} was used as selective medium. As high Ca^{2+}-concentrations have been reported to strongly reduce the phytotoxic effect of kanamycin (Joersbo and Okkels, 1996), kanamycin selection might be advantageous to use at low Ca^{2+} concentrations for legumes.

In order to enhance regeneration of the transformed tissue during selection, the growth regulator thidiazuron (TDZ) was added (Kathen and Jacobsen, 1995; Malik and Saxena, 1992) and at a concentration of 1 mg/L, the transformation frequency was enhanced 1.8-fold. The effect of TDZ was attributed to the observed improved vigour of the regenerated shoots (Joersbo *et al.*, 1999a). Silver thiosulfate was also added during selection, in order to suppress premature flowering of the regenerated shoots, and at 2.5–5.0 mM higher transformation frequencies were observed, due to better shoot vigour.

2.2. ROOTING

No spontaneous root formation was observed on the selected transgenic shoots and addition of various auxins did not stimulate this process, they only resulted in callus growth. Therefore, it was only possible to produce plants by grafting onto non-transgenic root stocks derived from decapitated seedlings (Joersbo *et al.*, 1999a). The seedlings were never observed to produce additional shoots. The lack of root formation might be related to the employed selection method as kanamycin has been reported to inhibit root formation in sugar beet (Konwar, 1994).

2.3. ANALYSIS OF TRANSFORMANTS

Transgenecity of the harvested shoots after selection was assessed by screening for GUS-positive leaf tips (Jefferson *et al.*, 1987). Grafted GUS-positive shoots were GUS-tested once more after 6–8 weeks of growth after transfer to soil. Only transformants displaying strong GUS-positive reactions in all leaf tips in both tests were cultivated further (Joersbo *et al.*, 1999a). The presence of the transgenes was confirmed by genomic DNA analysis in all analysed transformants (Dellaporta *et al.*, 1983) and the number of copies of these was estimated to be 1–3 for most transformants.

In a number of transformants, inheritance of the transgenes was studied and almost half of the transformants (41%) produced R_1 offspring where the GUS gene was expressed and stably inherited consistent with Mendelian genetics. For those transformants which were estimated to harbour only one copy of the transgenes, this type of predictable inheritance was also found for at least 2 more generations (Joersbo *et al.*, 1999a). However, a significant fraction of the transformants produced offspring which all were GUS-negative and the presence of the transgene(s) could not be detected, despite an unambiguously positive analysis on the parent plants.

The reason for the aberrant inheritance patterns observed in guar is not understood. Chimaerism of primary transformants has been reported for several transgenic species, including legumes (Bean *et al.*, 1997; McCabe *et al.*, 1988) but this results in segregational patterns covering a wide range of ratios between transgenic and non-transgenic offspring plants. In the study by Joersbo *et al.* (1999a), the authors observed a discontinuous phenomenon where either close-to-expected ratios were found or all of the offspring appeared to be non-transgenic. Instability of transgenes has been reported but generally at low frequencies (Cherdshewasart *et al.*, 1993; Risseeuw *et al.*, 1997).

3. Conclusions

It is now possible to produce transgenic guar plants of different varieties at reasonable transformation frequencies. These might be subject to improvement but considering the fact that guar belongs to the legumes which include a number of recalcitrant species, the transformation frequencies obtained so far may be quite acceptable for many purposes.

The transformation technology described here will enable the production of transgenic guar plants with improved agronomical characters, one of which may be guar gum with enhanced functional properties (Joersbo *et al.*, 1999b, 2000).

References

Bean S J, Gooding P S, Mullineaux P M and Davies D R (1997) A simple system for pea transformation. *Plant Cell Rep.*, **16**: 513–519.

Cherdshewasart W, Gharti-Chhertri G B, Saul M W, Jacobs M and Negrutiu I (1993) Expression instability and genetic disorders in transgenic *Nicotiana plumbaginifolia* L. plants. *Transgenic Res.*, **2**: 307–320.

Dellaporta S L, Wood J and Hicks J B (1983) A plant DNA mini-preparation: Version II. *Plant Mol. Biol. Rep.*, **1**: 19–21.

Gamborg O L, Miller R A and Ojima K (1968) Nutrient requirements of suspension cultures of soybean root cells. *Exp. Cell Res.*, **50**: 151–158.

Hinchee M A W, Connor-Ward D V, Newell C A, McDonnell R E, Sato S J, Gasser C S, Fischhoff D A, Re D B, Fraley R T and Horsch R B (1988) Production of transgenic soybean plants using *Agrobacterium*-mediated DNA transfer. *Bio/Technol.*, **6**: 915–922.

Jefferson R A, Kavanagh T A and Bevan M W (1987) GUS-fusions: ß-glucuronidase as a sensitive and versatile gene fusion marker in higher plants. *EMBO J.*, **6**: 3901–3907.

Joersbo M, Brunstedt J, Marcussen J and Okkels F T (1999a) Transformation of the endospermous legume guar (*Cyamopsis tetragonoloba* L.) and analysis of transgene transmission. *Mol. Breed.*, **5**: 521–529.

Joersbo M, Marcussen J and Brunstedt J (2001) *In vivo* modification of the cell wall polysaccharide galactomannan of guar transformed with cloned α-galactosidase gene from senna. *Mol. Breed.* (in press).

Joersbo M and Okkels F T (1996) Calcium reduces toxicity of amino-glycoside antibiotics in sugar beet explants *in vitro*. *Physiol. Plant.*, **97**: 245–250.

Joersbo M, Pedersen S G, Nielsen J E, Marcussen J and Brunstedt J (1999b) Isolation and expression of two cDNA clones encoding UDP-galactose epimerase expressed in developing seeds of the endospermous legume guar. *Plant Sci.*, **142**: 147–154.

Kathen A de and Jacobsen H J (1995) Cell competence for *Agrobacterium*-mediated DNA transfer in *Pisum sativum* L. *Transgenic Res.*, **4**: 184–191.

Konwar B K (1994) *Agrobacterium tumefaciens*-mediated genetic transformation of sugar beet (*Beta vulgaris* L.). *J. Plant Biochem. Biotechnol.*, **3**: 37–41.

Malik K A and Saxena P K (1992) Regeneration in *Phaseolus vulgaris* L.: High-frequency induction of direct shoot formation in intact seedlings by N^6-benzylaminopurine and thidiazuron. *Planta*, **186**: 384–389.

McCabe D E, Swain W F, Martinell B J and Christou P (1988) Stable transformation of soybean (*Glycine max*) by particle acceleration. *Bio/Technol.*, **6**: 923–926.

Morris E R (1990) Plant galactomannans. In: *Food Gels* (Ed Harris P), Elsevier Science Publishers, UK, 314–329.

Murashige T and Skoog F (1962) A revised medium for rapid growth and bioassay with tobacco tissue cultures. *Physiol. Plant.*, **15**: 473–497.

Puonti-Kaerlas J, Eriksson T and Engström P (1990) Production of transgenic pea (*Pisum sativum* L.) plants by *Agrobacterium tumefaciens*-mediated gene transfer. *Theor. Appl. Genet.*, **80**: 246–252.

Risseeuw E, Franke-van Dijk M E I and Hooykaas P J J (1997) Gene targeting and instability of *Agrobacterium* T-DNA loci in the plant genome. *Plant J.*, **11**: 717–728.

Schroeder H E, Schotz A H, Wardley-Richardson T, Spencer D and Higgins T J V (1993) Transformation and regeneration of two cultivars of pea (*Pisum sativum* L.). *Plant Physiol.*, **101**: 751–757.

Whistler R L and Hymowitz T (1979) *Guar: Agronomy, Production, Industrial Use and Nutrition*. Purdue University Press, Indiana, USA.

IN VITRO REGENERATION OF WINGED BEAN

S. DUTTA GUPTA
Department of Agricultural and Food Engineering,
Indian Institute of Technology, Kharagpur – 721 302, India
e-mail: sdg@agfe.iitkgp.ernet.in

Abstract

There is an increasing interest in the winged bean for its potential as a high protein crop, especially in the hot and humid tropics. Considerable efforts have been made to develop desirable traits through *in vitro* manipulations. Unlike other large seeded legumes, the plant regeneration in winged bean is still limited only to a few reports. Shoot formation directly from explants as well as from callus cultures have been obtained. The culture of thin cell layers has shown great promise for prolific shoot regeneration with less time. Such system is also amenable for gene transfer. In recent years attention has been focused on the development of embryogenic systems suitable to isolation of totipotent protoplasts and gene transfer. Somatic embryogenesis has been achieved both via a callus phase or directly without an intervening callus. Immature leaf segments are found to be a convenient, accessible, and efficient source of explants for somatic embryogenesis. The key to the successful accomplishment of somatic embryogenesis has been the selection of genotype, use of explants obtained at precise developmental stage and determination of critical concentration of growth regulators. Somatic embyrogenesis in winged bean appeared to consist of two distinct stage: induction and development. The induction of embryogenesis generally requires the application of low concentration of naphthalene acetic acid (NAA) along with high concentration of N^6-benzylaminopurine (BAP) in the primary culture medium. Somatic embryo development occurred on indole-3-acetic acid (IAA) and BAP containing secondary medium. Conversion ability of somatic embryos has been improved by inclusion of a maturation phase with abscisic acid (ABA). With the existing knowledge it is possible to optimize the regeneration systems for subsequent use in gene transfer and somatic hybridization experiments towards the genetic improvement of winged bean.

1. Introduction

The winged bean, *Psophocarpus tetragonolobus* (L.) DC. is a tropical multipurpose legume with unusually high protein levels. The genus *Psophocarpus* comprises 6 species.

P. K. Jaiwal and R. P. Singh (eds.), Applied Genetics of Leguminosae Biotechnology, 153–163.

It is widely distributed in Papua New Guinea and South East Asia. The center of origin is believed to be Papua New Guinea (Hymowitz and Boyd, 1977). The winged bean has been grown both as a perennial and annual crop, and is characterized by its tuberous roots and winged pod. An outstanding feature is that it contains a significant amount of proteins in all the parts of the plant, i.e. seeds, pods, leaves and tubes, all of which are edible. In its report on agricultural and nutritional potentials of the winged beans National Academy of Sciences of the United States (NAS) recognized this crop as 'A high protein crop for the tropics' (NAS, 1981). The total yield of tuber protein was estimated as 553–1175 kg/ha (Khan *et al.*, 1977). This compares with only 202 kg protein/ha for a sweet potato crop. The percentage of crude protein content (29.8–37.4) of the seeds is comparable to that of soybean seeds (35%). In addition the seeds are a source of edible oil. The oil content ranged from 15–20.4% with 30–40% saturation and 60–70% unsaturation (Khor and Chan, 1988). However, as an edible oil the winged bean oil is less preferred than peanut and soybean oil due to its high amount of C22:0 fatty acid behenic acid. The content of essential minerals particularly iron (11 mg/100 g) and calcium (230 mg/100 g) is very high in winged bean compared to other edible legumes (Jaffe and Korte, 1976).

Another point of interest of the winged bean is the formation of numerous nodules on the roots without artificial inoculation when the soil contains compost, peat and sand. It shows effective symbiotic association with a wide range of rhizobia. All these features made this crop an attractive candidate for when *in vitro* technologies have considerable potential. The development of *in vitro* culture techniques on winged bean thus offers an alternative method to conventional breeding programmes.

In recent years, biotechnological application using *in vitro* technology has offered a great potential for crop improvement. *In vitro* techniques can provide the means for obtaining somaclonal and induced variations for introducing agronomically important genes through *Agrobacterium*/particle bombardment-mediated transformation. However, the successful application of *in vitro* methods largely relies on a highly efficient regeneration system. Remarkable progress has been made in the development of regeneration system for legumes and subsequent genetic improvement with the production of transgenic plants (Parrott *et al.*, 1992, 1995; Christou, 1996). Unfortunately, winged bean is still difficult to manipulate because of its recalcitrance in culture. There is an increasing demand to develop methods which can facilitate high frequency regeneration in winged bean. The purpose of this review is to focus on the current status of regeneration of plants in winged bean emphasizing the modes of regeneration, with a view to facilitate the development and refinement of regeneration systems for research effort towards genetic improvement of this under exploited crop.

2. Plant regeneration

The first report on regeneration of plants in winged bean was announced by Venketeswaran and Huhtinen (1978). Since then, there are reports of regeneration both via organogenesis and embryogenesis either directly from the explants or indirectly via an intervening callus phase. The various studies on the regeneration of winged bean from several laboratories are summarized in Table 1.

Table 1. Summary of the plant regeneration studies on winged bean.

Genotype	Explant	Medium/growth regulator (mg/L)	Growth response	References
Not stated	Epicotyl, Cotyledon	MS + NAA (1 or 5) + KIN (0.1)	Shoot organogenesis, whole plants	Venketeswaran and Huhtinen, 1978
Acsn No. SU 110, SU 26	Leaf	MS + NAA (0.2) + BAP (2.0) MS + IAA (0.2–2.0) + BAP (1–10.0)	Callus Shoot organogenesis, whole plants	Gregory *et al.*, 1980
TPt 6	Stem thin cell layer (tcl)	MS + IAA (0.2 or 2.0) + BAP (0.2 or 2.0)	Direct shoot organogenesis, whole plants	Lie-Schricke and Tran Thanh Van, 1981
TPt 6	Epicotyl, Leaves, Cotyledons, Stem (tcl)	MS + IAA (0.2) + BAP (2.0)	Shoot organogenesis, whole plants	Tran *et al.*, 1986
TPt 1	Epicotyl, Cotyledon	MS (liquid medium) + 2,4-D (1 or 5) + NAA (1 or 5) + KIN (0.1 or 1)	Shoot organogenesis, whole plants, embryoids	Venketeswaran, 1981
Not stated	Seedlings	B_5 + BAP (1.12–6.7) $^1/_2$ B_5 + NAA (18.6) or IBA (1.61)	Shoot organogenesis, whole plants	Mehta and Mohan Ram, 1981
Not stated	Embryo, Cotyledons, Leaf, Stem explants	MS + 2, 4-D (1.0) + KIN (1.0) MS + BAP (1 or 2)/BAP (1.0) + KIN (1.0) MS + IAA (1.0) + IBA (1.0)	Callus Shoot organogenesis, whole plants	Venketeswaran *et al.*, 1985 Venketeswaran *et al.*, 1986
Su 623	Protoplast	UM + IAA (0.2) + BAP (1.0)	Shoot organogenesis, whole plants	Wilson *et al.*, 1985
LBNG 1, LBNG 3 TPt-1, TPt-2 UPS-122	Mature embryo Hypocotyl, Epicotyl, Cotyledon	MS + 2,4-D (0.5) MS or B_5 MS + 2,4-D (1.0) + NAA (1.0) or MS + 2,4-D (0.1–5.0), MS + BAP (1.0) MS or 1/2 MS (liquid culture)	Shoot organogenesis, whole plants Callus Shoot organogenesis Somatic embryogenesis	Venketeswaran, 1990 Venketeswaran *et al.*, 1992
EC-38825-2	Leaf	MS + NAA (0.5) + BAP (1.0) MS + IAA (0.1) + BAP (2.0) MS + IBA (0.1) + BAP (0.2)	Somatic embryo induction Somatic embryo development whole plants	Ahmed *et al.*, 1996; Dutta Gupta *et al.*, 1997

2.1. SHOOT ORGANOGENESIS

Shoot organogenesis has been reported from several types of explant, including epicotyls, leaves, cotyledons, thin cell layers and seeds. Among the various types of explant, considerable success was achieved with thin cell layer explants of the winged bean variety TPT-6. Development of buds and regeneration of complete plants were successful either directly from the explants or through a callus stage (Tran *et al.*, 1986). The thin cell layers initiated more buds in large number compared to fragment explants. *In vitro* bud formation on fragment explant was always accompanied by callus formation.

Shoot organogenesis was obtained by varying IAA and BAP ratio on MS (Murashige and Skoog, 1962) medium. Using thin cell layer explants nearly one hundred percent regeneration was obtained with 10^{-5} M BAP and 10^{-6} M IAA (Tran *et al.*, 1986).

Embryo segments from seeds sown on MS + 2,4-dichlorophenoxy acetic acid (2.4-D, 0.5 mg/L) for 5 days upon transfer to basal medium resulted in the formation of leafy shoots, shoot buds and differentiated plantlets (Venketeswaran, 1990). Callus could be induced from epicotyls, cotyledons, and stem segments on media containing 1 mg/L each of 2,4-D and kinetin (KIN) or 1 mg/L each of NAA and 2,4-D. Shoot regeneration occurred from such calluses after two to three subcultures on medium supplemented with 1 and 2 mg/L BAP. Complete plantlets were obtained by transferring the shoots on root induction medium containing mineral salts of MS or B_5 (Gamborg *et al.*, 1968) with 1 mg/L IAA or indole-3-butyric acid (IBA) or both IAA + IBA (Venketeswaran, 1990).

Whole seeds grown aseptically on B_5 medium containing various levels of BAP produced multiple shoot buds at the base of the epicotyl and cotyledonary node of primary leaves (Mehta and Mohan Ram, 1981). To obtain whole plant, transfer of shoots to rooting medium (modified B_5) containing 10^{-4} M NAA or 5×10^{-6} M IBA was necessary.

Leaf explants were also induced to form shoot buds via callus phase. Indeed, the first report of plantlet regeneration from callus tissues derived from the leaf of a grain legume came from the winged bean (Gregory *et al.*, 1980). Callus tissues were first induced from both mature and young leaves. A transfer from induction medium containing NAA and BAP to a regenerating medium containing IAA and BAP was essential to obtain shoot organogenesis. However, in our studies, combinations of IAA and BAP were found

*Table 2. Frequency of regeneration of buds/shoots on leaf callus**

Shoot regeneration medium BAP (mg/L)	Total no. of callus transferred	No. of callus in which buds/shoots produced	Percent regeneration	Mean no. of shoot/ responding callus ± SD
0	45	0	0	0 ± 0.0
2	45	9	20	2.6 ± 0.54
4	45	27	60	5.4 ± 0.89
6	45	30	66	9.0 ± 1.0
8	45	16	35	4.1 ± 0.75
10	45	4	8	1.4 ± 0.54

*Leaf callus was induced on MS medium containing 0.2 mg/L NAA and 2.0 mg/L BAP.

effective in somatic embryo development (Ahmed *et al.*, 1996; Dutta Gupta *et al.*, 1997). We obtained shoot organogenesis on BAP supplemented medium. Among the various concentrations tested the maximum response was obtained with 6 mg/L BAP (Table 2). Multiple shoots were originated adventitiously from leaf derived-callus (Fig. 1a) and also from the proliferation of axillary meristems (Fig. 1b). The shoots proliferated further into well developed leafy structures at the same culture conditions after 5 weeks of culture. Shoots regenerated on higher concentrations of BAP (8 mg/L and 10 mg/L) were fasciated and showed stunted growth (Fig. 1c). The promotive role of BAP in shoot regeneration has also been observed in other large-seeded legumes (Parrott *et al.*, 1992). The regenerated shoots were able to develop roots on MS medium supplemented with 0.2 mg/L IBA (Fig. 1d).

Histological examinations of shoot producing cultures revealed that the regenerated shoots had their origin in subepidermal tissue. Cells, 3–4 layers below the callus surface showed a significant high cell division activity and resulted in small protruberances and formation of typical shoots and leaf primordia within 6 weeks of culture. No direct vascular connections between the mother tissue and the regenerated shoots were found to exist, indicating *de novo* origin of shoot initiation. Apart from *de novo* organization of shoot meristems in callus cultures shoot regeneration also occurred through the development of axillary buds followed by inhibition of apical dominance.

2.2. SOMATIC EMBRYOGENESIS

The regeneration of plants via somatic embryogenesis is an important and essential component of biotechnology that is required for the genetic manipulation and improvement of plants. It overcomes many of the disadvantages of regeneration from adventitious shoot meristem of multicellular origin. Somatic embryogenesis may occur indirectly via a callus phase or directly without an intervening callus. Somatic embryogenesis in winged bean was first described by Venketeswaran (1990). Somatic embryos were induced by transferring newly initiated calluses maintained in a medium with 0.1–0.5 mg/L 2,4-D, 2 mg/L each of 2,4-D and NAA to a medium free of growth regulators, supplemented with full or half-strength MS salts. Somatic embryos were induced from epicotyl and hypocotyl cultures of the varieties TPt-1, TPt-2 and UPS-122 (Venketeswaran *et al.*, 1992). However, regeneration of plantlets was unsuccessful from these structures.

In our studies the protocol used to obtain somatic embryogenesis consisted of two distinct stages: induction and development. Initiation and development of somatic embryos from leaf segments occurred with a two step culture method.

2.2.1. Embryo induction

Somatic embryos direct as well as indirect were induced on primary culture medium containing NAA and BAP. No somatic embryo was detected on primary culture medium in winged bean as reported earlier in peanut (Baker and Wetzstein, 1992). Sequential development of somatic embryos directly or through a callus phase was observed on secondary medium (Fig. a and b). The ratio of NAA and BAP in the induction medium clearly affected the frequency of somatic embryogenesis. Apart from the NAA-BAP combination

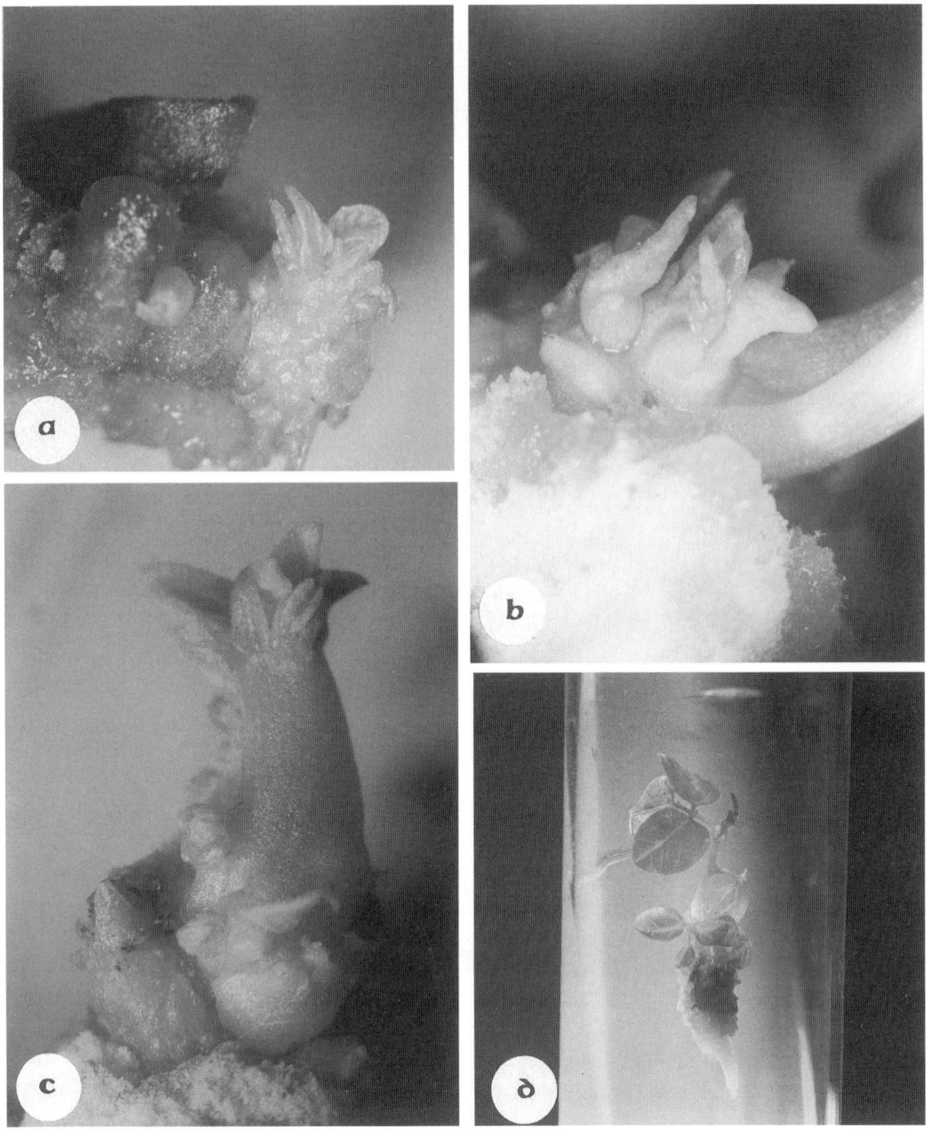

Figure 1. Shoot organogenesis and plant regeneration in winged bean. (a) Multiple shoot formation on leaf callus. (b) Proliferation of axillary meristem into multiple shoots. (c) Development of fasciated shoot with high concentration of BAP (8 mg/L). (d) Regeneration of a complete plantlet.

the duration of primary culture medium had a decisive influence on the type of somatic embryogenesis and embryogenic response. A narrow range of NAA concentration (0.1–0.5 mg/L) along with 1–2 mg/L BAP, was capable of inducing the somatic embryos. Leaf segments pre-cultured on these treatments for 28 days resulted in the formation of

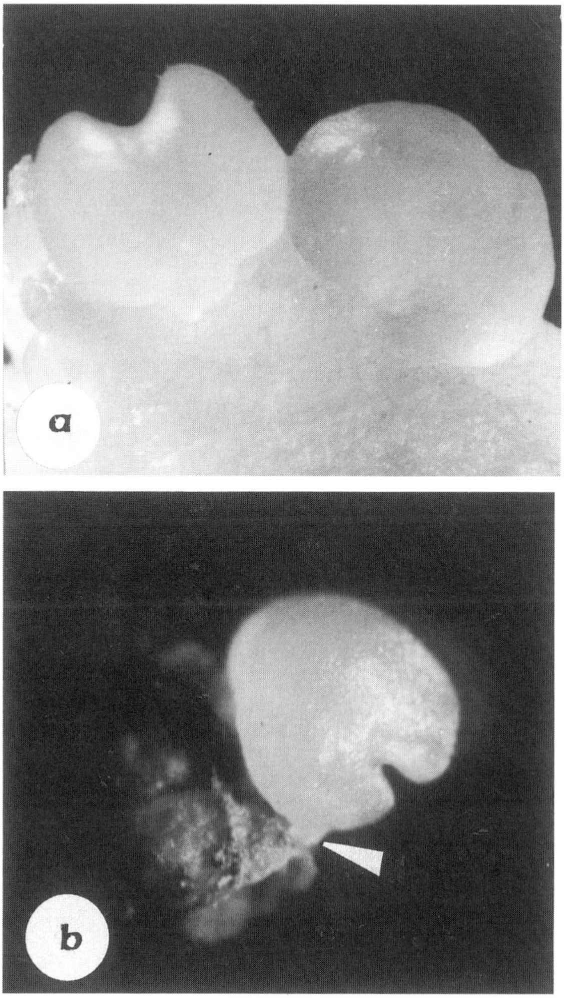

Figure 2. Somatic embryogenesis from leaf segments of winged bean. (a) A globular (right) and a heart shaped embryo on the surface of leaf callus. (b) A heart-shaped embryo with well developed suspensor (marked with arrow) formed directly on leaf surface.

direct somatic embryos. However, a prolonged exposure (48 days) of leaf segment to the induction produced embryogenic calluses. A clear trend for the effect of incubation time on embryo induction was noted and concurs with the finding of Baker and Wetzstein (1992).

The findings on NAA/BAP stimulated embryo induction is very striking, since in most of the reported examples of somatic embryogenesis in grain legumes, high auxins favoured embryo induction and gradual omission of auxin resulted in embryo development (Parrott *et al.*, 1992, p. 95). Conflicting results have been reported for the role of cytokinins in legume somatic embryogenesis. Inhibitory role of cytokinin has been noted in soybean (Lazzeri *et al.*, 1987), pea (Kysely and Jacobsen, 1990) and peanut (Baker and Wetzstein, 1994). Conversely, cytokinin induced somatic embryogenesis has been reported in *Phaseolus* (Malik and Sexena, 1992), chickpea (Murthy *et al.*, 1996) and peanut (Murthy *et al.*, 1995). Thus, it is difficult to formulate a generalized protocol for somatic embryogenesis in grain legumes as the growth regulator requirements appear to be very species and tissue specific.

Our studies on embryo induction also indicated the involvement of protein phosphorylation at an increased rate by protein kinase, as one of the components of transmembrane signalling during somatic embryogenesis.

2.2.2. Somatic embryo development

Somatic embryo development occurred on secondary media containing IAA and BAP. Somatic embryo development was found to be reduced when IAA levels were increased and BAP levels were decreased (Fig. 3). The apparent inhibition of somatic embryo development with IAA is analogous to what happens in peanut. In peanut embryo development occurred upon removal of the auxin from the medium (Parrott *et al.*, 1992). Combinations of IAA and BAP have also been found effective in shoot bud regeneration

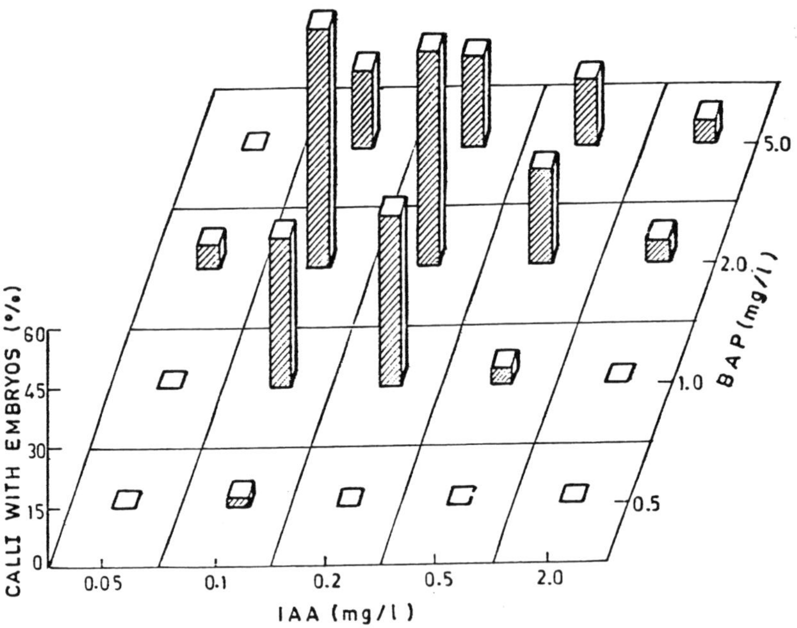

Figure 3. Influence of the growth regulator combination of the secondary medium on percent embryogenesis. Somatic embryos were induced from calluses initiated on primary medium with 0.5 mg/L NAA and 1.0 mg/L BAP. Number of callus pieces with somatic embryos were recorded 21 days after transfer to secondary medium containing the indicated IAA and BAP concentrations.

from leaf explants (Gregory *et al.*, 1980), and thin cell layer explants (Tran *et al.*, 1986) of winged bean. Cultivar differences may account for the different observations reported. Genotype has been shown to greatly influence regeneration potential in legumes (Bailey *et al.*, 1993; Parrot *et al.*, 1995; Brar *et al.*, 1999).

2.2.3. Somatic embryo germination and conversion

Embryo germination and successful conversion into plantlets was found to be associated with selective use of morphologically normal embryos. Approximately 42% conversion was noted on MS medium with 0.1 mg/L IBA and 0.2 mg/L BAP from well developed cotyledonary embryos. The development pattern of somatic embryo affecting conversion has been well documented (Wetzstein and Baker, 1993). Addition of ABA to the germination/conversion medium increased the conversion rate. It has been suggested that ABA initiates the synthesis of storage proteins, promotes desiccation tolerance, maturation and subsequent conversion of somatic embryos permitting plant recovery (Parrott *et al.*, 1995).

3. Conclusions and future prospects

Impressive progress has been made in recent years towards regeneration of plants via somatic embryogenesis as well as shoot organogenesis to meet the challenging demands of winged bean improvement. Various protocols manifesting the regeneration potential are currently available. A decision has to be made on the use of protocols considering the mode and type of regeneration for achieving the specific goals. The key to the successful development of regeneration systems has been the use of explants obtained at precise developmental stage and selection of growth regulator requirements at different cultural stages.

Although plant regeneration can be achieved with relative success, the process is yet to be maximized. There are many major problems that still remain. One of the major constrains is the lack of understanding of the factors that control embryogenic induction/competence. A detailed molecular understanding of the hormonal control of embryogenic induction is needed. Many of the somatic embryos germinate precociously to give rise to weak plants with abnormality. Poor embryo conversion also appeared to be under hormonal control. Thus optimization of growth regulators and other cultural parameters are also essential for the recovery of healthy plants. Further improvement in the efficiency of plant regeneration and subsequent recovery of plants may make the system highly amenable for transformation using *Agrobacterium* and/or particle bombardment-mediated DNA delivery. It is anticipated that through an efficient and reproducible regeneration system coupled with gene transfer technology new genotypes with desirable characteristics, such as improvement of the quality of proteins, vitamins and oils and erect short sized plants for easier cultivation, can be obtained.

References

Ahmed R, Dutta Gupta S and De D N (1996). Somatic embryogenesis and plant regeneration from leaf derived callus of winged bean (*Psophocarpus tetragonolobus* (L.) DC). *Plant Cell Rep.*, **15**: 531–535.

REGENERATION AND GENETIC TRANSFORMATION IN PEANUT: CURRENT STATUS AND FUTURE PROSPECTS

SUSAN EAPEN
Plant Biotechnology and Secondary Products Section, Nuclear Agriculture and Biotechnology Division, Bhabha Atomic Research Centre, Mumbai – 400085, India
e-mail: eapenhome@yahoo.com

Abstract

Peanut, *Arachis hypogaea* L. has long been the focus of conventional plant breeding efforts because of its importance as a source of high quality oil and protein. Recent techniques in genetic engineering coupled with developments in regeneration technology can be used for introduction of agronomically useful traits into established cultivars, which will supplement the conventional breeding programmes. Rapid strides have been made in the last two decades to develop a regeneration system for peanut. Highly reproducible regeneration systems through proliferation of axillary buds around a cultured meristem, *de novo* shoot organogenesis and somatic embryogenesis are available today as target explants for experiments on transformation. Success has been achieved in the development of transgenic plants using both *Agrobacterium*-mediated and direct DNA transfer using particle gun bombardment. Several useful genes have already been transferred to peanut using gene transfer techniques. Peanut production is severely limited by a number of diseases and pests and one of the most challenging needs of the day is to improve resistance to *Aspergillus* species which produce aflatoxins – which are potent carcinogenic metabolites. In addition, introduction of value added traits such as altered protein/oil composition today lies within the realms of biotechnology. Developing peanut plants with genes for abiotic stress or edible vaccines is not far away. The candidate genes which are currently available for transfer to peanut with possible implications in groundnut breeding programme is discussed. As developments in plant biotechnology unfolds, high frequency, disease resistant, drought tolerant and good tasting peanut, which are safe to eat, will continue to be the goals of peanut biotechnologists.

1. Introduction

Peanut or groundnut (*Arachis hypogaea* L.), a member of subtribe Stylosanthinae of tribe Aeschynomeneae of family Leguminosae, a native of South America, is at present an

P. K. Jaiwal and R. P. Singh (eds.), Applied Genetics of Leguminosae Biotechnology, 165–186.

important oil, food and forage crop grown in 80 countries from 40° N to 40° S in tropical and warm temperate regions of the world. The cultivated peanut is tetraploid and the species has been divided into two subspecies hypogaea and fastigata with several botanical varieties. About 13.5 million ha are grown in Asia, 5.3 million ha in Africa and 1.2 million ha in America and 0.1 million in other parts of the world (Carley and Fletcher, 1995). India and China are the largest producers of the crop (See Stalker, 1997; Isleib *et al.*, 1994). Average yields on a global scale has increased from 0.93 mt/ha in 1970s to 1.15 mt/ha in 1990s (Carley and Fletcher, 1995). However, when diseases and pests will be controlled, yields of 3 mt/ha or more can be achieved.

Peanut seed has from 41 to 59% oil (Savage and Keenan, 1994) and more than half of the global crop is grown as an oil seed. As major producers will become self sufficient for oil production, the large proportion of peanut crop will be used as a source of protein. The peanut protein is showing increasing potential as a food source, especially in the developing world where lack of adequate protein can be a very serious dietary problem. Protein makes up 12–36% of groundnut kernel (Savage and Keenan, 1994). The use of peanut as a source of peanut butter and snacks has boosted the popularity of the crop in Western countries. Besides seeds being used as a source of oil and protein for human consumption, the meal remaining after oil extraction is an important animal feed. The foliage is often used as a fodder for farm animals.

Peanut is unique among domesticated plants in that it flowers above ground, but produces seeds below the surface. It is a self fertilized crop. After pollination, the ovary elongates geotropically into a structure called peg (gynophore), which elongates and enters the soil, and after entering the soil, the apical end expands into a pod.

2. Crop improvement by breeding

The primary objective of groundnut breeders is to develop cultivars with high yield potential, adaptation to specific environments and production systems, resistance or tolerance to environmental stresses and resistance to diseases and insects. Breeding is a continuing process as the crop is introduced to new environments and production systems, as market demands change, and as diseases and pest populations shift in deployment of new cultivars (Isleib *et al.*, 1994). As the development of new cultivars adapted to local environments are produced, there will be a shift in breeding objectives to improve flavour and quality as desired by processors and consumers (Bunting *et al.*, 1985). Selection of short seasoned cultivars with drought resistance is a high priority in several programmes. Mass selection, pedigree method, recurrent selection, mutation breeding and hybridization have been used either alone or in combination for the improvement of the crop.

On a world-wide basis, the most important results of groundnut breeding in the past 10–20 years have been the identification of sources of resistance to foliar fungal pathogens and transfer of resistance into breeding populations with the locally appropriate agronomic attribute (Isleib *et al.*, 1994). Closely following the foliar diseases in importance is the aflatoxin problem. Breeding for resistance to insect pests and various diseases has not been emphasized to the same degree as breeding for resistance to foliar diseases. In Europe and the United States, peanut consumers are concerned about food quality and want to be certain that the food contains minimum amounts of pesticide residues and toxins. These consumer demands require cultivars resistant to biotic and abiotic stresses.

Until recently, the gene pool of cultivated groundnut comprised the global collection of the cultigen (about 12000 accessions) and the smaller collection of *Arachis* species of which genes only from species of section *Arachis* were accessible through sexual transfer. However, the whole scenario is being changed by the possibility of transfer of genes bypassing inter-specific, inter-generic and inter-kingdom barriers to produce transgenic peanut with desirable characters.

Developments in molecular and recombinant DNA techniques have opened up new opportunities for cloning of single genes from other organisms coupled with gene delivery techniques, and these techniques can now be applied to peanut, where successful regeneration and transformation techniques have been well documented.

3. Regeneration in peanut

Regeneration in peanut as in any other plant is of three different types:

(1) Axillary bud proliferation from area surrounding a meristem
(2) *De novo* shoot organogenesis
(3) Somatic embryogenesis.

3.1. AXILLARY BUD PROLIFERATION

When a meristem, shoot apex, an axillary bud or a cotyledon is cultured on a medium supplemented with a high cytokinin either alone or in combination with an auxin, axillary multiple shoots develop from the area surrounding the shoot or meristem. Shoot tips of peanut (Eapen *et al.*, 1998) and cultured cotyledons (Bhatia *et al.*, 1985) are known to produce plants by multiple axillary bud proliferation (Table 1). When cotyledons are cultured, normally shoot differentiation takes place only from the proximal end near the cotyledonary node and not from the distal end. This type of differentiation of shoot buds since it is restricted to cotyledonary nodal area can be considered as axillary bud proliferation from areas surrounding the meristem. Maximum number of multiple shoots was obtained on MS medium supplemented with 25 mgL^{-1} benzyladenine (BA) (McKently *et al.*, 1990). Twenty different genotypes were regenerated using the procedure and no phenotypic variant was observed among the regenerants (McKently, 1990).

Eapen *et al* (1998) found that BA (5 mgL^{-1}) in combination with naphthalene acetic acid (NAA, 0.1 mgL^{-1}) produced best results from cultured shoot tips of peanut. The regenerants showed a decrease in plant height, leaflet size, number of pegs, seeds and seed weight. They showed an increase in number of primary branches in comparison with seed derived control plants. No significant change in number of secondary branches and hundred seed weight was observed. In R_2, low percentage of variants ($< 1\%$) was observed, which were found to breed true in subsequent generations (Eapen *et al.*, 1998).

Table 1. Regeneration in Arachis hypogaea *by organogenesis or by adventitious meristem development*

Explant	Morphogenetic response	Reference
Shoot tips	Plants	Russo and Varnell, 1978 Kartha *et al.*, 1981 Eapen *et al.*, 1998 Ponsaamuel *et al.*, 1998
Cotyledon and cotyledonary node	Plants	Illingsworth, 1968 Illingsworth, 1974
	Roots	Guy *et al.*, 1978 Atreya *et al.*, 1984
	Plants	Sastri *et al.*, 1982
	Plants, flowers and pods	Narasimhulu and Reddy, 1984, 1985
	Plants	Bhatia *et al.*, 1985 Mhatre *et al.*, 1985 Rugman and Cocking, 1985 McKently *et al.*, 1990 Daimon and Mii, 1991 Venkatachalam and Jayabalan, 1997 Pestana *et al.*, 1999
Mature leaves	Shoot	Dunbar and Pitman, 1992
Immature leaves	Plants	Mroginski *et al.*, 1981
	Plants	Pitman *et al.*, 1983
	Plants	Rani and Reddy, 1996
	Plants	Venkatachalam *et al.*, 1999a
	Plants	Seitz *et al.*, 1987 McKently *et al.*, 1990, 1991 Cheng *et al.*, 1992 Eapen and George, 1993a
	Plants	Pestana *et al.*, 1999
Seedling explant Internode Petiole Hypocotyl Cotyledon Leaf	Plants	Kanyand *et al.*, 1994
Seed explants	Plants	McKently *et al.*, 1990
Epicotyl and hypocotyl	Plants	Venkatachalam *et al.*, 1998a
Leaf and petiole	Plants	Venkatachalam *et al.*, 1999a
Seeds	Plants Plants	Hisajima *et al.*, 1989 Gill and Ozias-Akins, 1999
Epicotyl and petiole	Plants	Cheng *et al.*, 1992

3.2. SHOOT ORGANOGENESIS

Organogenesis is the process by which a cell or a group of cells differentiate to form organs. Organogenesis refers to the formation of shoots and roots. Organogenesis can be induced by manipulations of exogenous phytohormones and can occur directly from the callus or from cultured explant. Primary leaves of peanut are known to regenerate shoots from the leaf lamina (Mroginski *et al.*, 1981; Seitz *et al.*, 1987; McKently *et al.*, 1990, 1991; Chang, 1992; Eapen and George, 1993a; Kanyanad *et al.*, 1994; Pestana *et al.*, 1999) (Fig. 1). A simple medium with a cytokinin such as BA either alone or in combination with an auxin or auxin-aminoacid conjugate is known to induce shoot organogenesis in peanut leaf discs (Eapen and George, 1993a). Pestana *et al.* (1999) used BA in combination with $AgNO_3$ to induce shoot differentiation. Immature leaflets are more responsive to regeneration than mature leaflets (Venkatachalam *et al.*, 1999). Epicotyl (Cheng *et al.*,

Figure 1. Development of multiple shoots from cultured cotyledon.

1992; Venkatachalam, *et al.*, 1998a) and petiole (Cheng *et al.*, 1992) also regenerated plantlets on culture. Kanyand *et al.* (1994) used thidiazuron (0.5 to 30 mg/L) to induce shoots from seedling explants such as leaf lamina, whole leaf, whole leaf without petiole, hypocotyl and internode segment. When cotyledons are used, if shoot buds originate from distal end, it can be considered as organogenesis. However, many authors confuse shoot induction from proximal end of cotyledon as *de novo* shoot regeneration, which is actually development of multiple shoots around the cotyledonary area. Normally shoot development from cotyledon is restricted to the proximal end and hence they can be considered as multiple bud development and not true *de novo* shoot organogenesis.

3.3. SOMATIC EMBRYOGENESIS

Somatic embryogenesis is a process whereby a cell or a group of cells from somatic tissues form an embryo. The development of somatic embryos parallels that of zygotic embryo and similar stages as globular, heart shaped, torpedo and cotyledonary stages can be observed. As with organogenesis, somatic embryogenesis in peanut is very similar to that of other legumes and other plants. A somatic embryo will be distinct from shoot organogenesis, in that it should have a root-shoot meristem at opposite poles, should have a closed vascular system and should be capable of germination when detached from mother explant, with simultaneous development of shoots and roots at opposite poles (Fig. 2a). Often secondary somatic embryos develop from the primary somatic embryos (Fig. 2b) and the process can be continuous when subcultured in the same medium. Durham and Parrot (1992) obtained repetitive somatic embryogenisis in liquid medium. Many researchers confuse shoot organogenesis as somatic embryogenesis without providing substantial evidence, just by visual morphogenic appearance. Somatic embryogenesis can occur directly from the cultured explants or develop from callus initiated from the explant.Somatic embryogenesis can be induced in peanut using a variety of explants such as immature embryo axes (Hazra *et al.*, 1989; Ozias-Akins, 1992; Ramdev Reddy and Reddy, 1993; Eapen *et al.*, 1993), whole immature embryos (Sellars *et al.*, 1990), immature cotyledons (Ozias-Akins, 1989; Ozias-Akins *et al.*, 1992; Durham and Parrott, 1992; Baker and Wetzstein, 1992, 1994; Eapen *et al.*, 1993), embryonal axes of mature seeds (McKently, 1991, 1995) and leaves (Baker *et al.*, 1992; Venkatachalam *et al.*, 1999 b) (Table 2.). Out of the different explant sources studied, immature embryonal axes proved to be the best explant source followed by immature cotyledon, mature embryonal axis and whole immature embryo (George and Eapen, 1993).

For induction of somatic embryogenesis, an auxin alone at a high concentration is sufficient (Eapen *et al.*, 1993). Auxins such as 2,4-dichlorophenoxyacetic acid (2,4-D), NAA, picloram, dicamba 2,4,5-trichlorophenoxyacetic acid (2,4,5-T), indole propionic acid, 2,4,5 trichlorophenoxypropionic acid and α-napthoxyacetic acid could induce somatic embryogenesis in peanut (Eapen and George, 1993b), but 2,4-D was the best. Cytokinins are not essential for induction of somatic embryos, but incorporation of zeatin to the medium slightly enhanced the frequency of responding cultures (Eapen and George, 1993b). Kinetin, 6-γ-γ-dimethylallyl amino purine (2-iP) and BA reduced the frequency of response (Eapen and George, 1993b). Among the different sugars, sucrose at 6% gave the best results followed by glucose and fructose, while maltose did not

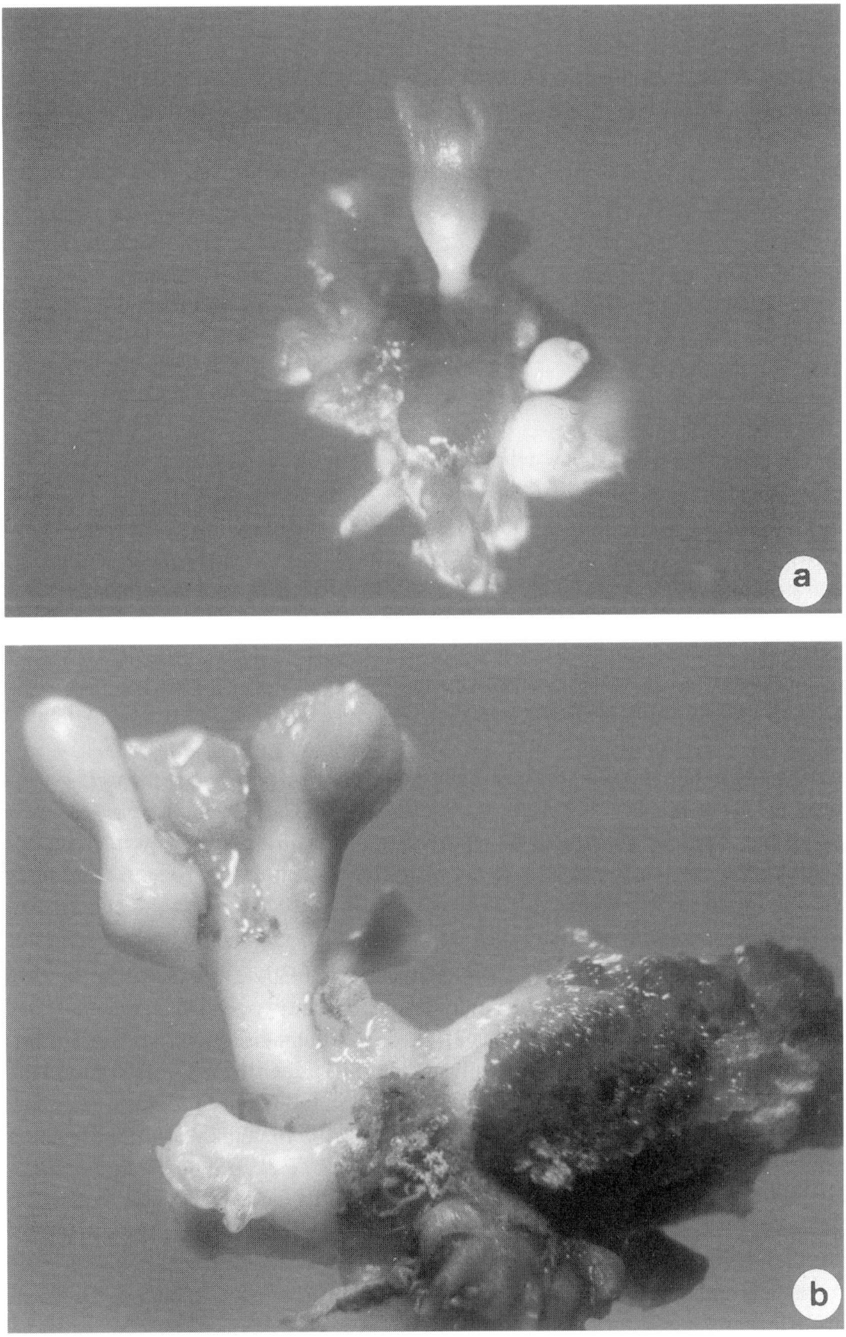

Figure 2. (a) Development of a somatic embryo from cultured immature zygotic embryo; (b) Development of secondary somatic embryos from primary somatic embryo.

Figure 3. GUS assay of peanut leaf explants after co-cultivation with A. tumefaciens *LBA-4404 with pB1121 plasmid. (a) One week after co-cultivation; (b) 45 days after culture.*

Table 3. Selected examples of genetic transformation in peanut using Agrobacterium tumefaciens

Gene transferred	Explant used	Transformant	Reference
nptII	—	—	Dong *et al.*, 1990
	—	Tumor	Lacorte *et al.*, 1991
	Hypocotyl	Callus	Mansur *et al.*, 1993
	—	Callus	Franklin *et al.*, 1993
	Whole plant	Plants	McKently *et al.*, 1995
uidA	Cotyledon	Plant	Freitas *et al.*, 1997
uidA, nptII	Leaf	Plants	Eapen and George, 1994
	—	Plants	Cheng *et al.*, 1994
nptII, uidA	Cotyledon	Plants	Sharma *et al.*, 1993
	Leaf	Plants	Ping *et al.*, 1996
uidA, nptII	Leaf	Plants	Cheng *et al.*, 1996
	—	T_0,T_1,T_2 generations	Cheng *et al.*, 1997
	—		
uidA, nptII and	Epicotyl and leaf	Plants	Egnin *et al.*, 1998
Edible vaccine - Rinderpest for animals	—		Khandelwal *et al.*, 1999
uidA, nptII	Cotyledon	Plants	Venkatachalam *et al.*, 1998b
uidA	Embryo with one cotyledon	Plants	Rohini and Shankara, Rao 2000
Nucleocapsid gene of tomato wilt virus	—	Plants	Jarret and Demski, 1997
Indian peanut clump virus (IPCVcp), *nptII, uidA*	Cotyledon	T_1 plant	Sharma and Anjaiah, 2000

Among the different genes that have been introduced by particle gun bombardment are 2S albumin gene from Brazil nut (Lacorte *et al.*, 1997), *cry1Ac* gene from *Bacillus thuringensis* (Singsit *et al.*, 1997) and nucleocapsid protein gene of tomato spotted wilt virus (Yang *et al.*, 1998a), improved nutritional protein gene (Lewis *et al.*, 1998) and synthetic antifungal peptide gene (Yang *et al.*, 1998b).

Invasion of peanut pods and seeds by aflatoxin forming species of *Aspergillus* is linked to injury by the lesser cornstalk borer. Introduction of a codon modified *cry1Ac* gene into peanut resulted in complete larval mortality to 66% reduction in larval weight (Singsit *et al.*, 1997).

Transgene copy number ranged from 1 to 20 with multiple integration (Livingtone *et al.*, 1999). Gene silencing operating in primary transgenic plants with multiple insertions have been observed (Yang *et al.*, 1998a). When a single copy of tomato spotted wilt

Table 4. Selected examples of particle gun mediated gene transfer in peanut

Gene transferred	Explant used	Transformant	Reference
	Embryos	—	Schnall and Weissinger, 1993
	Leaflet	—	Clementale *et al*., 1992
uidA, bar, tomato spotted wilt nucleocapsid protein gene	Embryonic axes	Plant	Brar *et al*., 1994
	Embryonic callus	Plant	Ozias-Akins *et al*., 1993
uidA and Luciferase gene	Leaf	Plant	Livingstone and Birch, 1995
GUS and 2 S albumin gene	Cotyledon with or without embryonal axes	—	Lacorte *et al*., 1997
Bt *Cry1Ac* gene	Somatic embryo	Plant	Singsit *et al*., 1997
uidA	Somatic embryo	Plant	Wang *et al*., 1998
uidA	Embryogenic tissues from leaf		Kim *et al*., 1999
Synthetic antifungal peptide gene			Yang *et al*., 1998b
Tomato anionic peroxidase gene	Embryos	Plant	Gill *et al*., 1998
Nucleocapsid protein gene of Tomato spotted wilt virus		Plant	Yang *et al*., 1998a

virus nucleocapsid protein (N) gene of the lettuce isolate (TSWV) was inserted into peanut, the gene segregated in 3:1 ratio in the progeny (Yang *et al*., 1998a). Livingstone and Birch (1999) used diverse cultivars of peanut to develop transgenics using particle gun technology. However, much effort needs to be put in to develop transgenic plants with modified agronomic traits.

5. Traits to be altered and candidate genes for peanut improvement

The primary objective of peanut improvement is to develop cultivars with high yield potential, adaptation to specific environments and production system, resistance or tolerance to environmental stresses and resistance to diseases and insects. Improvement of the crop is a continuing process since the crop is introduced to new environments and production systems as market demands change and as disease and insect population shift in deployment of new cultivars (Isleib *et al*., 1994). As the primary constraints for production improves, requirements will be shifted towards improving quality parameters such as flavour.

Molecular methodologies have opened up new opportunities in isolation of single genes from other *Arachis* species/plants/organisms for introduction into peanut. While yield is a complex character with many components, it is easier to introduce single genes for resistance to diseases, pests and environmental stresses.

Peanuts are plagued by foliar and seed borne diseases and insect pests. Leaf spots caused by *Cercospora arachidicola* Hori and *Cercosporidium personatum* are the most widespread diseases, but other diseases cause serious losses on a regional basis (Table 5). Production can be severely hampered by a number of diseases caused by viruses (Peanut stripe, tomato wilt spot, etc.). Substantial losses also occur due to boring and pod damage caused by borers and many other insects (Table 6.). Perhaps the most pressing challenge to peanut industry is the susceptibility of seeds to the toxigenic fungus – *Aspergillus flavus* or *parasiticus* which invades peanut seeds, particularly those exposed to improper storage and conditions such as mechanical damage, to produce aflatoxins which are potent carcinogenic metabolites. Wild *Arachis* species exist which are superior to *A. hypogaea* in diseases and insect resistance, but differences in ploidy level and other incompatibility barriers have hampered efforts to introduce these useful traits into peanut using conventional breeding techniques.

Table 5. Common diseases in peanut

Bacteria
Bacterial wilt by *Pseudomonas solanacearum*
Bacterial wilt by *Xanthomonas sp. (eye-spot)*
Viruses and mycoplasmas
Peanut mottle virus (PMV)
Peanut stripe virus (PStV)
Peanut stunt virus (PSV)
Peanut clump virus (PCV)
Groundnut rosette virus (GRV)
Tomato spotted wilt virus (TSWV)
Fungal diseases
Early leafspot by *Cercospora arachidicola*
Late leafspot by *Cercosporidium personatum*
Rust by *Puccina arachidis*
Web blotch by *Phoma arachidicola*
Pepper spot and leaf scorch by *Leptosphaerulina crassiasca*
Anthracnose by *Colletotrichum* sp.
Scab by *Sphaceloma arachidis*
Phoma blight by *Phoma microspora*
Powdery mildew by *Oidium arachidis*
Stem rot by *Sclerotium rolfsii*
Afflatoxin production in seeds by *Aspergillus* spp.
Verticillium wilt by *Verticillium dahliae*
Rhizoctonia disease by *Rhizoctonia solani*
Nematodes
Pod and root knot disease by *Meloidogyne* spp.
Groundnut chlorosis by *Aspasmatylenchus* spp.
Yellow patch disease by *Scutellonema cavenessi*

Taken from Middleton *et al.*, 1994.

5.1. PEST/INSECT RESISTANCE

The insects that live in the soil are responsible for higher levels of yield loss than foliage feeders in groundnut. They attack pods and roots and foliage via roots (Table 6.) Peanut is also subjected to a reduction of yield and quality due to feeding by insects on leaves, pegs, pods and seeds. In addition to causing damage directly, some insects serve as vectors of viral diseases. Insects of global importance include aphids, thrips, jassids and *Spodoptera*. Leaf miner, *Helicoverpa, Hilda* and other lepidopteran insects present problems in specific regions. Damage from pod borer not only reduces yield, but also permits entry into the pod of soil-borne pathogenes such as *Aspergillus flavus* (Isleib *et al*., 1994). Post harvest pests also cause damage adding to the risk of infection by *Aspergillus flavus*, the fungal source of afflatoxin.

Table 6. Selected examples of some of the important insect pests of groundnut

Soil Insects

Coleoptera
- *Sphenoptera indica* – Jewel beetles (Root borer)
- *Scarabaeidae* – White grubs as larvae

Lepidoptera
- *Agrotis* spp. – Crown borer
- *Feltia* spp. – Crown borer
- *Spodoptera litura* – Pod borer
- *Elasmopalpus lignosellus* – The lesser cornstalk borer

Isoptera
- Termites

Dermaptera – *Earwigs*
- *Euborella annulipes, E. plebeja*

Hemiptera
- Mealy bugs – *Pseudococcus solani*
- *Hilda* bug – *Hilda partruelis*

Hymenoptera

Ants – *Dorylus orientalis*

Insect that live on leaves and flowers

Orthopteroid order
- Grasshopper, Locusts, Crickets, Mantids

Thrips as leaf eaters
- *Thrips palmi*

Homoptera
- Aphids – *Aphis craccivora*
- Jassids or leaf hoppers – *Empoasca kerri*
- Whiteflies

Lepidoptera
- Many leaf eating caterpillars
- Armyworms – *Spodoptera* spp.
- *Helicoverpa armigera*

Post-harvest pests

Bruchid – *Carvedon serratus*

Taken from Wightman and Ranga Rao (1999).

5.1.1. Bacillus thuringiensis *δ-endotoxin gene*

Bacillus thuringiensis, a gram positive bacterium produces many crystalline proteins (Cry proteins) which are specific for specific groups of insects. *cryI* genes are specific for lepidopteran insects and *cryII* specific for lepidoptera and deptera and *cryIII* for coleoptera. Since peanut plants are infested by lepidopteran and coleopteran type of insects (Table 6), *cryI, II* and *III* are the candidate genes for transfer to peanut to develop plants resistant to these specific group of insects. Since native genes do not perform well in transgenic plants, codon-optimized synthetic gene is the gene of choice for development of transgenic plants with insect resistance.

Lesser corn-stalk borer is a lepidopteran insect which causes damage in North America. Singsit *et al.* (1997) successfully developed transgenic peanut plants with *cry1Ac* gene which showed resistance to lesser cornstalk borer. In spite of concerns about the practical implementation and sustainability of genes from *Bacillus thuringiensis Bt* used in transgenic cultivars, it is clear that *Bt* genes have potential for controlling a number of insect pests of peanut.

5.1.2. *Proteinase inhibitor gene*

Proteolysis is an essential metabolic process required for protein processing and turn over. Proteases function in pathogen infection and are essential for the digestion of plant proteins by herbivores. Four types of proteases have been identified such as serine, cysteine, aspartic and metallo-proteases based on active amino acid in the reaction centre. Protease inhibitors are important elements of plant defense response to insect predator and may also act to restrict infection by some nematodes. Insects mainly use one or a combination of serine, cysteine or aspartic proteases as major proteolytic enzymes. Inhibitors of these enzymes are produced by plants and modulate the growth of insects and pests (see Koiwa *et al.*, 1997). Serine protease inhibitors like soybean trypsin inhibitor, potato protease inhibitor I and II are active against herbivores.

Plant derived proteinase inhibitors are of particular interest because they are part of plant's natural defense system against insect predators. No information is available on impact of proteinase inhibitors on insect pests of peanut. Protease inhibitors I and II from potato and trypsin inhibitors need be tested in peanut before producing transgenic plants.

5.1.3. *Lectins*

Lectins are carbohydrate binding proteins and may be involved in plant defense against fungi and insects (Chrispeels and Raikhel, 1991). Lectins with specificity for GlcNAc residues appear to be insecticidal to cowpea weevil. *Galanthus nivalis* agglutinin (GNA) and wheat germ agglutinin (WGA) are some of the important lectins which have significant effects on insects when fed. Effects of these lectins on aphids of peanut have to be studied to determine the utility of the corresponding gene for development of transgenic plants with aphid resistance.

Legume lectins such as PHA, α – amylase inhibitor and arcelin appear to function against insects. Legume lectin-like proteins interact with seed beetles of family Bruchidae against the host family Leguminosae.

Bruchid is one of the important post-harvest pests of peanut (Table 6). α-amylase inhibitor of kidney bean origin is known to be active against certain bruchids such as cowpea weevil from digesting the starch granules. It is the deprivation of this major nutrient source that probably accounts for the effectiveness of α-amylase inhibitors in preventing cowpea weevil growth, development and survival. An effort by scientists at Purdue University of California at San Diego and the CSIRO, Canberra, Australia transferred the bean gene into garden pea *Pisum sativum* and expressed it in the pea seeds at levels comparable to those naturally occurring in common bean (Shade *et al.*, 1994). Pea seeds expressing the bean α-amylase inhibitor gene were immune or highly resistant to adzuki bean weevil. Seeds from the same plant were found immune, highly resistant or moderately resistant to cowpea weevil depending on the level of α-amylase inhibitor or expression. In case of peanut, work has to be carried out to test whether α-amylase from different sources are effective for bruchids before attempting development of transgenic plants.

Many wild species of Leguminosae possess lectin group of chemicals which are toxic to weevils. Screening of wild species will lead to detection of new lectins which may find use in development of transgenic plants for insect control.

5.2. FUNGAL AND BACTERIAL RESISTANCE

Genes encoding hydrolytic enzymes such as chitinase and glucanase which degrade fungal cell wall components are attractive candidates for development of transgenic disease resistant plants. Peanut itself contains class II chitinases, which are active against certain fungi (Kellmann *et al.*, 1996). Wild species of *Arachis* may contain chitinases which are candidate genes for transfer to cultivated peanut for imparting fungal resistance.

Some microorganisms produce phytotoxic substances which play an important role in causing the disease. For example, *Pseudomonas syringae* produce phytotoxic tabtoxin. A tabtoxin resistance gene cloned from *Pseudomonas* and introduced into plants showed resistance to that bacteria. If *Pseudomonas* species causing bacterial wilt in peanut produces similar toxins, genes from the bacteria itself can be introduced into peanut to develop resistance to the bacteria.

Besides tabtoxin acetyl transferase gene, stilbene synthase gene, ribosome inactivating protein gene, glucose oxidase gene and human lysozyme gene are known to impart disease resistance in plants.

On a global scale, leaf spots and leaf rust together can cause upto 70% yield losses in peanut. Leaf rust caused by *Puccinia arachidis* is also a major problem in peanut. It is of interest to identify disease resistance genes present in wild species of *Arachis* so that they can be cloned and used for transformation of cultivated species. Breeding for *Aspergillus* infection of seeds should be a major thrust area in *Arachis hypogaea*. Synthetic antifungal peptide gene has been introduced into peanut by transformation (Yang *et al.*, 1998b).

Aspergillus species cause major problems in peanut producing areas of the world since it produces the mycotoxin – aflatoxin. Development of genetic resistance by breeding is difficult. Since *A. flavus* is a week pathogen, prevention of its initial infection into peanut plant may be an important strategy for developing resistance. Pod on a developing plant is predisposed to *A. flavus* infection when plant is undergoing drought stress and high temperature stress. Hence, besides transfer of chitinase and other genes for fungal resistance, transfer of genes for drought and temperature tolerance is desirable.

5.3. VIRAL RESISTANCE

Virus disease cause serious problems in peanut in many parts of the world. Viruses such as peanut mottle virus, peanut stripe virus, peanut stunt virus, tomato spotted wilt virus, etc. cause considerable damage to the crop (Table 5). The genes which are normally transferred to plants to impart virus resistance are:

(1) Coat protein genes from the virus.
(2) DNA sequences corresponding to small satellite RNAs.
(3) cDNA sequences that are transcribed into virus specific antisense RNAs.
(4) Genes for whole or a portion of viral replicase.
(5) Sequence of a non-structural viral gene.

Peanut plants resistant to tomato spotted wilt virus was produced by incorporation of nucleocapsid gene (Jarret and Demski, 1997; Yang *et al.*, 1998a). Recently coat protein gene of Indian peanut clump virus (IPCVcp) has been stably introduced into peanut using *Agrobacterium* to induce resistance to this virus (Sharma and Anjaiah, 2000). Other strategies for developing virus resistance in peanut needs much concentrated effort.

5.4. CHANGING THE QUALITY PARAMETERS

Deficiency in sulphur containing aminoacids can be alleviated by introducing genes encoding proteins such as Brazil nut *2S albumin*, zein or nodulin 21. Attempts have been made for producing transgenic plants with improved protein quality by transfering *asp-1* gene (Lewis *et al.*, 1999) and Brazil nut *2S albumin* gene into peanut (Lacorte *et al.*, 1997). However Brazil nut *2S albumin* gene is known to be allergenic to humans.

Peanut oil, despite its unsaturated fatty acid components, has been found to be extremly atherogenic to rhesus monkeys. Genes encoding many key enzymes of fatty acid biosynthesis and modification have been isolated and available and can be transferred in peanut also.

90% of peanut oil is composed of palmitic acid, oleic acid and linoleic acid. Reducing the palmitic acid content of seeds results in better keeping quality of peanut oil and preliminary evidence suggests association of the trait with reduced aflatoxin contamination (Holbrook *et al.*, 1994). Peanut also contains four forms of antioxidant tocopherol and it should be possible to improve the antioxidant content of peanut by genetic engineering.

Peanuts are allergenic foods for people, especially for children. Allergen activity is present in two major proteins in peanut – arachin and conarachin (Burks *et al.*, 1991). It should be possible to reduce the content of these by genetic manipulation.

Peanut industry is interested in improving roasted flavour of peanut. As more information is available, it will be possible to change the flavour of seeds by introduction of useful genes.

5.5. TOLERANCE TO ABIOTIC STRESS

Peanuts are exposed to a range of temperature during growth. In many parts of the world, cool temperature exists during the planting season. Besides this, peanut is also subjected to drought and salinity. Compatible solutes (osmoprotectants) accumulate in the cytoplasm. Some of the important osmoprotectants are proline, mannitol and glycine betajine. Genes for these osmoprotectants can be transferred to plants to develop stress tolerance. *CodA* gene for choline oxidase from *Arthrobacter globiformis* (Sakamoto *et al.*, 1998) can be transferred to peanut too to obtain stress tolerant plants.

5.6. PEANUT AS A SOURCE OF EDIBLE VACCINE

Genes for edible vaccines can also be introduced into peanut. Khandelwal *et al.* (1999) transferred the edible vaccine gene for Rinderpest in animals to peanut.

5.7. HERBICIDE RESISTANCE

On a worldwide basis, yield losses of peanut from weeds are significant. Molecular manipulation for development of herbicide tolerant lines in peanut is the need of the day. Several genes for herbicide tolerance have been identified in microorganisms and transferred to plants to produce transgenic plants. A similar approach can be applied to peanut also.

6. Conclusions and future prospects

Rapid progress has been achieved in the last few years to develop reproducible regeneration and transformation methodology in peanut. Ever since the development of the first transgenic plant in tobacco in 1983, it is indeed remarkable that in less than two decades the tools of recombinant DNA technology and cell biology are at the disposal of the plant breeders. We are now in a position to embark on ambitious programmes of improvement of our major crops including peanut surpassing the inter-generic and inter-kingdom barriers for directed gene flow. This effectively converts the gene pool from a portion of genetic information in the genus *Arachis* to virtually all genes in the planetary biosphere. Research should now concentrate on production of disease resistant, aflatoxin free peanut with improved flavour and quality parameters. The problems of developing drought tolerant, temperature tolerant, salinity tolerant peanut by genetic manipulation will improve the level of prosperity and well-being of farmers and the community they serve. The expression level of the inserted genes and interactions of the new genotypes with different environments deserve special attention.

Acknowledgement

This paper is dedicated to the late Dr. Leela George, a long-time colleague and friend who has contributed to regeneration and transformation in peanut. I thank Dr. R. Mitra for critical evaluation of the manuscript.

References

Atreya C D, Rao J P and Subrahmanyam N C (1984) *In vitro* regeneration of peanut (*Arachis hypogaea* L.) plantlets from embryo axes and cotyledon segments. *Plant Sci. Lett.*, **34**: 379–383.

Baker C M, Durham R E, Burns J A, Parrott W A and Wetzstein H Y (1995) High frequency somatic embryogenesis in peanut (*Arachis hypogaea* L.) using mature dry seed. *Plant Cell Rep.*, **15**: 38–42.

Baker C M and Wetztein H Y (1992) Somatic embryogenesis and plant regeneration from leaflets of peanut, *Arachis hypogaea. Plant Cell Rep.*, **11**: 71–75.

Baker C M and Wetztein H Y (1994) Influence of auxin type and concentration on peanut somatic embryogenesis. *Plant Cell Tiss. Org. Cult.*, **36**: 361–368.

Bhatia C R, Murty G S S and Mathews V H (1985) Regeneration from "de-embryonated" peanut cotyledons cultured without nutrients and agar. *Zeitschrift für Pflanzenzuchtung*, **94**: 149–155.

Brar G S, Cohen B A, Vick C L and Johnson G W (1994) Recovery of transgenic peanut (*Arachis hypogaea*) plants from elite cultivars using ACCELL technology. *Plant J.*, **5**: 745–753.

Bunting G A H, Wynne J C and Gibbons R W (1985) Groundnut (*Arachis hypogaea* L.) In: *Grain Legume Crops* (Eds Summerfield R J and Roberts E H), Collins Professional and Technical Books, London, 747–800.

Burks A W, Williams L W, Helm R M, Connaughton C, Cockrell G and O'Brien T (1991) Identification of a major peanut allergen, Ara h1 in patients with atopic dermatitis and positive peanut challenge. *J. Allergy Clin Immunol.*, **90**: 962–969.

Carley D H and Fletcher S M (1995) An overview of world peanut markets. In: *Advances in Peanut Science* (Eds. Pattee H E and Stalker H T), Peanut Res. and Educ. Soc. Inc., Stillwater, UK, 554–557.

Cheng M, Hsi D C H and Phillips G C (1992) *In vitro* regeneration of Valencia type peanut (*Arachis hypogaea* L.) from cultured petioles, epicotyl sections and other seedling explants. *Peanut Sci.*, **19**: 82–87.

Cheng M, Hsi D C H and Phillips G C (1994) Recovery of primary transformants of Valencia type peanut using *Agrobacterium tumefaciens. Peanut Sci.*, **22**: 82–88.

Cheng M, Jarret R L, Li Z, Xing A and Demski J W (1996) Production of fertile transgenic peanut (*Arachis hypogaea* L.) plants using *Agrobacterium tumefaciens. Plant Cell Rep.*, **15**: 653–657.

Cheng M, Jarret R L, Li Z and Demski J W (1997) Expression and inheritance of foreign genes in transgenic peanut plants generated by *Agrobacterium* mediated transformation. *Plant Cell Rep.*, **16**: 541–544.

Chengalrayan K, Mhaske V R and Hazra S (1998) Genotype control of peanut somatic embryogenesis. *Plant Cell Rep.*, **17**: 522–525.

Chengalrayan K, Sathaye S S and Hazra S (1994) Somatic embryogenesis from mature embryo derived leaflets of peanut. *Plant Cell Rep.*, **13**: 578–581.

Chrispeels M J and Raikhel N V (1991) Lectins, lectin genes and their role in plant defense. *Plant Cell*, **3**: 1–9.

Clementale T E, Robertson D, Isleib T G, Beute M K and Weissinger A K (1992) Evaluation of peanut (*Arachis hypogaea* L.) leaflets from mature zygotic embryos as recipient tissue for biolistic gene transfer. *Transgenic Res.*, **1**: 275–284.

Daimon H and Mii M (1991) Multiple shoot formation and plant regeneration from cotyledonary node in peanut (*Arachis hypogaea* L.). *Japan J. Breed.*, **41**: 461–466.

Dong J D, Bi Y P, Xia L S, Sun S M, Song Z H, Guo B T, Jiang X C and Shao Q Q (1990) Teratoma induction and nopaline synthase gene transfer in peanut. *Acta. Genet. Sin.*, **17**: 13–16.

Dunbar K B and Pittman R N (1992) Adventitious shoot formation from mature leaf explants of *Arachis* species. *Crop Sci.*, **32**: 1353–1356.

Durham R E and Parrot W A (1992) Repetitive somatic embryogenesis from peanut culture in liquid medium. *Plant Cell Rep.*, **11**: 122–125.

Eapen S and George L (1993a) Plant regeneration from leaf discs of peanut and pigeonpea: influence of benzyladenine, indole acetic acid and indole acetic acid aminoacid conjugates. *Plant Cell Tiss. Org. Cult.*, **35**: 223–227.

Eapen S and George L (1993b) Somatic embryogenesis in peanut: Influence of growth regulators and sugars. *Plant Cell Tiss. Org. Cult.*, **35**: 151–156.

Eapen S, George L and Rao P S (1993) Plant regeneration through somatic embryogenesis in peanut (*Arachis hypogaea* L.). *Biol. Plant.*, **35**: 499–504.

Eapen S and George L (1994) Agrobacterium mediated gene transfer in peanut (*Arachis hypogaea* L.). *Plant Cell Rep.*, **13**: 582–586.

Eapen S, Kale D M and George L (1998) Embryonal shoot tip multiplication in peanut. Clonal fidelity and variation in regenerated plants. *Tropical Agr. Res. Extension*, **1**: 23–27.

Egnin M, Mora A and Prakash C S (1998) Factors enhancing *Agrobacterium* tumefaciens mediated gene transfer in peanut (*Arachis hypogaea* L.) *In Vitro Cell. Dev. Biol. Plant.*, **34**: 310–318.

Franklin C I, Shorrosh K M, Trieu A N, Cassidy B G and Nelson R S (1993) Stable transformation of peanut callus via *Agrobacterium* mediated DNA transfer. *Transgenic Res.*, **2**: 321–324.

Freitas Valeria G De, Lacorte C, Sachetto M A, Krul William R, Olivera D F De, Neves I J and Mansur E (1997) Identification of competent cells for *Agrobacterium* transformation and *in vitro* regeneration in peanut leaf and cotyledon explants. *Rev. Bras. de Fisol. Veget.*, **9**: 157–167.

George L and Eapen S (1993) Influence of genotype and explant source on somatic embryogenesis in peanut. *Oleagineaux*, **48**: 361–364.

Gill R and Ozias-Akins P (1999) Thidiazuron-induced highly morphogenic callus and high frequency regeneration of fertile peanut (*Arachis hypogaea*) plants. *In Vitro Cell Dev. Biol. Plant.*, **35**: 445–450.

Gill R and Saxena P K (1992) Direct somatic embryogenesis and regeneration of plants from seedling explant of peanut (*Arachis hypogaea*) promotive role of thidiazuron. *Can J. Bot.*, **70**: 1186–1192.

Gill R, Yang H and Ozias-Akins P (1998) Plant regeneration from transgenic peanut lines (*Arachis hypogaea* L.) transformed with a tomato anionic peroxide gene. In Congress on *in vitro* Biol., Israel, 52.

Guy A L, Heins J L and Pancholy S K (1978) Induction of biochemical parameters of callus growth from three peanut cultivars. *Peanut Sci.*, **5**: 78–82.

Hazra S, Sathaye S S and Mascarenhas A F (1989) Direct somatic embryogenesis in peanut (*Arachis hypogaea*). *Bio/Technology*, **7**: 749–751.

Hisajima S, Paek K Y, Namwongprom K, Subhadrabandhu S and Ishizuka K (1989) Mass propagation of peanut (*Arachis hypogaea* L.) plant through cultured seeds *in vitro. Japan J. Trop. Agr.*, **33**: 237–242.

Holbrook C C, Hunter J E, Knauft D A, Wilson D M and Matheron M E (1994) Fatty acid composition as a possible mechanism for resistance to preharvest aflatoxin contamination of peanut. In: *Proc. Am. Peanut Res. Ed. Soc.*, **26**: 224–226.

Illingsworth J E (1968) Peanut plants from single de-embryonated cotyledons. *Hortscience*, **3**: 238–276.

Illingsworth J E (1974) Peanut plants from single de-embryonated cotyledons or cotyledonary fragments. *Hortscience*, **9**: 462.

Isleib T G, Wynne J C and Nigam S N (1994) Groundnut Breeding. In: *The Groundnut crop – a scientific basis for improvement* (Ed Smartt J), Chapman and Hall, London.

Jarret R I and Demski J W (1997) Engineered resistance to tomato spotted wilt virus in transgenic peanut expressing nucleocapsid gene. *Transgenic Res.*, **6**: 297–305.

Kanyand M, Dessai A P and Prakash C S (1994) Thidiazuron promotes high frequency regeneration of peanut (*Arachis hypogaea*) plants *in vitro. Plant Cell Rep.*, **14**: 1–5.

Kartha K K, Pahl K, Leung N L and Mroginski L A (1981) Plant regeneration from meristems of grain legumes: soybean, cowpea, peanut, chickpea and bean. *Can. J. Bot.*, **59**: 1671–1679.

Kellmann J W, Kleinow T, Engelhardt K, Philipp C, Wegner P, Schell J and Schreier P H (1996) Characterization of two class II chitinase genes from peanut and expression studies in transgenic tobacco plants. *Peanut Mol. Biol.*, **30**: 351–358.

Khandelwal A, Geetha N, Venkatachalam P, Shaila M S and Lakshmi Sita G (1999) Generation of transgenic plants as a source of edible vaccine for Rinderpest, an animal disease. In: *NCL Golden Jubilee National Seminar on Emerging Frontiers in Plant Biotechnology*, NCL, Pune, 39–40.

Kim T, Chowdhury M K U and Wetztein H Y (1999) A quantitative and histological comparison of GUS expression with different promoter constructs used in microprojectile bombardment of peanut leaf tissue. *In Vitro Cell Dev. Biol. Plant.*, **35**: 51–56.

Klein T M, Wolf E D, Wu R and Sanford J C (1987) High velocity microprojectiles for delivering nucleic acids into living cells. *Nature*, **327**: 70–73.

Koiwa H, Bressan R A and Hasegawa PM (1997) Regulation of protease inhibitors and plant defense. *Trends Plant Sci.*, **2**: 379–384.

Lacorte C, Aragu F J L, Almeida E R, Mansur E and Rech E L (1997) Transient expression of GUS and 2S albumin gene from Brazil nut in peanut (*Arachis hypogaea* L.) seed explants using particle bombardment. *Plant Cell Rep.*, **16**: 628.

Lacorte C, Mansur E, Timmerman B and Corderio A R (1991) Gene transfer into peanut (*Arachis hypogaea*) by *Agrobacterium tumefaciens. Plant Cell Rep.*, **10**: 354–357.

Lewis J, Egnin M, Walker M, Jaynes J and Prakash C S (1998) Introduction and expression of an improved nutritional protein gene in peanut. In *In Vitro* Biol Congress.

Livingstone D M and Birch R G (1995) Plant regeneration and microprojectile mediated gene transfer in embryonic leaflets of peanut (*Arachis hypogaea* L.). *Australian J. Plant Physiol*, **22**: 585–591.

Livingstone D M and Birch R G (1999) Efficient transformation and regeneration of diverse cultivars of peanut (*Arachis hypogaea* L.) by particle bombardment into embryogenic callus produced from mature seeds. *Mol. Breed.*, **5**: 43–51.

Lurquin P F (1987) Foreign gene expression in plant cells. *Progress in Nucleic acid Research and Molecular Biology*, **34**: 143–188.

McKently A, Moore G A and Gardner F P (1990) *In vitro* plant regeneration of peanut from seed explants. *Crop Sci.*, **30**: 192–196.

McKently A H (1991) Direct somatic embryogenesis from axes of mature peanut embryos. *In Vitro Cell Dev. Biol. Plant*, **27**: 197–200.

McKently A H, Moore G A and Gardner F P (1991) Regeneration of peanut and perennial peanut from cultured leaf tissue. *Crop Sci.*, **31**: 833–837.

McKently A H (1995) Effect of genotype on somatic embryogenesis from axes of mature peanut embryos. *Plant Cell Tiss. Org. Cult.*, **42**: 251–254.

McKently A H, Moore G A, Doostdar H and Niedz R P (1995) *Agrobacterium* mediated transformation of peanut (*Arachis hypogaea* L.) embryo axes and the development of transgenic plants. *Plant Cell Rep.*, **14**: 699–703.

Mansur E, Lacorte C and De Freitas V G (1993) Regulation of transformation efficiency of peanut (*Arachis hypogaea* L.) explants by *Agrobacterium tumefaciens*. *Plant Sci.*, **99**: 89–91.

Mhatre M, Bapat V A and Rao P S (1985) Micropropagation and protoplast culture in peanut (*Arachis hypogaea* L.). *Curr. Sci.*, **54**: 1052–1056.

Middleton K J, Pande S, Sharma S B and Smith D H (1994) Diseases. In: *The Groundnut Crop* (Ed Smartt J), Chapman and Hall, London, 336–394.

Mroginski L A, Kartha K K and Shyluk J P (1981) Regeneration of peanut (*Arachis hypogaea*) plantlets by *in vitro* culture of immature leaves. *Can. J. Bot.*, **59**: 826–830.

Murch S J and Saxena P K (1997) Modulation of mineral and fatty acid profiles during thidiazuron mediated somatic embryogenesis in peanut (*Arachis hypogaea* L.). *J. Plant Physiol.*, **151**: 358–361.

Murthy B N S, Murch S H and Saxena P K (1995) Thidiazuron induced somatic embryogenesis in intact seedlings of peanut (*Arachis hypogaea*) endogenous growth regulation levels and significance of cotyledons. *Physiol. Plant.*, **94**: 268–276.

Narasimhulu S B and Reddy G M (1984) *In vitro* flowering and pod formation from cotyledons of groundnut (*Arachis hypogaea*). *Theor. Appl. Genet.*, **61**: 87–91.

Narasimhulu S B and Reddy G M (1985) Callus induction and morphogenesis in *Arachis hypogaea* L. In: *Proc. of an International Workshop on Cytogenetics of Arachis* (Eds Morr J P and Feakin S D), ICRISAT, Patancheru, A.P., India, 159–163.

Ozias-Akins P (1989) Plant regeneration from immature embryos of peanut. *Plant Cell Rep.*, **8**: 217–218.

Ozias-Akins P, Anderson W F and Holbrook CC (1992) Somatic embryogenesis in *Arachis hypogaea* L. genotype comparison. *Plant Sci.*, **83**: 103–111.

Ozias-Akins P, Schnall J A, Anderson W F, Sinsit C, Clemente T E, Adang M J and Weissinger A K (1993) Regeneration of transgenic peanut plants from stably transformed embryonic callus. *Plant Sci.*, **93**: 185–194.

Pestana M C, Lacorte C, De Freitas V G, De Oliveria D E and Mansur E (1999) *In vitro* regeneration of peanut (*Arachis hypogaea* L.) through organogenesis. Effect of culture temperature and silver nitrate. *In Vitro Cell Dev. Biol. Plant*, **35**: 214–216.

Ping Y, Yong X Z, Yi Z Y, Lining Y and Kunrong C (1996) Plant regeneration and *Agrobacterium* mediated gene transformation in leaflets of groundnut (*Arachis hypogaea* L.). *Oil Crops of China*, **18**: 52–56.

Pittman R N, Banks D J, Kirby J S, Mitchell E D and Richardson (1983) *In vitro* culture of immature peanut (*Arachis* spp.) leaves: morphogenesis and plantlet regeneration. *Peanut Sci.*, **10**: 21–25.

Ponsamuel J, Huhman D V, Cassidy R A and Post-Reittenmiller D (1998) *In vitro* regeneration via caulogenesis and brassin induced shoot conversion of dormant seeds from plumular explants of peanut (*Arachis hypogaea* L.). *Plant Cell Rep.*, **17**: 373–378.

Ramdev Reddy L and Reddy G M (1993) Factors affecting direct somatic embryogenesis and plant regeneration in groundnut, *Arachis hypogaea* L. *Indian J. Exp. Biol.*, **31**: 57–60.

Rani A S and Reddy G M (1996) Multiple shoot regeneration from de-embryonated cotyledons of cultivated and wild species of *Arachis*. *J. Genet. Plant Breed.*, **50**: 351–355.

Rohini V K and Shankara Rao K (2000) Transformation of peanut (*Arachis hypogaea*) a non tissue culture approach for generating transgenic plants. *Plant Sci.*, **150**: 41–49.

Rugman E E and Cocking E C (1985) The development of somatic hybridization technique for groundnut improvement. In: *Proc. of an International Workshop on Cytogenetics of Arachis* (Eds Moss J P and Feakin S D), ICRISAT Patencheru, A.P., India, 167–174.

Russo S L and Varnell R J (1978) *In vitro* responses of peanut shoot tips. *Proc. Soil and Crop. Sci. Soc.*, **37**: 34–36.

Sanford J C, Klein T M, Wolf E D and Allen N (1987) Delivery of substances into cells and tissue using a particle gun bombardment process. *Particle Gun Technol.*, **5**: 27–37.

Sastri D C, Nalini M S and Moss J P (1982) Tissue culture and prospects for improvement of *Arachis hypogaea* and other oil seed crops. In: *Tissue Culture of Economically Important Plants* (Ed Rao A N), *Proc. Costed Symp., Singapore*, 42–57.

Savage G P and Keenan J I (1994) The composition and nutritive value of groundnut kernels. In: *The Groundnut Crop – Scientific Basis for Improvement* (Ed Smartt J), Chapman and Hall, London, 173–213.

Saxena P K, Malik K A and Gill R (1992) Induction by thidiazuron of somatic embryogenesis in intact seedlings of peanut. *Planta*, **187**: 421–424.

Schnall JA and Weissinger A K (1993) Culturing peanut (*Arachis hypogaea* L.) zygotic embryos for transformation via microprojectile bombardment. *Plant Cell Rep.*, **12**: 316–319.

Seitz M H, Stalker H T and Green G C (1987) Genetic variation for regenerative response in immature leaflet cultures of the cultured peanut *Arachis hypogaea. Plant Breed.*, **98**: 104–110.

Sellars R M, Southward G M and Phillips G C (1990) Adventitious somatic embryogenesis and cultured immature zygotic embryos of peanut and soybean. *Crop Sci.*, **30**: 408–414.

Shade R E, Shroeder H E, Pueyo J J, Table L M, Murdock L L, Higgins T J V and Chrispeels M J (1994) Transgenic pea seeds expressing α-amylase inhibitor of the common bean are resistant to bruchid beetles. *Bio/Technology*, **12**: 793–796.

Sharma K K and Anjaiah V (2000) An efficient method for the production of transgenic plants of peanut (*Arachis hypogaea* L.) through *Agrobacterium tumefaciens* mediated genetic transformation. *Plant Sci.*, **159**: 7–19.

Sharma K K, Anjaiah V and Moss J P (1993) Production of transgenic plants of groundnut by *Agrobacterium*-mediated genetic transformation. *Groundnut Newsletter*, 23–25.

Singsit C, Arlang M, Lynch R F, Anderson W F, Wang A, Cardineau A and Ozias-Akins P (1997) Expression of a *Bacillus thuringiensis cry1Ac* gene in transgenic peanut. *Transgenic Res.*, **6**: 169–176.

Stalker H T (1997) Peanut (*Arachis hypogaea* L.). *Field Crop Res.*, **53**: 205–207.

Varsha R L, Dubey R K, Srivastava A K and Kumar S (1997) Microprojectile plant transformation: A bibliographic search. *Euphytica*, **95**: 269–294.

Venkatachalam P and Jayabalan N (1997) Effect of auxins and cytokinins on efficient plant regeneration and multiple shoot formation from cotyledons and cotyledonary node explants of groundnut (*Arachis hypogaea* L.) by *in vitro* culture technology. *Applied Biochem. and Biotech.*, **67**: 237–247.

Venkatachalam P, Geetha N and Jayabalan N (1998a) Influence of growth regulators on plant regeneration from epicotyl and hypocotyl cultures of two groundnut (*Arachis hypogaea* L.) cultivars. *J. Plant Biol.*, **41**: 1–8.

Venkatachalam P, Geetha N, Jayabalan N, Saravana Babu and Lakshmi Sita G (1998b) *Agrobacterium* mediated genetic transformation of groundnut (*Arachis hypogaea* L.). An assessment of factors affecting regeneration of transgenic plants. *J. Plant Res.*, **111**: 565–577.

Venkatachalam P, Geetha N, Khandelwal A, Shaila M S and Lakshmi Sita G (1999c) Induction of direct somatic embryogenesis and plant regeneration from mature cotyledon explants of *Arachis hypogaea* L. *Curr. Sci.*, **77**: 269–273.

Venkatachalam P, Geetha N, Sankara Rao K and Jayabalan N (1999a) Rapid and high frequency in vitro plant regeneration from leaflet and petiole explants of groundnut (*Arachis hypogaea*) L. *Applied Biochem. and Biotech.*, **80**: 193–203.

Venkatachalam P, Kavikishor P B, Geetha N, Thangavelu M and Jayabalan M (1999b) A rapid protocol for somatic embryogenesis from immature leaflets of groundnut (*Arachis hypogaea* L.). *In Vitro Cell Dev. Biol. Plant*, **35**: 409–412.

Venkatachalam P and Jayabalan N (1996) Efficient callus induction and plant regeneration via somatic embryogenesis from immature leaf derived protoplasts of groundnut (*Arachis hypogaea* L.). *Israel J. Plant Sci.*, **44**: 387–394.

Victor J M R, Murch S J, Krishnaraj S and Saxena P K (1999) Somatic embryogenesis and organogenesis in peanut: The role of thidiazuron and N6 benzylaminopurine in the induction of plant morphogenesis. *Plant Growth Reg.*, **28**: 9–15.

Wang A, Fan H, Singsit C and Ozias-Akins P (1998) Transformation of peanut with a soybean VspB promoter – *uidA* chimeric gene. 1. Optimization of a transformation system and analysis of GUS expression in primary transgenic tissues and plants. *Physiol. Plant.*, **102**: 38–48.

Wightman J A and Ranga Rao G V (1994) Chapter 11. Groundnut pests. In: *The Groundnut Crop* (Ed Smartt J), Chapman and Hall, London and Glasgow, 336–394.

Yang H, Singsit C, Wang A, Gonsalves D and Ozias-Akins P (1998a) Transgenic peanut plants containing nucleocapsid protein gene of tomato spotted wilt virus and divergent levels of gene expression. *Plant Cell Rep.*, **17**: 693–699.

Yang H Y, Gill R and Ozias-Akins P (1998b) Transformation of peanut with a synthetic antifungal peptide gene. In: Congress on *in vitro* Biology, Israel, 65.

MICROPROJECTILE-MEDIATED TRANSFORMATION OF PEANUT

D. MALCOLM LIVINGSTONE
CSIRO Tropical Agriculture, 120 Meiers Rd., Indooroopilly, Queensland 4068, Australia
e-mail: Malcolm.Livingstone@tag.csiro.au

Abstract

Like many legumes peanut has been recalcitrant to genetic transformation. Here I describe some of the parameters that must be considered when establishing a transformation system with reference to legumes in general and peanut in particular. Firstly, I discuss the relative advantages and disadvantages of organogenic and embryogenic tissue culture systems. A procedure for the efficient conversion of peanut somatic embryos into fertile plants is described. Secondly, consideration is given to the important parameters involved in the optimisation of gene transfer using microprojectiles. Finally, I give a detailed description of a protocol to efficiently transform cultivars in both botanical types of peanut (Spanish and Virginia), by particle bombardment into embryogenic callus derived from mature seeds, followed by single-step selection for hygromycin B resistance. This method produces three to six independent non-chimeric transgenic plants per bombardment of 10 cm^2 embryogenic callus. Copy number of integrated transgenes ranged from one to twenty with a mean of four and 57% coexpression of *hph* and *luc* or *uidA* genes coprecipitated on separate plasmids. Potted transgenic plant lines can be regenerated within 9 months of callus initiation or 7 months after bombardment.

1. Introduction

The exact origin of peanut or groundnut (*Arachis hypogaea* L.) is not known but it is likely to have originated in Brazil, Peru or Bolivia. The cultivation of peanut has been shown to have occurred in Peru as early as 600 AD and by approximately 1550 AD cultivation had reached eastern South America (Bajaj, 1983). Peanut is now grown globally in sub-tropical regions such as India, China, USA, Sudan, Indonesia, Nigeria and Australia. Peanut is one of the world's major oilseed crops, cultivated on over 20 million ha, in over 100 tropical and subtropical countries. It is now the 13th most widely grown crop in the world with an annual yield estimated to be 28 million tons (FAO, 1996). Peanut is a member

P. K. Jaiwal and R. P. Singh (eds.), Applied Genetics of Leguminosae Biotechnology, 187–204.

of the Leguminoseae family and fixes nitrogen from the atmosphere with the help of *Rhizobium* bacteria. Peanut is therefore a plant that does not require the application of exogenous fertilisers in the form of nitrates and nitrites and is an important provider of fixed nitrogen in crop rotations. It is also useful for preventing soil erosion and the leaves can be used as forage for livestock (Bajaj, 1983). As a major food legume, it is also an important source of protein and carbohydrates in human and animal diets.

Current commercial peanut cultivars are allotetraploids, apparently derived from a single hybridisation event between diploid parents (*Arachis duranensis* and *Arachis ipaensis*) (Kochert, 1996). There are 2 botanical types of peanut: Virginia and Spanish, and these can be further classified into 4 market types, i.e. Virginia and runner (Virginia) and Spanish and Valencia (Spanish) (Amernick, 1986). The crop has a relatively narrow germplasm base, without satisfactory resistance to several major pathogenic fungi and viruses. The most severe of these are fungal pathogens such as *Cercospora arachidicola* (leaf spot), *Aspergillus flavus* (source of aflatoxin), and *Puccinia arachidis* (rust). However, insect pests such as aphids, jassids, and thrips as well as viruses such as tomato spotted wilt virus, peanut mottle virus and peanut stripe virus also have the potential to cause serious crop losses. Wild *Arachis* species do have resistance to a number of these diseases but interspecific hybridisation is difficult due to incompatibility barriers such as differences in ploidy levels (Pattee *et al.*, 1988). Novel genes have been cloned for resistance to such diseases in peanut (Teycheney and Dietzgen, 1994). Expression of viral coat protein or coat protein mRNA in plants confers immunity from infection by that and closely related viruses (Nejidat *et al.*, 1988). Within the *Arachis* gene pool no resistance to peanut stripe virus was found in over 8000 accessions. However the coat protein genes of peanut stripe virus (PStV), peanut mottle virus and tomato spotted wilt virus have been cloned and may confer resistance if expressed in transgenic peanut plants (Teycheney and Dietzgen, 1994). Thus a strong practical incentive exists for the development of an efficient genetic transformation system for peanut. Before genetic manipulation can be used for peanut improvement, it is necessary to establish:

- An efficient system for plant regeneration from transformed peanut cells;
- A method to deliver foreign genes at high frequencies to peanut cells suitable for transformation; and
- Genetic constructs with promoters and reporter genes allowing detection of both transient and stable expression of an introduced gene in peanut cells.

2. Peanut tissue culture

The procedures of plant tissue culture have developed to such a level in the last decade that most plant species can be regenerated *in vitro* through several methodologies. Plant regeneration is the cornerstone of tissue culture methodology. However, the rate and efficiency of plant regeneration varies greatly from one species to another (Tisserat, 1985). With the advent of microprojectile bombardment, gene transfer to most plant cells is now possible with careful attention to optimisation of the bombardment conditions. In most transformation-recalcitrant species the limiting factor is not gene transfer but regeneration of fertile plants from isolated tissues or callus. Some species are quite susceptible to *Agrobacterium* infection but tissue culture is problematic. An example of this is sunflower, where resort to meristem bombardment has been used to overcome this deficiency, resulting in a reasonable efficiency of transformation (Knittel *et al.*, 1994). In general, however, the ideal tissue culture system is one in which:

- Tissue can be easily isolated from readily available plant material (e.g. leaves or seeds).
- Cells can be induced to multiply in tissue culture in an undifferentiated state.
- Cells can be readily induced to regenerate into whole, fertile plants without somaclonal variation.

Plants regenerated after long periods in tissue culture often lead to changes in chromosome number, single gene nuclear mutation and cytoplasmic changes (epigenetic effects) and, consequently, time in tissue culture should be minimised (Pugliesi *et al.*, 1990). Epigenetic effects are modifications in gene expression that are brought about by heritable, but potentially reversible, changes in chromatin structure and/or DNA methylation. Epigenetic effects can be a serious impediment to successful practical application of plant transformation. Frequently explants that can be regenerated quickly in tissue culture are not good candidates for transformation and vice versa.

2.1. ARE ALL SOMATIC CELLS TOTIPOTENT?

Unlike animals, where differentiation is generally irreversible, mature and differentiated plant cells retain the ability to regress to a meristematic state as long as they have an intact membrane system, chloroplasts, mitochondria, a viable nucleus and if the cell wall has not lignified (Bhojwani, 1983). However, this does not mean that all cell types (or even individual cells) in a particular organ can be induced to undergo dedifferentiation at the same time, under the same conditions. Callus derived from a multicellular explant (e.g. leaflet) will be heterogeneous with respect to the ability of its component cells to form a whole plant (Bhojwani, 1983). This may mean in effect, that very few cells are regenerable, *in vitro*, under any particular set of conditions (a tissue culture system). The ideal target for transformation then is a mass of:

- Undifferentiated cells all equally capable of regeneration; or
- Differentiated cells (e.g. the parenchyma of a leaf) that are all equally capable of forming regenerable callus.

There may be several processes of plant regeneration for each species but it is the most efficient system with regard to plant transformation that is desired by the molecular biologist (as long as somaclonal variation is minimal). Most plants can be regenerated *in vitro* from excised meristems but in most instances this approach is not efficient at producing non-chimeric transformed plants. Callus derived from plant material can be placed into one of two types:

- Embryogenic.
- Organogenic.

2.1.1. Somatic embryogenesis

The potential to form embryos from somatic cells was unique to the plant kingdom until recently when cloned sheep were produced from somatic animal cells (Wilmut *et al.*, 1997). Somatic embryos can develop into whole plants through stages that are remarkably similar to those found in zygotic embryos. Conditions have now been developed for *in vitro* somatic embryogenesis of many plant species, including the formerly transformation intransigent cereals and legumes. Somatic embryogenesis can occur directly from differentiated tissues without an intervening embryogenic callus phase (e.g. *Citrus*, Tisserat, 1985) but this is a rare event. Most examples of somatic embryogenesis occur by secondary or indirect embryogenesis where the inducible tissue first forms embryogenic callus. The particular advantage of a somatic embryogenic system is that somatic embryos are produced from a single cell and this facilitates the application of selection to transformed tissue. The greatest impediment to successful tissue culture of somatic embryos is successful regeneration. It is necessary to induce, maintain and regenerate somatic embryos quickly to avoid unwanted somaclonal variation. Usually a loss of regeneration potential occurs over time in tissue culture (Wochok, 1972). However, the decline in morphogenic potential of somatic embryos can often be overcome by altering chemical and physical factors.

Somatic embryogenesis has been induced from immature peanut embryos by the use of various auxin-like compounds. Hazra *et al.* (1989) induced somatic embryogenesis from immature embryos of the cultivar SB-11 by supplementing growth media with 2,4-dichlorophenoxy acetic acid (2,4-D). Ozias-Akins *et al.* (1989 and 1992) induced somatic embryogenesis in eight peanut cultivars by culturing immature embryos on medium containing 0.5–1 mg/L picloram. Eapen and George (1993) also induced somatic embryogenesis from immature embryos of the cultivar JLM-1 using the auxins 2,4-D, dicamba, picloram, indolepropionic acid, and α-naphthaleneacetic acid (NAA). Durham and Parrott (1992) isolated somatic embryos formed from immature embryos (cultivar AT127) cultured on solid medium supplemented with 40 mg/L 2,4-D. Rapid growth, secondary and tertiary embryogenesis of these somatic embryos was achieved by culturing them in a liquid form of the same medium. Other explants have also been shown to be a source of embryogenic callus and somatic embryos. Chengalryan *et al.* (1994) obtained embryogenic masses and somatic embryos from immature leaflets of the cultivar JL24 by incorporating 20 mg/L 2,4-D. Baker and Wetzstein (1992) induced somatic embryos from leaflets of germinating seedlings using 40 mg/L 2,4-D (cultivar AT127) and more recently from mature zygotic embryos (Baker *et al.*, 1995).

Although many of these reports show that somatic embryos can be obtained from various peanut tissues, the most useful (in terms of efficiency and genotype range) are immature embryos cultured on either picloram or 2,4-D. Unfortunately immature embryos are not easy to harvest and in many places are available only seasonally.

However, plant regeneration, and not somatic embryo induction, is the most limiting factor in the successful tissue culture of any of the above explants. Plants have been regenerated in all of the published methods but with poor or variable responses (Table 1).

2.1.2. Organogenesis

The production of adventitious shoots *in vitro* is more common and easier to control than the development of somatic embryos from cultured explants (Tisserat, 1985). Organogenic

Table 1. Variation in media used for the regeneration of peanut somatic embryos

Author	Medium[1]	Regeneration (%)	Cultivars	Liquid Phase
Ozias-Akins (1989)	MSO	0–18	8 genotypes used but no information on which regenerated	–
Hazra *et al.* (1989)	MSO + 0.25% charcoal	50	SB-11	–
Durham and Parrott (1992)	MSO + dehydrating step	15	AT127	+
Ozias-Akins (1992)	MSO + 0.5% activated charcoal → MSO + 25 mg/L BAP → MSO	5–71	7 genotypes used but with large variation in response	–
Ozias-Akins (1993)	MSO (3% sucrose) + 0.5 mg/L BAP + 0.5 mg/L kinetin + 0.5 mg/L zeatin + 0.1 mg/L NAA + 500 mg/L casein → MSO (3% sucrose) + 0.1 mg/L BAP + 0.1 mg/L NAA + 25 mM $AgNO_3$ → MSO (2% sucrose) + 3 mg/L BAP + 1 mg/L gibberellic acid	No data	Toalson	+
Eapen (1993)	MSO + 0.1% activated charcoal	10	JLM-1	–
Eapen (1993)	MSO + 1% mannitol + 0.1% activated charcoal → 5 mg/L BAP → 0.5 mg/L NAA → 1 mg/L NAA	16		–

[1]MSO: MS Salts (Murashige and Skoog, 1962), MS vitamins, 3% sucrose and 0.8% agar.

callus may arise from single cells (Link and Eggers, 1942) under tissue culture conditions favouring unorganised growth. Organogenic callus is composed of cells, which have the capacity, under certain tissue culture conditions, to form *de novo* meristems (meristemoids). These meristemoids may be induced to regenerate into whole plants. The end of the induction process however, is not when the shoots or buds can first be seen but the time when groups of cells are committed to the differentiation and growth of shoots (Christianson, 1987). This time is difficult to determine. Ideally the callus should have the capacity to remain as undifferentiated, rapidly multiplying cells for as long as required (for example, to allow sufficient time for the selection of genetically transformed cells) with the change to organised meristematic tissue being undertaken only when the tissue is given an appropriate stimulus. This stimulus is usually chemical, but physical factors such as light intensity and quality, agar concentration and whether the tissue is cultured in solid or liquid medium may also be involved. Practically however, it is often hard to achieve regeneration. It is also difficult to differentiate between organogenic callus and a closely related phenomenon called direct shoot organogenesis. In direct shoot organogenesis, cells in an isolated explant quickly form a meristem without an associated callus phase (Bhojwani, 1983). Often callus and direct shoot organogenesis occur simultaneously on the same explant and this can complicate a transformation system.

The earliest reports on the tissue culture and regeneration of peanut explants *in vitro* were based on direct shoot regeneration or organogenic callus. Atreya *et al.* (1984) regenerated peanut plants at a frequency of 10–18% from excised cotyledon segments on medium containing 2 mg/L N6-benzylaminopurine (BAP). Bajaj *et al.* (1983) cultured peanut callus derived from mesocotyl tissues of 3 to 5 week old seedlings on a medium containing 2 mg/L 2,4-D. This callus could be induced to form shoots occasionally when placed on a medium containing 2 mg/L kinetin. Mroginski *et al.* (1981) produced shoots from immature leaflet explants derived from 3 to 5 day old seedlings. Although buds were obtained on 5 out of 10 cultivars tested only one cultivar developed shoots at a low frequency and none of these grew larger than 5 cm. Clemente *et al.* (1992) evaluated peanut embryonic leaflets excised from 4 to 8 day old seedlings (cultivars UPL PN 4, Tamnut 74 and NC-7) as targets for gene transfer. Twenty-three growth regulator ratios were tested for bud and shoot production. Organogenic callus formed on media with a BAP concentration of between 2 and 4 mg/L and NAA concentrations of between 1 and 2 mg/L. An average of 25% of leaflets of the cultivar UPL PN 4 produced shoots on medium containing 4 mg/L BAP and 0.8 mg/L NAA, but cultivar NC-7 produced shoots from only 5.6% of leaflets. Fertile plants were recovered from cultivars UPL PN 4 (12 plants per leaflet) and Tamnut 74 (six plants per leaflet) but none from NC-7. The leaflets produced a chlorophyllous cell mass (callus) containing buds, but it was not clear whether the plants regenerated via a callus phase or via direct shoot organogenesis. Regeneration from callus was reported to be as quick as 3 weeks.

3. Peanut transformation

Peanut, like many other legumes, is recalcitrant to transformation (Mansur *et al.*, 1995) with few confirmed transgenic plants following particle bombardment or *Agrobacterium* treatment. Low efficiency, cultivar specificity, chimeric transformants, or availability of explants limit reported transformation methods.

3.1. *AGROBACTERIUM*-MEDIATED TRANSFORMATION

Agrobacterium-mediated transformation of peanut has been reported by several groups (Eapen and George, 1994; McKently *et al.*, 1995; Cheng *et al.*, 1996) at frequencies of 0.2–9% of treated explants. As with other recalcitrant legumes, *Agrobacterium*-mediated transformation of peanut appears to be cultivar and strain specific (see chapter by Susan Eapen in this volume).

3.2. DIRECT GENE TRANSFER

Direct gene transfer has been accomplished by several methods. For example, by the electroporation of protoplasts (Wang *et al.*, 1992), the electroporation of intact tissues (D'Halluin *et al.*, 1992; Kloti *et al.*, 1993), microinjection of protoplasts or meristems (Lusardi, 1994) and polyethylene glycol mediated transformation of protoplasts (Larkin,

1990). However, microprojectile-mediated transformation of intact plant cells has proved the most useful and practical method.

It has been known for many years that high velocity microparticles can allow entry of virus particles or nucleic acids into intact plant cells (McKenzie *et al.*, 1966). In 1987 Sanford *et al.*, demonstrated that the transfer of non-infectious nucleic acids to intact plant cells was possible. To achieve this it was necessary to design a device to accelerate microparticles (in this case tungsten particles with a diameter of 1–4 μm) to velocities capable of penetrating plant cell walls (~250 ms^{-1}) (Franks and Birch, 1991b). Devices used successfully have employed electric discharge, gunpowder, or gas pulse to propel the microparticles towards the target tissue (Klein *et al.*, 1987 and 1988a, b, c) (Fig. 1).

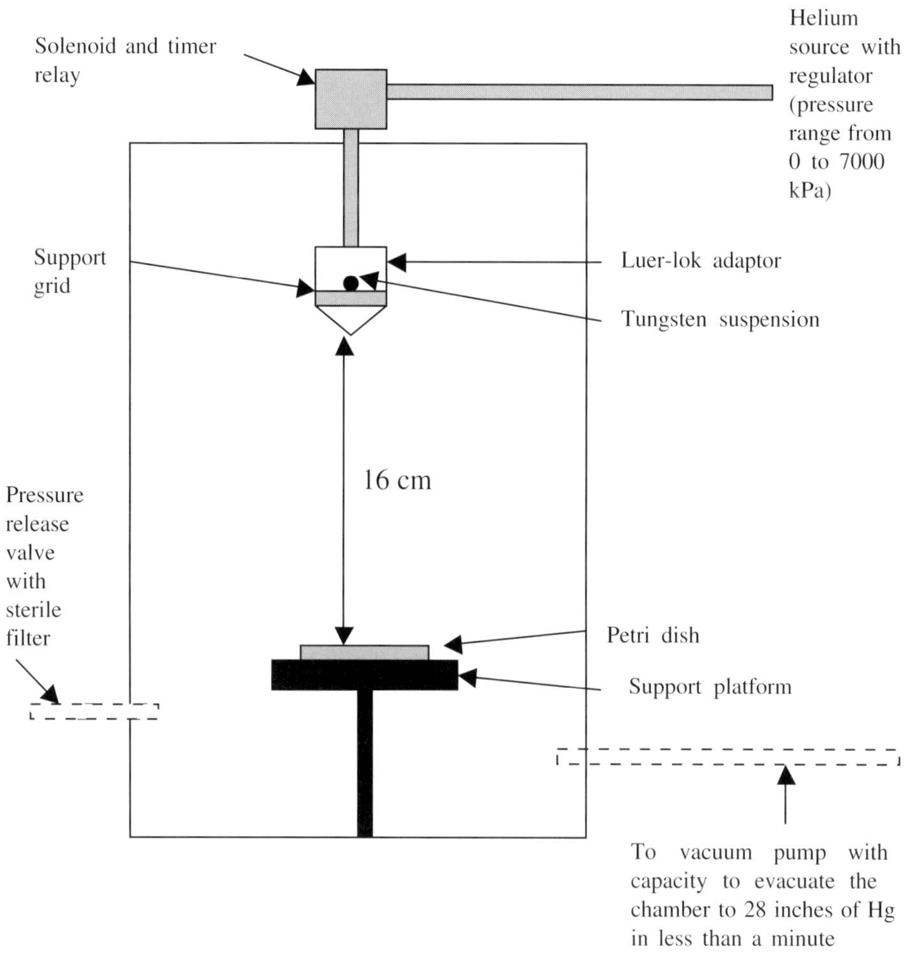

Figure 1. Diagram of a device used for accelerating microparticles with a pulse of helium.

3.3. CONDITIONS THAT INFLUENCE GENE TRANSFER BY MICROPROJECTILE BOMBARDMENT

A number of physical and biological factors can affect the transfer of genes by particle bombardment to intact plant cells. These parameters are commonly optimised by the detection of transient expression of an introduced reporter gene construct in the target tissue of interest. For example, the *uidA* gene codes for the enzyme β-glucuronidase for which a simple histochemical assay allows for easy detection. The number of cells which survive and integrate the DNA into their genome (stably transformed cells) is estimated to be no more than 5% of transiently expressing cells (Franks and Birch, 1991a). Unfortunately some bombardment conditions may lead to the death of most transiently expressing cells within a few days. Alternative genes are available for use as screenable markers, such as the *R1/C1* gene from maize (Ludwig *et al.*, 1990), the *luc* gene isolated from firefly (Ow *et al.*, 1986) or the *gfp* gene isolated from *Aequoria victoria* (Chalfie *et al.*, 1994).

3.3.1. Precipitation

The DNA containing the gene of interest is coated onto the surface of microparticles by precipitation with $CaCl_2$ and spermidine, or by treatment with ethanol (Finer *et al.*, 1992). Tungsten or gold microparticles with diameters ranging from 0.5 to 4 μm are commonly used. Other particles such as platinum, iridium, glass needles and dried bacteria have been used but with little success (Rasmussen, 1994). Experience shows that several factors, which influence precipitation, must be carefully controlled to achieve reproducible results:

- The condition of the microparticles (e.g. oxidation of tungsten);
- The quality of the reagents (e.g. spermidine and DNA are easily degraded); and
- Adding and thoroughly mixing the reagents in the correct sequence at each stage of the precipitation process.

3.3.2. Velocity of particles

In order to achieve sufficient momentum to penetrate plant cell walls, dense microparticles such as tungsten and gold must be propelled towards the target at velocities of $> 250\ ms^{-1}$. The optimum velocity of the microparticles for each tissue must be determined empirically (Russell *et al.*, 1992).

If the device uses gunpowder charges, then one must alter the position of the target with respect to the stop plate or alter the position of the macroprojectile in the gun barrel. This is a crude way of adjusting the velocity of the microparticles and combined with the variability inherent in blank gunpowder charges, makes optimising velocity with this system difficult. When using electric discharge and gas pulse devices particle velocity can be more finely controlled and a voltage or gas pressure can be found which will result in the highest number of transiently expressing cells.

3.3.3. Size and type of microparticles

Transgene expression in plant cells has been demonstrated following microprojectile bombardment using microparticles ranging in size from 0.5 to 4 μm in diameter. Which size is optimal depends on the type and size of the target cell and can only be determined empirically. However, it is likely that smaller targets (e.g. chloroplasts) will show higher transient gene expression when bombarded with microparticles with diameters at the lower end of this range. Larger particles might be required if the target cells are several cell layers deep or if the target tissue has thicker cell walls. Tungsten particles show greater variation in size within any one batch (0.5 to 2 μm) and a greater degree of aggregation of particles when compared to gold (Kausch *et al.*, 1995) but they are inexpensive and easier to coat with DNA. The range of sizes also means that tungsten particles are useful over a wide range of plants. The average size of particles may be important. Hébert *et al.* (1993) found that tungsten particles with an average size of 1.07 μm gave higher transient gene expression in grape suspension cells than particles with an average size of 0.77 μm. Kausch *et al.* (1995) reported that tungsten particles with an average diameter of 1 μm diameter or less penetrated only the first few layers of embryonic suspension cells of maize, and average particle diameters of 2.0–3.5 μm were found two to three layers deep. Taylor and Vasil (1991) found that particles could be detected as deep as 12 layers in pearl millet embryos. However, such deep penetration resulted in a large amount of damage to the first few cell layers.

3.3.4. *Multiple bombardments*

Multiple bombardments of the target tissue may result in higher levels of transient gene expression (Wang *et al.*, 1988), or no increase (Castillo *et al.*, 1994) or a decrease (Reggiardo *et al.*, 1991). The difference in results reported by different experimenters may reflect differences in the amount of time between bombardments. A longer rest period between bombardments may allow injured cells to recover somewhat before the shock of the second blast.

3.3.5. *Bombardment damage*

Taylor and Vasil (1991) showed that damage to *Pennisetum glaucum* embryos was caused by the shock wave of the bombardment and not the DNA/microprojectile suspension and that this damage extended to cells several layers from the surface. Both a gunpowder driven and a helium driven device caused injury to tobacco suspension cells attributed to gas blast and acoustic shock. The damage was less using the helium device and was further reduced by the use of a baffle or screen over the tissues, which also increased the conversion from transient to stable expression (Russell *et al.*, 1992). Kausch *et al.* (1995) showed that maize suspension cells were not damaged by the acoustic shock of bombardment.

3.3.6. *Osmoticum*

The use of an osmoticum to increase transient expression frequencies, first reported by Ye *et al.* (1990), has become one of the universally accepted means of increasing transient and stable expression rates. Osmotic treatment to reduce the turgor of cells may reduce cytoplasmic extrusion following bombardment, allowing time for repair of the cell membrane (Iglesias *et al.*, 1993). Vain *et al.* (1993) found this not to be the case following transformation of maize embryogenic suspension cells. Yamashita *et al.* (1997) found that DNA-coated gold particles were preferentially delivered into the vacuole and cytoplasm of tobacco suspension cells. Cells, which contained a particle in the nucleus, had a 912 times higher transient gene expression rate than those which had particles in the vacuole. Desiccation of the target cells by the use of an osmoticum might ensure that the microparticles are delivered to the nucleus and result in higher numbers of 'hit' cells expressing the introduced DNA. This approach will not result in higher transient expression frequency as claimed by Yamashita (1997) as the target area is reduced in plasmolysed cells. However, plasmolysed cells may be able to tolerate higher numbers of particles and hence an unwanted increase in transgene copy number.

3.3.7. *Stage of tissue development*

The state of the target tissue at the time of bombardment is also crucial for the achievement of high transient expression rates and stable transformation frequencies (Klein *et al.*, 1992). For example, changing cell wall thickness and metabolic patterns will affect tissue penetration and relative expression strengths of different constructs. Cells may be more likely to express introduced DNA at specific stages of the cell cycle, for example when the nuclear membrane has been dissolved or when the cell is particularly metabolically active (Bower and Birch, 1990). This factor may be even more important in achieving high levels of stable, integrative transformation – a process not well understood at this time. The stage of development of explants can also be important for accessibility of specific target cell types, or for subsequent regenerability of transformants. For example, Iglesias *et al.* (1993) found that wheat embryos at 11 days old were optimal for microprojectile bombardment into the meristem.

3.3.8. *Vacuum*

In general, evacuation of a gun chamber increases the number of transient expression events by minimising the loss of microparticle velocity due to air friction, and by increasing dispersal of the accompanying gas pulse to reduce percussive shock. However, vacuum can damage target tissues. *Phaseolus vulgaris* mature embryos subjected to a vacuum of 16.6 kPa absolute pressure were undamaged but at 9.14 kPa absolute pressure (commonly used in microprojectile bombardment) development was blocked (Birch and Bower, 1994). Kausch *et al.* (1995) showed that maize suspension cultures exposed to vacuum alone did not show any signs of damage.

Takeuchi *et al.* (1992) demonstrated transient GUS expression in soybean seedlings, potato and carrot root slices and tobacco leaves without the use of a vacuum. Subsequently an evacuated version of this device (Finer *et al.*, 1992) was developed and no further instances of transformation without a vacuum has been published.

3.3.9. *Toxicity of particles*

Heavy metals are toxic to most organisms and bombardment with tungsten microparticles might kill or damage cells in the target tissue. Russell *et al.* (1992) found evidence that tungsten is toxic to some cells in tobacco suspension cultures and that gold particles substituted for tungsten reduced cell death. This finding may be particularly important for obtaining stable transformants, although the authors acknowledged that even using tungsten particles many transformants were obtained.

3.3.10. Gene silencing

One of the characteristics of direct gene transfer is the frequent integration of repeated copies and rearrangements of transgenes in the plant genome (Hansen and Chilton, 1996). *Agrobacterium*-mediated transformation often results in lower transgene copy numbers but plants transformed by *Agrobacterium* still exhibit gene silencing.

Numerous reports demonstrate that multiple gene insertions can affect the activity of introduced or homologous genes, but others show that gene activity is not well correlated with transgene copy number (Barcelo *et al.*, 1994), and even when present at low copy numbers, transgenes can still be silenced. The reason for this is not known, but the effect is termed 'position effect'. Transgenes are thought to integrate into the plant genome randomly. However, some regions of the genome may be very active and others generally silenced or active only when certain stimuli are received. It has been shown that T-DNA is able to target such active regions. T-DNA targeting provides an explanation for the observation that *Agrobacterium*-mediated transformation tends to be less susceptible to gene silencing than direct gene transfer methods. It has been recently shown that T-DNA delivered by *Agrobacterium* may in fact prefer one chromosomal locus (De Neve *et al.*, 1997). Nevertheless plants transformed by Agrobacterium also show 'position effect' silencing even if at a reduced rate. In an attempt to unequivocally demonstrate position effect Qin *et al.* (1994) and Albert *et al.* (1995) used the P1 bacteriophage *cre-lox* site-specific recombination system in tobacco to integrate transgenes at five specific chromosomal sites. This work demonstrated that even when the transgene is integrated at the same chromosomal site there is variability between individual transformants (Gelvin, 1998). Silencing attributed to 'position effect' is likely to be a complex phenomenon and comparison of five sites may be inadequate. Another approach to demonstrate 'position effect' has been targeted at the microenvironment of the transgene. The attachment of DNA sequences to the 5′ and 3′ ends of transgenes (MARs) has sometimes reduced variation in transgene expression. Intragenic regions can also be important for correct temporal and spatial expression. For example, Sieburth and Meyerowitz (1997) showed that a 3.8 kb intragenic sequence of the *AG* gene was required in addition to 5′ flanking regions to obtain *uidA* expression in *Arabidopsis*. Mlynarova *et al.* (1996) showed that variation observed amongst genetically identical individuals with an attached MAR region was no greater than that which is caused by environmental factors.

Recently an interesting gene (*MOM*) has been found in *Arabidopsis* that maintains transcriptional silencing in highly methylated transgenes (Amedeo *et al.*, 2000). *Arabidopsis* mutants with defective *MOM* continue to transcribe transgenes even when highly methylated. If elite crops containing *MOM* mutations (homozygous) are found, then these mutant lines could be used as a base for introducing transgenes to valuable crops (Chandler and Jorgenson, 2000).

3.3.11. Other factors

Hébert *et al.* (1993) reported that the addition of 3 g/L activated charcoal to the post-bombardment medium resulted in a four-fold increase and the use of Parafilm (instead of porous venting tape) to wrap plates resulted in a three-fold increase in long-term *uidA* expression in embryogenic grape cell suspensions. This effect may be due to increased growth and health of the cultured cells and not to an increased transformation frequency. However, no further reports of this effect have been published.

3.4. FEATURES OF MICROPROJECTILE-MEDIATED TRANSFORMATION

- Plasmid vectors are small and easy to manipulate.
- The integration of multiple copies of the transgene and the vector DNA is common, and may contribute to gene silencing.
- Rearrangement and concatermerisation of the introduced genes is common.
- There is no host range limitation (cf. *Agrobacterium*). However, it may difficult to target a particular cell type.

3.5. SELECTION

An important difference between organogenesis and somatic embryogenesis is that a shoot primordium or meristemoid is dependent on the surrounding callus tissue and a somatic embryo may not be. This dependence arises because a shoot primordium establishes a connection with pre-existing vascular tissue dispersed within the callus. A somatic embryo is a bipolar structure with a closed radicular end and may not require physical connection to surrounding cells. This difference may complicate the use of selective agents in organogenic regeneration systems.

Chia *et al.* (1994) used selective subculture of protocorm-like-bodies of *Dendrobium* transformed with the *luc* gene to recover transformed plants after only 3 rounds of screening using a high-power dissecting microscope. Twenty-seven lines were recovered in three experiments. However, no indication of transformation efficiency or evidence of germline transformation was reported.

Bower *et al.* (1996) were able to successfully recover transformed sugarcane by the application of selective subculture to embryogenic callus transformed with the *luc* gene, but found that selective subculture was inefficient when compared to selection with the antibiotic geneticin (G-418).

Vasil *et al.* (1992) failed to recover any transformed wheat callus when full selection (10 or 20 mg/L PPT) was applied immediately after bombardment of embryogenic callus with pBARGUS (35SP-*adh1-bar*). However, when PPT concentration was increased every 2 weeks after bombardment (0, 1, 5, 10, 20, and 40 mg/L PPT), transgenic callus was obtained at a frequency of 0.6%.

Many plants are naturally resistant to particular selective agents, and an effective antibiotic/herbicide for a particular species can only be determined empirically. It is important to design an experiment to test the frequency of ‘escapes’ in any particular explant and for each selective condition. Selective agents can act to effectively inhibit growth in a large proportion of particular explants but allow the ‘escape’ of a few percent. In a large transformation experiment much work can be wasted discovering this fact. In general a ‘kill curve’ should be generated to determine the lowest concentration of selective agent that will kill 100% of explants. If however, the proposed transformation protocol is expected to generate transformants at a very low frequency then caution must be exercised when interpreting a ‘kill curve’ based on lower numbers of untransformed explants.

3.5.1. Selection strategies

Because few cells in any treated explant are initially transformed, the application of a selection step for the recovery of transformed tissues is essential. For many monocotyledonous plants the lack of an effective selective agent may be the principal obstacle to transformation (Gordon-Kamm *et al.*, 1990). The application of selection can be approached in one of two ways. It is possible to transform tissues with a gene that allows for the non-toxic detection of tissues expressing the transgene (e.g. *luc*, and *gfp*) and then to selectively subculture these regions. Alternatively it is possible to introduce a gene that will confer a selective advantage to those cells expressing the introduced gene (e.g. antibiotic (*neo*) or herbicide resistance (*bar*)). In addition, the application of selective agents can be applied in gradual increments (step-wise selection) or in lethal concentrations soon after transformation. All these approaches have been used successfully in the transformation of a number of species and a variety of explants.

3.6. MICROPROJECTILE-MEDIATED TRANSFORMATION OF PEANUT

Clemente *et al.* (1992) bombarded embryonic leaflets excised from 4 to 8 day old seedlings (cultivars UPL PN 4 and Tamnut 74) with a plasmid containing the *neo* gene with a 35S promoter, and recovered kanamycin resistant callus lines on 50 mg/L kanamycin. Of ten shoots regenerated no transformed plants were recovered (NC-7 and Tamnut). Of 875 leaflets of the cultivar UPL PN 4 bombarded, 202 kanamycin resistant calli were recovered but only 1 untransformed shoot was produced. Regeneration from these explants was reported to be rapid and efficient (Mroginski *et al.*, 1981) but selection with kanamycin appeared to be problematic. Despite prolonged culture of the transformed callus on 50 mg/L for 75 to 100 days prior to excising any shoots, only 10 plants from 10 callus lines were regenerated and none of these was transformed (NPTII ELISA and Southern analysis). Livingstone and Birch (1995) excised leaflets from mature embryos of peanut cultivars Robut, Gajah, McCubbin and NC-7 and produced multiple shoot primordia via organogenesis on Murashige and Skoog medium supplemented with 3 mg/L BAP and 1 mg/L NAA. Plants were regenerated by elongation of shoot primordia on transfer to MS medium supplemented with 5 mg/L BAP. The regeneration frequency of five plants per embryonic leaflet was ten to forty-fold higher than previously reported from seedling leaflets. Microprojectile bombardment parameters were optimised for high transient expression rates of the β-glucuronidase (*uidA* or *gus*) reporter gene in peanut embryos ($>$ 2000 cells per bombardment) and in excised embryonic leaflets ($>$ 1200 cells per bombardment). The firefly luciferase reporter gene (*luc*) allowed repeated non-toxic assays of callus grown without selection for 8 weeks after bombardment of embryonic leaflets. The LUC assays revealed the transition from hundreds of transiently expressing cells soon after bombardment to a few stably expressing regions in organogenic callus at 8 weeks. No transformed plants were recovered from 2500 leaflets bombarded with pGN1 (35S-*neo*) and cultured for 6 months on medium containing 50 mg/L kanamycin. Also no transformed plants were produced from 1000 leaflets bombarded with pOsaPAT (Osa-*bar*) and cultured for 2 months on medium containing 5 mg/L phosphinothricin (PPT).

Brar *et al.* (1994) subjected peanut embryos to a pre-bombardment treatment to increase the size of the meristem (and probably the number of meristematic cells).

The treated embryos were then bombarded with microprojectiles containing the *bar* gene and resulted in a surprisingly high rate of transformation with at least 1% of bombarded embryos giving rise to transformed plants. Livingstone and Birch (unpublished) optimised conditions for gene transfer into peanut meristems (cultivars Florunner, Gajah and McCubbin) and obtained transient *uidA* expression frequencies of up to 23 ± 1.4 GUS expression events per meristem with 82% of bombarded meristems showing transient *luc* expression. Stable expression of both the *uidA* and *luc* genes was demonstrated for up to 3 weeks. However, no transgene expression could be detected beyond this time. Twenty-six percent of peanut meristems were damaged by the bombardment conditions optimal for gene transfer and failed to germinate.

Ozias-Akins *et al.* (1993) transformed 1 to 2 year old embryogenic callus derived from immature embryos (cultivar Toalson) by microprojectile bombardment, followed by stepwise selection for resistance to hygromycin B in solid and liquid media. This approach resulted in about 0.3 hygromycin B resistant callus lines per 10 cm^2 of embryogenic callus bombarded. Five callus lines and one plant were confirmed as independent transformants by Southern analysis, but the fertility of transformants was not reported. Further to this work, Singsit *et al.* (1997) increased the efficiency for the cultivar Florunner to 0.85 to 2.3 transgenic lines per bombardment. One progeny plant was characterised, but sterility of transformants was prevalent, possibly associated with chromosomal abnormalities. The most recent report shows further increased transformation efficiencies, ranging from 2.6 ± 3.5 to 19.8 ± 18.5 hygromycin B resistant lines per bombardment (5 cm^2) with plant fertility rates of 32% (Wang *et al.*, 1998).

Building on the work of Ozias-Akins *et al.* (1993), Livingstone and Birch (1998) efficiently transformed both botanical types of peanut (Spanish and Virginia) by particle bombardment into embryogenic callus derived from mature seeds, followed by single-step selection for hygromycin B resistance. Somatic embryos were efficiently converted into fertile, non-chimeric transgenic plants, by brief osmotic desiccation followed by sequential incubation on charcoal and cytokinin-containing media. This method produces three to six independent transformants per bombardment of 10 cm^2 embryogenic callus. The copy number of integrated transgenes ranged from one to twenty with a mean of four and 57% coexpression of *hph* and *luc* or *uidA* genes coprecipitated on separate plasmids. Potted transgenic plant lines were regenerated within 9 months of callus initiation or 7 months after bombardment. Transgene expression was confirmed in T_1 progeny from each of six tested independent transformants.

4. Conclusions and future prospects

Microprojectile-mediated transformation of peanut has been demonstrated by several independent sources and is clearly a practical means of introducing agronomically important genes to this important crop. The obstacle of infertility has been overcome but doubts remain about the long term stability of transgene expression. This problem might be addressed by careful selection of promoters. Plant promoters with tissue and developmental specificity already exist and modification of promoters with matrix attachment regions (MARS) is now routinely increasing the stability of expression *in vivo*. Even without increased gene stability this technique is a useful and practical tool to transform the major commercial peanut varieties in Australia (cultivar NC-7, a Virginia market type) and Indonesia (cultivar Gajah, a Spanish market type).

Important features of this protocol are :

i. Use of embryos from mature seeds as a readily available explant for initiation of highly regenerable embryogenic callus, suitable for bombardment within 3 months.
ii. 'Escape-free' selection for hygromycin B resistance on solid media after a short recovery period. This is a simpler approach than stepwise selection methods and produces non-chimeric transformants.
iii. A high efficiency of production of transgenic plant lines of both Virginia (cultivar NC-7, six lines per 10 cm^2 of bombarded callus) and Spanish (cultivar Gajah, three lines per bombardment) botanical types of peanut.
iv. Fertility of 53% of transgenic plant lines, which is likely to be increased by reduced time in tissue culture.
v. Coexpression in 44% of plant lines of an unselected reporter gene introduced on a separate plasmid from the selected hygromycin B resistance gene.

This research also provided the first Southern analysis showing transgene integration patterns in multiple peanut lines, and expression of the transgene in T_1 progeny. The procedure is suitable for the efficient transfer of agronomically important genes into major commercial peanut cultivars.

References

Albert H (1995) Site-specific integration of DNA into wild-type and mutant lox sites placed in the plant genome. *Plant J.*, **7**: 649–659.

Amadeo P, Habu Y, Afsar K, Mittelsten Scheid O and Paszkowski J (2000) Disruption of the plant gene *MOM* releases transcriptional silencing of methylated genes. *Nature*, **405**: 203–206.

Amernick B A (1986) *Patent Law for the Nonlawyer: A Guide for the Engineer, Technologist and Manager*. New York, Van Nostrand Reinhold.

Atreya C D, PapaRao J and Subrahmanyam N C (1984) *In vitro* regeneration of peanut (*Arachis hypogaea* L.) plantlets from embryo axes and cotyledon segments. *Plant Sci. Lett.*, **34**: 379–383.

Bajaj Y P S (1983) Peanut. In: *Handbook of Plant Cell Culture* (Ed Evans D A), Vol. 3, Macmillan, New York, 193–225.

Baker C M, Durham R E, Burns J A, Parrott W A and Wetzstein H Y (1995) High frequency somatic embryogenesis in peanut (*Arachis hypogaea* L.) using mature, dry seed. *Plant Cell Rep.*, **15**: 38–42.

Baker C M and Wetzstein H Y (1992) Somatic embryogenesis and plant regeneration from leaflets of peanut, *Arachis hypogaea. Plant Cell Rep.*, **11**: 71–75.

Barcelo P, Hagel C, Becker D, Martin A and Lörz H (1994) Transgenic cereal (*tritordeum*) plants obtained at high efficiency by microprojectile bombardment of inflorescence tissue. *Plant J.*, **5**: 583–592.

Bhojwani S S and Razdan M K (1983) *Plant Tissue Culture: Theory and Practice*. Elsevier, Amsterdam.

Birch R G and Bower R (1994) Principles of gene transfer using particle bombardment. In: *Particle Bombardment Technology for Gene Transfer* (Eds Yang N S and Christou P), Oxford University Press, New York, 3–37.

Bower R and Birch R G (1990) Competence for gene transfer by electroporation of protoplasts from uniform carrot cell suspension cultures. *Plant Cell Rep.*, **9**: 386–389.

Bower R, Elliott A R, Potier B A M and Birch R G (1996) High-efficiency, microprojectile-mediated cotransformation of sugarcane, using visible or selectable markers. *Mol. Breed.*, **2**: 239–249.

Brar G S, Cohen B A, Vick C L and Johnson G W (1994) Recovery of transgenic peanut (*Arachis hypogaea* L.) plants from elite cultivars utilizing ACCELL technology. *Plant J.*, **5**: 745–753.

Castillo A M, Vasil V and Vasil I K (1994) Rapid production of fertile transgenic plants of rye (*Secale cereale* L.). *Bio/Technology*, **12**: 1366–1371.

Chalfie M, Tu Y, Euskirchen G, Ward W W and Prasher D C (1994) Green fluorescent protein as a marker for gene expression. *Science*, **263**: 802–805.

Chandler V L and Jorgenson R (2000) Silencing Morpheus awakens transgenes. *Nature Bio/Technology*, **18**: 602–603.

Cheng M, Jarret R L, Li Z and Demski J W (1997) Expression and inheritance of foreign genes in transgenic peanut plants generated by *Agrobacterium*-mediated transformation. *Plant Cell Rep.*, **16**: 541–544.

Cheng M, Jarret R L, Li Z, Xing A and Demski J W (1996) Production of fertile transgenic peanut (*Arachis hypogaea* L.) plants using *Agrobacterium tumefaciens. Plant Cell Rep.*, **15**: 653–657.

Chengalrayan K, Sathaye S and Hazra S (1994) Somatic embryogenesis from mature embryo-derived leaflets of peanut (*Arachis hypogaea* L.). *Plant Cell Rep.*, **13**: 578–581.

Chia T F, Chan Y S and Chua N H (1994) The firefly luciferase gene as a non-invasive reporter for *Dendrobium* transformation. *Plant J.*, **6**: 441–446.

Christianson M L (1987) Causal Events in Morphogenesis. In: *Plant Tissue and Cell Culture* (Eds Green C E, Somers D A, Hackett W P and Biesboer D D), Alan R. Liss Inc., N.Y., 45–55.

Clemente T E, Robertson D, Isgleib T G, Beute M K and Weissinger A K (1992) Evaluation of peanut (*Arachis hypogaea* L.) from mature zygotic embryos as recipient tissue for biolistic gene transfer. *Transgenic Res.*, **1**: 275–284.

D'Halluin K (1992) Transgenic maize plants by tissue electroporation. *Plant Cell*, **4**: 1495–1505.

Durham R E and Parrot W A (1992) Repetitive somatic embryogenesis from peanut cultures in liquid medium. *Plant Cell Rep.*, **11**: 122–125.

Eapen S and George L (1994) *Agrobacterium tumefaciens* mediated gene transfer in peanut (*Arachis hypogaea* L.). *Plant Cell Rep.*, **13**: 582–586.

FAO (1996) *Quarterly Bulletin of Statistics*. Vol. 8 (3/4). FAO, Rome.

Finer J J, Vain P, Jones M W and McMullen M D (1992) Development of the particle inflow gun for DNA delivery to plant cells. *Plant Cell Rep.*, **11**: 323–328.

Franks T and Birch R G (1991a) Gene transfer into intact sugarcane cells using microprojectile bombardment. *Australian J. Plant Physiol.*, **18**: 471–480.

Franks T and Birch R G (1991b) Microprojectile techniques for direct gene transfer into intact plant cells. In: *Advanced Methods in Plant Breeding and Biotechnology* (Ed Murray D A). CAB International, Wallingford, UK.

Gordon-Kamm W J, Spencer T M, Mangano M L, Adams T R, Daines R J, Start W G, O'Brien J V, Chambers S A, Adams W R Jr., Willetts N G, Rice T B, Mackey C J, Krueger R W, Kausch A P and Lemaux P G (1990) Transformation of maize cells and regeneration of fertile transgenic plants. *Plant Cell*, **2**: 603–618.

Hansen G and Chilton M D (1996) Agrolistic transformation of plant cells: integration of T-strands generated in planta. *Proc. Natl. Acad. Sci. USA*, **93**: 4978–4983.

Hazra S, Sathaye S S and Mascarenhas A F (1989) Direct somatic embryogenesis in peanut (*Arachis hypogea*). *Bio/Technology*, **7**: 949–951.

Hébert D, Kikkert J R, Smith F D and Reisch B I (1993) Optimization of biolistic transformation of embryogenic grape cell suspensions. *Plant Cell Rep.*, **12**: 585–589.

Iglesias V A, Gisel A, Bilang R, Leduc N, Potrykus I and Sautter C (1993) Transient expression of visible marker genes in meristem cells of wheat embryos after ballistic micro-targeting. *Planta*, **192**: 84–91.

Kausch A P, Adams T R, Mangano M, Zachwieja S J, Gordon-Kamm W, Daines R, Willetts N G, Chambers A A, Adams W Jr., Anderswon A, Williams G and Haines G (1995) Effects of microprojectile bombardment on embryogenic suspension cell cultures of maize (*Zea mays* L.) used for genetic transformation. *Planta*, **196**: 501–509.

Kiernan J M, Goldberg M J, Young M J, Schoelz J E and Shepard R J (1989) Transformation and regeneration of *Nicotiana edwardsonii. Plant Sci.*, **64**: 67–78.

Klee H, Horsch R and Rogers S (1987) *Agrobacterium*-mediated plant transformation and its further applications to plant biology. *Annu. Rev. Plant Physiol.*, **38**: 467–486.

Klein T M, Arentzen R, Lewis P A and Fitzpatrick-McElligott S (1992) Transformation of microbes, plants and animals by particle bombardment. *Bio/Technology*, **10**: 286–291.

Klein T M, Fromm M E, Weissinger A, Tomes D, Schaff S, Sletten M and Sanford J C (1988a) Transfer of foreign genes into intact maize cells with high velocity microprojectiles. *Proc. Natl. Acad. Sci. USA*, **85**: 4305–4309.

Klein T M, Gradziel T, Fromm M E and Sanford J C, (1988b) Factors influencing gene delivery into *Zea mays* cells by high velocity microprojectiles. *Bio/Technology*, **6**: 559–563.

Klein T M, Harper E C, Svab Z, Sanford J C, Fromm M E and Maliga P (1988c) Stable genetic transformation of intact Nicotiana cells by the particle bombardment process. *Proc. Natl. Acad. Sci. USA*, **85**: 8502–8505.

Klein T M, Wolf E D, Wu R and Sanford J C (1987) High velocity microprojectiles for delivering nucleic acids into living cells. *Nature*, **327**: 70–73.

Kloti A (1993) Gene transfer by electroporation into intact scutellum cells of wheat embryos. *Plant Cell Rep.*, **12**: 671–675.

Knittel N, Gruber V, Hahne G and Lénée P (1994) Transformation of sunflower (*Helianthus annuus* L.): a reliable protocol. *Plant Cell Rep.*, **14**: 81–86.

Kochert G (1996) Molecular markers and genome mapping. In: *Current Status of Agricultural Biotechnology in Indonesia* (Ed Darussamin A). AARD, Jakarta, 89–108.

Kokko H I and Kärenlampi S O (1998) Transformation of arctic bramble (*Rubus arcticus* L.) by *Agrobacterium tumefaciens*. *Plant Cell Rep.*, **17**: 822–826.

Larkin P J (1990) Direct gene transfer to protoplasts. *Australian J. Plant Physiol.*, **17**: 291–302.

Link G K K and Eggars V (1942) Hyperauxing in crown gall of tomato. *Botanical Gazette*, **103**: 87–106.

Livingstone D M and Birch R G (1995) Plant regeneration and microprojectile-mediated gene transfer in embryonic leaflets of peanut (*Arachis hypogaea* L.). *Australian J. Plant Physiol.*, **22**: 585–591.

Livingstone D M and Birch R G (1999) Efficient transformation and regeneration of diverse cultivars of peanut (*Arachis hypogaea* L.) by particle bombardment into embryogenic callus produced from mature seeds. *Mol. Breed.*, **5**: 43–51.

Ludwig S R, Bowen B, Beach L and Wessler S R (1990) A regulatory gene as a novel visible marker for maize transformation. *Science*, **247**: 449–450.

Lusardi M C (1994) An approach towards genetically engineered cell fate mapping in maize using the Lc gene as a visible marker: transactivation capacity of Lc vectors in differentiated maize cells and microinjection of Lc vectors into somatic embryos and shoot apical meristems. *Plant J.*, **5**: 571–582.

McKently A H, Moore G A, Doostar H and Niedz R P (1995) *Agrobacterium*-mediated transformation of peanut (*Arachis hypogaea* L.) embryo axes and the development of transgenic plants. *Plant Cell Rep.*, **14**: 699–703.

MacKenzie D R, Anderson P M and Wernham (1966) A mobile air blast inoculator for plot experiments with maize dwarf mosaic virus. *Plant Disease Rep.*, **50**: 363–367.

Mansur E, Lacorte C and Krul W R (1995) Peanut transformation. In: *Agrobacterium Protocols: Methods in Molecular Biology* (Eds Gartland K M A and Davey M R). Humana Press, Totowa, NJ, 87–100.

Mlynarova L, Keizer L C P, Stiekema W J and Nap J P (1996) Approaching the lower limits of transgene variability. *Plant Cell*, **8**: 1589–1599.

Mroginski L A, Kartha K K and Shyluk J P (1981) Regeneration of peanut (*Arachis hypogaea*) plantlets by *in vitro* culture of immature leaves. *Can. J. Bot.*, **59**: 826–830.

Nejidat R S, McCormick S M, Delennay X, Dupe P and Layton J (1988) Transgenic tobacco plants expressing a tobacco virus coat protein are resistant to some tobamoviruses. *Mol. Plant-Microbe Interact.*, **3**: 247–251.

Ow D W, Wood K V, DeLuca M, deWet J R, Helinski D R and Howell S H (1986) Transient and stable expression of the firefly luciferase gene in plant cells and transgenic plants. *Science*, **234**: 856–859.

Ozias-Akins P (1989) Plant regeneration from immature embryos of peanut. *Plant Cell Rep.*, **8**: 217–218.

Ozias-Akins P, Anderson W F and Holbrook C C (1992) Somatic embryogenesis in *Arachis hypogaea* L.: genotype comparison. *Plant Sci.*, **83**: 103–111.

Ozias-Akins P, Schnall J A, Anderson W F, Singsit C, Clemente T E, Adang M J and Weissinger A K (1993) Regeneration of transgenic peanut plants from stably transformed embryogenic callus. *Plant Sci.*, **93**: 185–194.

Pugliesi C, Cecconi F, Mandolfo A and Baroncelli S (1990) Plant regeneration and genetic variability from tissue cultures of sunflower (*Helianthus annuus* L). *Plant Breed. Z. Planzenzucht*, **106**: 114–121.

Rasmussen J L (1994) Biolistic transformation of tobacco and maize suspension cells using bacterial cells as microprojectiles. *Plant Cell Rep.*, **13**: 212–217.

Reggardio M I, Arana J L, Orsaria L M, Permingeat H R, Spitteler M A and Vallejos R H (1991) Transient transformation of maize tissues by microprojectile bombardment. *Plant Sci.*, **75**: 237–243.

Russell J A, Roy M K and Sanford J C (1992) Physical trauma and tungsten toxicity reduce the efficiency of biolistic transformation. *Plant Physiol.*, **98**: 1050–1056.

Sanford J C, Klein T M, Wolf E D and Allen N (1987) Delivery of substances into cells and tissues using a particle bombardment process. *Particulate Science and Technology*, **5**: 27–37.

Sieburth L E and Meyerowitz E M (1997) Molecular dissection of the AGAMOUS control region shows that *cis* elements for spatial regulation are located intragenically. *Plant Cell*, **9**: 355–365.

Singsit C, Adang M J, Lynch R E, Anderson W F, Wang A, Cardineau G and Ozias-Akins P (1997) Expression of a *Bacillus thuringiensis cryIA*(c) gene in transgenic peanut plants and its efficacy against lesser cornstalk borer. *Transgenic Res.*, **6**: 169–176.

Takeuchi Y, Dotson M and Keen N T (1992) Plant transformation: a simple particle bombardment device based on flowing helium. *Plant Mol. Biol.*, **18**: 835–839.

Taylor M G and Vasil I K (1991) Histology of, and physical factors affecting, transient GUS expression in pearl millet (*Pennisetum glaucum* (L.) R. Br.) embryos following microprojectile bombardment. *Plant Cell Rep.*, **10**: 120–125.

Teycheney P Y and Dietzgen R G (1994) Cloning and sequence analysis of the coat protein genes of an Australian strain of peanut mottle and an Indonesian 'blotch' strain of peanut stripe potyviruses. *Virus Research*, **31**: 235–244.

Tisserat B (1985) Embryogenesis, Organogenesis and Plant Regeneration. In: *Plant Cell Culture: a practical approach* (Ed Dixon R A), IRL Press, Oxford, 79–105.

Vain P, McMullen M D and Finer J J (1993) Osmotic treatment enhances particle bombardment-mediated transient and stable transformation of maize. *Plant Cell Rep.*, **12**: 84–88.

Vasil V, Castillo A M, Fromm M E and Vasil I K (1992) Herbicide resistant fertile transgenic wheat plants obtained by microprojectile bombardment of regenerable embryogenic callus. *Bio/Technology*, **10**: 667–674.

Wang Y, Klein T M, Fromm M, Cao J, Sanford J C and Wu R (1988) Transient expression of foreign genes in rice, wheat and soybean cells following particle bombardment. *Plant Mol Biol.*, **11**: 433–439.

Wang Z Y, Takamizo T, Iglesias V A, Osusky M, Nagel J, Potrykus I and Spangenberg G (1992) Transgenic plants of tall fescue (*Festuca arundinacea* Schreb.) obtained by direct gene transfer to protoplasts. *Bio/Technology*, **10**: 691–696.

Wilmut I, Schnieke A E, McWhir J, Kind A J and Campbell K H S (1997) Viable offspring derived from fetal and adult mammalian cells. *Nature*, **385**: 810–813.

Wochok Z S and Weterell D F (1972) Restoration of declining morphogenetic capacity in long-term cultures of *Daucus carota* by kinetin. *Experientia*, **28**: 104–105.

Ye G N, Daniell H and Sanford J C (1990) Optimization of delivery of foreign DNA into higher-plant chloroplasts. *Plant Mol. Biol.*, **15**: 809–819.

TRANSFORMATION OF LUPINS

CRAIG A. ATKINS* AND PENELOPE M. C. SMITH
Department of Botany, University of Western Australia,
Nedlands WA 6907, Australia
**e-mail: catkins@cyllene.uwa.edu.au*

Abstract

Effective methods for routine transformation of narrow-leafed lupin (*Lupinus angustifolius* L.) that rely on *Agrobacterium tumefaciens* have been developed. Stable transformants that express high level resistance to broad spectrum herbicides have been the initial outcomes but genetically modified (GM) lines with potentially enhanced disease resistance and altered growth/yield traits have also been generated. The methodology has been transferred and modified to suit a range of other pulse crops and leguminous pas-ture species and techniques that extend this potential to other lupin species are being developed.

1. Introduction

Recombinant DNA technology, together with newly developed procedures for cell and tissue culture, has permitted direct genetic manipulation of plants. Members of the Solanaceae and Brassicaceae, including tobacco, tomato, potato, canola and *Arabidopsis*, have proved to be the most easily engineered. However, recently a number of cereals and leguminous species have also been transformed to express novel genes. Among grain legumes most attention has been given to soybean, pea and *Vigna* species, while of the pasture legumes, clovers and medics, including alfalfa (*Medicago sativa*), and the model species *M. truncatula*, have been successfully transformed (Christou, 1994). The genes transferred have in many cases only conferred expression of a selectable marker, such as resistance to an antibiotic, or of an enzyme that is easily assayed *in vivo*, such as β-glucuronidase (the 'blue' or GUS gene). However, genes are now being transferred which introduce novel traits with agronomic utility or which increase the value of the crop or pasture. These include tolerance to broad-spectrum herbicides, resistance to insect pests and diseases, altered grain quality, plant growth and architecture and features of reproductive development. Strategies to exploit plants by promoting synthesis of

P. K. Jaiwal and R. P. Singh (eds.), Applied Genetics of Leguminosae Biotechnology, 205–211.

a particular organic molecule, effectively using the plant as a 'bioreactor' or as a source of pharmaceutically valuable proteins, have also been proposed (Poirier *et al.*, 1992).

Routine procedures for transforming narrow-leafed lupin (*Lupinus angustifolius* L.), by *Agrobacterium tumefaciens*-mediated gene transfer, have now been developed (Atkins *et al.*, 1994; Molvig *et al.*, 1994; Pigeaire *et al.*, 1997) and applied to improvement of the species by genetic engineering. Initially transformations to confer herbicide tolerance were made but subsequently a number of other traits have been investigated (Atkins *et al.*, 1998). The strategies developed may also apply to other lupins and effective methods for their transformation and regeneration are being sought.

2. Early studies with lupins

Protocols for organogenesis and regeneration from tissue culture have been described for a number of *Lupinus* species. Lee (1955a, b) showed that explants from mature embryos of *L. hartwegii* could be induced to callus or regenerate rooted plants while Sator (1985) extended these protocols to other species (*L. angustifolius*, *L. luteus* and *L. polyphyllus*). Callus cultures for *L. angustifolius*, *L. albus*, *L. luteus*, *L. hartwegii* and *L. polyphyllus* from embryogenic tissue as well as from mature vegetative and cotyledonary tissues were described by Sroga (1987). Calli from embryonic axes and hypocotyls for all species except *L. albus* and *L. luteus* were demonstrated with high frequencies of multiple shoot induction but with lower frequencies from leaves and cotyledons, and no calli from roots. As many as 96% of shoots formed roots in culture and of these 33–65% formed viable plants. In general *L. angustifolius* was more amenable to the methods compared to *L. polyphyllus*. Daza and Chamber (1993) reported regeneration of viable *L. luteus* plants following shoot induction from hypocotyl explants in culture and Rahim and Caligari (1998) have described an *in vitro* protocol for shoot regeneration in *L. mutabilis* using explants from developing seeds. Wetten *et al.* (1994) have described preparation of protoplasts from leaf tissue of *L. angustifolius*, *L. albus*, *L. luteus*, *L. mutabilis*, *L. hispanicus*, *L. pilosus*, *L. polyphyllus* and *L. palestinus* using enzymic digestion of cell walls. The yields of protoplasts were suitable for interspecific crosses using dielectrophoresis (Wetten *et al.* 1996), and potentially could also be used for direct transformation using electroporation, but their exploitation has yet to be reported.

Two Australian groups independently described *Agrobacterium*-mediated transformation and regeneration of *L. angustifolius*. In one case slices of the cotyledonary node of young seedlings or of the developing embryonic axis were used (Molvig, 1992; Molvig *et al.*, 1993, 1994). The technique had proved successful for transformation and regeneration of *Pisum sativum* (Schroeder *et al.*, 1993) and was applied to lupins more or less directly. However, the frequency with lupins (*cv.* Warrah) was very low (<0.01%) compared with peas (2–4%) (Molvig *et al.* 1993). Pigeaire *et al.* (1994, 1995) developed a method that used explants from germinating seeds and relied on gene transfer to the shoot apex with development of shoots from transformed axillary buds. Transformation frequencies ranged from as low as 1.5 to 6%. Both protocols led to the generation of transgenic lines but because of the much greater utility of that developed by Pigeaire *et al.* (1994, 1997) the details of their methods are outlined below. The protocols have been reviewed in the context of methods for legume transformation elsewhere (Atkins and Smith, 1997, 2000; Atkins *et al.*, 1998). The method described is reproducible and is now in routine use to genetically engineer novel lines for potential commercialization.

3. *Agrobacterium*-based transformation of lupins

3.1. SELECTION OF LUPIN CULTIVAR AND *AGROBACTERIUM* STRAIN

Following studies that relied on transient GUS expression of one *Agrobacterium tumefaciens* strain, AgL0 (Lazo *et al.*, 1991), was chosen from a screen of nineteen as being superior with a number of different lupin lines. Initially a rather early lupin cultivar, *cv.* Unicrop, was used. However, the methods have been tested successfully with a wide range of cultivars including the most recently released cultivars and advanced breeding lines. Strain AgL0 has also proved superior for transformation of *L. luteus* (Somsap *et al.*, 1993, 1994).

The constructs used to develop the lupin transformation protocol were pCGP963 or pCGP1258 (both in *Agrobacterium* strain AgL0). The plasmid pCGP963 was derived from pGA492 (An, 1986) and contained the *bar* gene (encoding an enzyme that acetylates and thus detoxifies glufosinate, the active ingredient in the herbicide Basta®) controlled by the CaMV35S promoter and an *ocs* (octopine synthase) terminator. pCGP1258 contains a *uidA* gene from pKIWI101 (Janssen and Gardner, 1989) also under control of the 35S promoter.

3.2. LUPIN EXPLANTS AND INOCULATION

The best explant for transformation has been found to be the shoot apex of the embryonic axis excised from a germinated seed. These explants are prepared from surface sterilized seed that has germinated overnight at 25°C or until the root was at least 2 mm but not more than 10 mm long. The seed coat, cotyledons and the two pairs of primordial leaves are excised to leave a plumule with an apical dome and the third pair of leaf primordia. Before being plated upright (i.e. with the root in solid medium) on a co-cultivation medium containing MS salts (Murashige and Skoog, 1962) with B_5 vitamins, 10 mg/L BAP and 1 mg/L NAA, adjusted to pH 5.8 and solidified with 0.25% (w/v) gelrite (Sigma) the apical dome is stabbed with a 30G needle to facilitate infection by the inoculum (*Agrobacterium*). Addition of glucose (10 mM) and acetosyringone (20 μM) in some cases has been found to increase the frequency of transformation.

A drop (~5 μL) of *Agrobacterium* culture (resuspended in MS salts at 5×10^8 cells per mL) is then placed on top of the wounded apex and the explants co-cultivated with the bacterium for 2 days at 25°C in continuous low intensity white light after which they are transferred to a regeneration medium (MS medium supplemented with one tenth the level of growth regulators and 150 mg/L timentin to inhibit growth of the *Agrobacterium*). A solution containing glufosinate (2 mg/mL) and 0.1% Tween 20 is applied to the apex and the explants are grown in the same culture medium under low intensity white light at 25°C with an 8 h photoperiod.

3.3. SELECTION AND REGENERATION OF TRANSGENIC AXILLARY SHOOTS

Depending on the cultivar and the nature of the transforming *Agrobacterium* strain as many as 100% of explants co-cultivated will regenerate shoots (10–15 mm). However, in controls where explants are co-cultivated with a host *Agrobacterium* strain without the *bar* construct there is no regeneration. After 1–2 weeks on the selection medium axillary shoots are excised and placed on a 'micropropagation' medium (as above but with a further tenth dilution of the growth regulators, 150 mg/L timentin, 20 mg/L glufosinate and solidified with 0.9% (w/v) agar) to encourage development of transformed axillary buds. When green axillary shoots are produced they are transferred to fresh medium. Control shoots (when grown on media without glufosinate), will usually produce axillary shoots on all four sides but the transformed shoots are chimeric and often produce a secondary shoot on only one side. The secondary shoots are excised and cultured with glufosinate and will in turn produce axillary shoots (usually on 2 sides) which are again excised and cultured. Frequently the primary and secondary shoots do not survive a long period of selection (3 months) and it is usually tertiary and quaternary shoots that come through the transformation procedure and are eventually transferred to the glasshouse.

3.4. ROOTING TRANSGENIC SHOOTS

To induce root formation, the base of a shoot is dipped in a solution of 1 mg/mL indolebutyric acid (IBA) and then grown on micropropagation medium containing 3mg/L IBA but no other hormones. Roots develop after approximately 2 weeks in culture and the plantlets can be transferred to the glasshouse after a further 3 weeks. They are most easily grown in aerated liquid culture (equivalent to 1/4 strength Hoagland's solution with 5 mM KNO_3). However, difficulty may be experienced in rooting transformed lupin shoots and a grafting protocol has been developed. Initially, grafting was essential (Pigeaire *et al.*, 1997). This protocol involved attaching the transformed material to a slit at the cotyledonary node of 10–14 day old seedlings of the corresponding cultivar growing in sand with a complete nutrient solution (equivalent to Hoagland's solution with 5 mM KNO_3), or potting mix supplemented with nutrients. The recovery of shoots is relatively high but to ensure success several shoots (clones) representing each transgenic event are grafted. The strength of the stem is adequate to hold the grafted shoot in place without the need for tape or support and as a result there is little problem with infection at the wound. The grafted plants are covered in a plastic vial itself wrapped in aluminium foil for 3–4 days and then the foil is removed. Over this period a graft union is formed and new shoot tissue is evident. After a further 1–2 day(s) the vial is removed and the plants grown on under normal glasshouse conditions to obtain viable seed. The extent to which the transgenic shoots grow depends on the time to flowering. If shoots are held in tissue culture for extended periods (more than 4 months) before grafting they flower so soon that the flowers abort and no seed is set. This problem may be overcome to some extent if the floral organs on the main axis are removed once the graft is established. The axillary shoots then flower later when sufficient vegetation has been formed to support pod set and seed filling.

3.5. ASSESSMENT FOR GENETIC TRANSFORMATION

The *bar* gene offers a highly sensitive assay for transformation. Axillary shoots which develop on glufosinate-containing medium (20 mg/mL) are tolerant of a 2 mg glufosinate per ml solution (in 0.1% Tween 20) painted on excised leaflets. Grafted shoots and foliage which express the *bar* gene in subsequent generations are tolerant to 0.1 mg/mL glufosinate in a leaf paint assay and for a whole plant spray treatment Basta® at 0.1 mg/mL glufosinate. Resistance of leaflets or whole plants can be assessed after 6–14 days. To further confirm stable integration of the *bar* gene Southern analysis of genomic DNA following PCR amplification and by northern analysis of RNA isolated from leaf tissue is essential. Genetic analysis of inheritance and expression of the *bar* gene in subsequent generations of transgenic *L. angustifolius* lines indicates that inheritance patterns in the first generation of transgenic seed (T_1) equate roughly to a 3 : 1 Mendelian ratio, i.e. segregation expected for a dominant gene at a single locus in a self-pollinating species.

3.6. TRANSFORMATION FREQUENCY

Although newer methods for transformation will undoubtedly be developed for lupins the best at present continue to rely on *Agrobacterium*. Overall the transformation frequency is about 0.4% but there is considerable variation between cultivars and with the nature of the construct used. Frequencies of up to 6% have been found for an older variety, *cv.* Unicrop, and with current methodology newer varieties are returning frequencies of about 2.5%. With constructs that do not interfere with the growth and differentiation of transformed shoots or roots, T_1 seed is available 5–6 months after rooting. Sufficient seed is generated in the T_3 generation for field and screen house trials. From the initial explant to the point where sufficient transgenic seed had been generated for controlled field site trials at a number of locations requires just over three years. While this approach may save some time in the breeding of lupins the time required to meet the regulatory requirements for field trials and release is significant.

4. Transformation and regeneration of other lupin species

There have been no published reports of routine protocols for the stable transformation of any other species of lupin. Mugnier (1988) demonstrated transformation of *L. albus* and *L. polyphyllus* with *A. rhizogenes* and established a hairy root culture for each species but did not regenerate transformed plants. The current protocols for narrow-leafed lupin have been applied with success to a number of other pulses (peas, lentils and chickpeas) but have not yet been successful with *L. albus*. Some progress has been made with *L. luteus* (Somsap *et al.*, 1993, 1994) using a binary plasmid system in *A. tumefaciens* AGL1 carrying a GUS-*bar* fusion (similar to that used above for *L. angustifolius*). The transformation and regeneration protocols for *L. luteus* are similar to those outlined above for narrow-leafed lupin except that an *in vitro* grafting technique has been used to increase the frequency of successful grafts in the species reducing possible losses in the glasshouse due either to desiccation or disease. The transformations have been confirmed

NUTRITIONAL IMPROVEMENT OF LUPIN SEED PROTEIN USING GENE TECHNOLOGY

L. MOLVIG[1], L. M. TABE[1], J. HAMBLIN[2], V. RAVINDRAN[3], W. L. BRYDEN[3], C. L. WHITE[4] and T. J. V. HIGGINS[1,*]

[1]*CSIRO Plant Industry, GPO Box 1600, Canberra ACT 2601, Australia*

[2]*Export Grains Centre Ltd, 219 Canning Hwy, South Perth WA 6151, Australia*

[3]*Faculty of Veterinary Science, The University of Sydney, Camden NSW 2570, Australia*

[4]*CSIRO Animal Production, Private Bag, PO Wembley WA 6014, Australia*

**e-mail: tj.higgins@pi.csiro.au*

Abstract

The nutritive value of lupin, an important grain legume for animal production, has been improved by the transformation of a chimeric gene specifying seed-specific expression of a sulfur-rich sunflower seed albumin (SSA). This gene encodes a protein which contains 16% methionine and 8% cysteine residues and the protein was expressed in transgenic seeds at a level of 5% of total soluble seed protein. The expression of the SSA gene in lupin seeds caused an overall increase in sulfur containing amino acids of 19%, while methionine levels were doubled. The improved nutritive value of these transgenic lupin seeds was demonstrated in feeding trials with rats, chickens, and sheep.

1. Introduction

Improvement of the nutritive value of seeds, in particular amino acid profiles, is now an achievable goal using gene technology. Examples include significant increases in lysine levels in transgenic canola and soybean seeds by manipulating key control points in the biosynthetic pathway of the aspartate family of amino acids (Falco *et al.*, 1995). The introduction of a Brazil nut albumin gene has increased the levels of methionine in the seeds of transgenic canola (Altenbach *et al.*, 1992) and narbon bean (Saalbach *et al.*, 1995).

Narrow leaf lupin (*L. angustifolius* L.) is the most widely grown grain legume crop in Australia and is an important source of protein in animal diets. However, the seed protein of this legume is deficient in the essential sulfur containing amino acids for optimum

P. K. Jaiwal and R. P. Singh (eds.), Applied Genetics of Leguminosae Biotechnology, 213–219.

animal growth, which necessitates the addition of synthetic methionine to commercial diets containing lupin seed. A chimeric gene coding for a sunflower seed albumin (SSA), which contains 16% methionine and 8% cysteine, was stably transformed into narrow leaf lupin (Molvig *et al.*, 1997) to improve nutritive value. A homozygous transgenic line, in which the new protein constituted about 5% of the seed protein, was selected for chemical analysis, field trials and animal feeding studies. The level of sulfur containing amino acids was increased in the transgenic line and this was reflected in its enhanced nutritive value in animal feeding trials.

2. Transgenic lupin seed with enhanced nutritive value for monogastric animals

2.1. TRANSFORMATION AND REGENERATION

Two methods for lupin transformation were developed independently. One method is briefly described here (for more detail see Molvig *et al.*, 1997). The other is described in detail by Pigeare *et al.* (1997). Both methods are reproducible, although both yield fewer than 1% of the original explants as transformants.

Pods were harvested from glasshouse-grown lupin (*Lupinus angustifolius* cv. Warrah) plants when seeds were at a stage corresponding to maximum fresh weight. The pods were surface sterilised by immersion for 1 minute in 70% (v/v) ethanol followed by 20 minute sterilisation in a solution of commercial bleach (10 g/L of sodium hypochlorite) and rinsed several times in sterile distilled water. Seeds were removed from pods, and explants prepared by slicing the embryonic axis longitudinally into two halves while still attached to the cotyledons. The root pole was removed 2 to 3 mm above the root tip and discarded, and thin longitudinal slices taken from the halves of the embryonic axis (three to five slices per axis).

The sliced embryonic axes were incubated for 30 to 90 minutes in a suspension of *Agrobacterium tumefaciens* containing a binary plasmid consisting of three genes encoding phosphinothricin acetyl transferase, sunflower seed albumin and β-glucuronidase (GUS), respectively. Following incubation, explants were removed from the *Agrobacterium* suspension and transferred to solid B_5 medium (Gamborg *et al.*, 1968) containing 3 mg/L NAA, 0.5 mg/L BAP, 100 mM coniferyl alcohol (Sigma) for co-cultivation in a horizontal position. After three days' co-cultivation at 24°C with a 16 hour photoperiod, explants were washed several times in sterile distilled water and blotted dry on sterile filter paper before plating on regeneration medium.

Shoot formation was initiated on modified MS medium (Murashige and Skoog, 1962; Shetty *et al.*, 1992) containing 1 mg/L each of NAA and BAP, 0.3% w/v sucrose, 0.8% Difco Bacto agar at pH 5.8 (before autoclaving). Timentin (SmithKline Beecham Australia Pvt. Ltd.) was filter-sterilised and added to cooled media (150 mg/L) to control *Agrobacterium* growth. Phosphinothricin (5 mg/L) was used for selection.

After 14 days of culture at 24°C with a 16 hour photoperiod, shoots were transferred to MS medium containing 0.1 mg/L NAA and 1 mg/L BAP. Phosphinothricin and timentin levels were kept constant throughout the regeneration and selection procedures. Once shoots reached 10 mm in height, the clumps of shoots were divided and cultured every 2 weeks on MS medium containing 0.5 mg/L BAP for further shoot growth, or

*Table 1. Levels of cysteine and methionine in non-transgenic and transgenic lupin seeds**

Amino acid	Non-transgenic lupin (g/16g N)**	Transgenic lupin (g/16g N)**	Difference
Cysteine	1.36	1.2	−12%
Methionine	0.55	1.07	+94%

*Modified and reprinted with permission from Molvig *et al.*, 1997.
**In the non-transgenic and transgenic lupin seeds, the protein levels were 34.3 and 35.8 (% DW), respectively.

individual shoots were transferred to B_5 medium without phosphinothricin but containing 1 mg/L indolebutyric acid to induce root growth. Regenerated plantlets were transferred to soil in the glasshouse with an inverted polycarbonate jar over the plant for the first few weeks until they became acclimated to the glasshouse conditions.

The SSA gene was under the control of a seed-specific promoter and the new protein was expressed at a level of 5% of the extractable protein in transgenic seeds. The expression of the SSA gene in lupin seeds at this level resulted in a 94% increase in methionine content accompanied by a 12% reduction in cysteine content (Table 1). This translated into a 19% increase in sulfur amino acid levels in grain protein.

The expression of this transgene has been stable through seven generations, including four generations in the field. The effect of the increased methionine level in transgenic lupin seed has been evaluated in feeding trials with rats, chickens, sheep, gilthead seabream (*Sparus aurata*) and black tiger prawn (*Penaeus monodon*).

2.2. RAT FEEDING TRIALS WITH TRANSGENIC, HIGH METHIONINE LUPIN SEED

Small-scale rat feeding trials tested the effect of elevated methionine levels in transgenic lupin seeds on animal nutrition (Molvig *et al.*, 1997). Two groups of rats were fed a diet balanced for N content containing 22.6% non-transgenic or transgenic lupin seed meal, a nitrogen-free mixture and vitamin-mineral mixture. The transgenic diet appeared to be more palatable to the rats; at the end of each day, an average of 45% of the non-transgenic diet was left uneaten, whereas only 21% remained in the case of the transgenic diet. The biological value and true protein digestibility were both increased in the transgenic lupin. Therefore, net protein utilisation, a derived value, was also increased. As expected, there was no change in digestible energy (Table 2). Significant increases in biological value and in net protein utilisation have also been observed in earlier rat feeding trials when lupin diets were supplemented with synthetic methionine (Eggum *et al.*, 1993) although true protein digestibility was not affected by methionine supplementation in that study.

2.3. AVIAN DIETS CAN ALSO BE ENHANCED WITH TRANSGENIC LUPIN SEED

In the poultry industry, feed mixes are routinely supplemented with synthetic methionine to compensate for the low level of the sulfur-containing amino acids in feed ingredients.

*Table 2. Nutritional evaluation of non-transgenic and transgenic lupin seeds in a rat feeding trial**

	True protein digestibility (%)	Biological value (%)	Net protein utilisation (%)	Digestible energy (%)
Non-transgenic lupin	89.39	63.18	56.37	77.74
Transgenic lupin	95.66	73.00	69.78	78.40
% Change	7.0	15.5	23.8	0.85
Significance	P = 0.03	P = 0.03	P = 0.002	NS

*Modified and reprinted with permission from Molvig *et al.*, 1997.

*Table 3. Nutritional evaluation of non-transgenic and transgenic lupin seeds compared to a basal diet in a chicken feeding trial (Day 6 to 20 post hatching). Synthetic methionine was added to meet the recommended level required for young broiler chickens. Differences in weight gain, feed intake and feed/gain were not significant ($P > 0.05$)**

Diet	Added methionine (g/kg diet)	Weight gain (g)	Feed intake (g)	Feed/gain (g/g)
Maize/soybean	2.2	454	743	1.64
Non-transgenic lupin	2.8	442	803	1.82
Transgenic lupin	2.2	441	767	1.74

*Modified and reprinted with permission from Ravindran *et al.*, 1998.

Transgenic grain legumes with elevated levels of methionine would reduce the addition of this essential amino acid to poultry diets. Chicken feeding trials were conducted with non-transgenic and transgenic lupin seeds that were incorporated at a level of 25% into a maize-soybean meal diet and compared to a non-lupin supplemented maize-soybean meal diet (Ravindran *et al.*, 1998). All diets were balanced to contain similar levels of apparent metabolisable energy (AME), protein, lysine and sulfur-containing amino acids (SAA). Synthetic methionine was added to all diets to meet the SAA requirements for broiler starter chickens. The diet containing the transgenic lupin needed less supplemental methionine than the non-transgenic lupin diet.

The production response of the chickens fed the three diets was similar. Weight gain and feed intake were not influenced by dietary treatments, but the feed conversion ratio (feed/gain) tended to be higher in chickens fed lupin diets compared to those on the control diet (Table 3). Feed conversion ratio of the birds fed the non-transgenic lupin diet were numerically higher than those fed the trangenic lupin diet. In a follow-up trial, the AME of non-transgenic and trangenic lupins was determined to be 9.42 and 10.18 MJ/kg dry matter, respectively. The higher AME values, and the lower feed conversion ratios observed in the feeding trial, with trangenic lupin may be related to their lower content of total non-starch polysaccharides (369 vs 417 g/kg dry matter).

Methionine supplementation of poultry diets containing 25% lupin can be lowered by 0.6 g/kg diet (600 g/ton) when using high methionine transgenic lupin. This represents

*Table 4. Gains in wool growth and liveweight show improved feed value of transgenic lupin seed compared to non-transgenic lupin seed in a sheepfeeding trial**

Diet: 65% hay+	Wool growth ($g/cm^2/day$)	Liveweight gain (g/day)
35% Non-transgenic lupin	0.74	90
35% Transgenic lupin	0.80	96
Difference	+8%	+7%
Significance	$P < 0.001$	$P < 0.05$

*Modified and reprinted with permission from White *et al.*, 2001.

a 21% reduction in the addition of synthetic methionine and therefore a cost saving and a nutritional benefit in the use of these seeds. There may also be a cost saving in substituting lupin seed for the more expensive soybeans in poultry diets.

2.4. INCREASED WOOL GROWTH AND LIVE WEIGHT GAIN IN RUMINANT ANIMALS ON A TRANSGENIC LUPIN DIET

The supply of sulfur-containing amino acids in the diet is a major limiting factor for wool growth in sheep (Reis, 1979). To test the effect of increased levels of methionine in transgenic lupin seeds, sheep feeding trials were conducted where non-transgenic or transgenic lupin seeds were incorporated into a hay diet at a level of 35% w/w (White *et al.*, 2000). For the first 6 weeks of the trial all the sheep were fed the same hay/lupin diet containing control lupins. The sheep were then divided into two groups and each group fed a restricted amount per day. This consisted of either non-transgenic or transgenic lupin seed incorporated into the hay diet. Sheep ate all they were offered, so intake was constant for the two diets. Measurements of wool growth and liveweight gain were taken at two 3-week intervals. Over the period of the 6-week trial, a significant gain of 8% in the growth of wool and 7% in live weight gain was detected in sheep fed the transgenic lupin seed (Table 4). Since feed intakes were constant, the improvements in wool and body growth were due to an increased efficiency of feed conversion.

The additional methionine in transgenic lupin seed is rumen-resistant (McNabb *et al.*, 1994) and it has now been shown to be a good source of protected amino acids for ruminant animals, such as sheep.

3. Conclusions and future prospects

Grain legumes are an important, protein-rich feed source, but their deficiency in sulfur amino acids for animal diets has limited their widespread use. The results of animal feeding trials have shown that it is possible to improve the nutritive value of a feed grain using gene technology. Sunflower seed albumin is a source of high levels of sulfur amino acids and expression of the transgene resulted in a doubling of the lupin seed methionine content. This level of methionine was sufficient to yield significant improvements in growth rate in rats, chickens and sheep when transgenic lupin meal was included at levels ranging from 20 to 40% of the diet. It was also noted in the rat trial that consumption of transgenic lupin seeds was greater than non-transgenic seeds suggesting improved palatability. With the use of high methionine transgenic lupin seed in animal diets there would be a cost saving by reduced need for addition of synthetic methionine. In some animal diets, such as those for aquaculture, there could also be a cost saving from the use of lupin seeds as they are cheaper than other protein sources, e.g. soybeans or fishmeal. In aquaculture systems it was shown (Molvig *et al.*, 1999) that there was no significant benefit of the elevated methionine levels in the transgenic lupin seed diets at the levels of dietary inclusion. Higher proportions of lupin seed in the diet may be able to test the efficacy of the high methionine lupin seed although inclusion levels of lupin will be limited by their fibre content.

The *bar* gene was initially used to select transgenic lupin, but other selectable markers will need to be found because of freedom to operate issues and consumer reluctance to accept herbicide tolerance genes in genetically modified crops. The neomycin phosphotransferase II gene, which confers resistance to aminoglycoside antibiotics such as kanamycin and paromomycin, has been successfully used to select transformed plants in tissue culture. The current lupin transformation procedure was modified using this selection system. Antibiotic resistance is thus a viable strategy for selecting high methionine transgenic lupin, but some consumers also have objections to antibiotic-resistance genes in feeds and foods. Further research is therefore required to produce a high methionine transgenic lupin line for commercialisation that does not also contain any selectable marker gene. Although lupin seeds are rarely, if ever, used as a human food, there is interest in developing this grain legume for human consumption. The safety of transgenic lupin seed containing the sunflower albumin is being assessed in various ways using animal feeding trials and biochemical and chemical analyses prior to release as a product.

The improvement of seed protein quality is not the only characteristic being modified in lupin by the techniques of gene transformation. Fungal and viral diseases are a major threat to both yield and grain quality of lupin. Genes for resistance to these diseases are being transformed into lupin (Hamblin *et al.*, 1998). Constructs containing genes for resistance to bean yellow mosaic virus (BYMV) and cucumber mosaic virus (CMV) have been individually inserted into transgenic lupin. Suitable fungal resistance genes for transformation into lupin are currently being sought.

References

Altenbach S B, Kuo C C, Staraci L C, Pearson K W, Wainwright C, Georgescu A and Townsend J (1992) Accumulation of a Brazil nut albumin in seeds of transgenic canola results in enhanced levels of seed protein methionine. *Plant Mol. Biol.*, **18**: 235–246.

Eggum B O, Tomes G, Beames R M and Datta F U (1993) Protein and energy evaluation with rats of seed from 11 lupin cultivars. *Anim. Feed Sci. Technol.*, **43**: 109–119.

Falco S C, Guida T, Locke M, Mauvais J, Sanders C, Ward R T and Webber P (1995) Transgenic canola and soybean seeds with increased lysine. *Bio/Technology*, **13**: 577–582.

Gamborg O L, Miller R A and Ojima K (1968) Nutrient requirements of suspension cultures of soybean root cells. *Exp. Cell. Res.*, **50**: 151–158.

Hamblin J, Barton J, Atkins C, Jones M, Smith P, Wylie S, Schroeder H, Molvig L, Tabe L, Higgins T J V (1998) The development and status of transgenic pulses in Australia. In: *3rd European conference on grain legumes. Opportunities for high quality, healthy and added-value crops to meet European demands*, 70–71.

McNabb W C, Spencer D S, Higgins T J and Barry T N (1994) *In vitro* rates of rumen proteolysis of ribulose-1, 5-biphosphate carboxylase (rubisco) from lucerne leaves, and of ovalbumin, vicilin and sunflower seed albumin 8 storage proteins. *J. Sci. Food Agric.*, **64**: 53–61.

Molvig L, Tabe L M, Eggum B O, Moore A E, Craig S, Spencer D and Higgins T J V (1997) Enhanced methionine levels and increased nutritive value of seeds of transgenic lupins (*Lupinus angustifolius* L.) expressing a sunflower seed albumin gene. *Proc. Natl. Acad. Sci. USA*, **94**: 8393–8398.

Molvig L, Tabe L M, Hamblin J, Ravindran V, Bryden W L, Smith D M, Kissil G W M, Lupatsch I, White C L and Higgins T J V (1999) Increased nutritional quality of seeds of transgenic lupins (*Lupinus angustifolius* L.) expressing a sunflower seed albumin gene. In: *Plant tissue culture at the edge of the new millennium*, Proceedings of the International Association for Plant Tissue Culture and Biotechnology (Australian Branch), VIth National Meeting, 109–114.

Murashige T and Skoog F (1962) A revised medium for rapid growth and bioassays with tobacco tissue cultures. *Physiol. Plant.*, **15**: 473–497.

Pigeare A, Abernethy D, Smith P M, Simpson K, Fletcher N, Lu C Y, Atkins C A and Cornish E (1997) Transformation of a grain legume (*Lupinus angustifolius* L.) via *Agrobacterium tumefaciens*-mediated gene transfer to shoot apices. *Mol. Breed.*, **3**: 341–349.

Ravindran V, Bryden W L, Tabe L M, Higgins T J V and Molvig L (1998) Evaluation of high-methionine transgenic lupins in broiler diets. *Proc. Aust. Poult. Sci. Symp.*, **10**: 208.

Reis P J (1979) Effects of amino acids on the growth and properties of wool. In: *Physiological and environmental limitations to wool growth* (Eds Black J L and Reis P J), University of New England Publishing Unit, Australia, 223–242.

Saalbach I, Pickardt T, Waddell D R, Hillmer S, Scheider O and Muntz K (1995) The sulphur-rich Brazil nut 2S albumin is specifically formed in transgenic seeds of the grain legume *Vicia narbonensis*. *Euphytica*, **85**: 181–192.

Shetty K, Asano Y and Oosawa K (1992) Stimulation of *in vitro* shoot organogenesis in *Glycine max* (Merrill.) by allantoin and amides. *Plant Sci.*, **81**: 245–251.

White C L, Tabe L M, Dove H, Hamblin J, Young P, Phillips N, Taylor R, Gulati S, Ashes J and Higgins T J V (2001) Increased efficiency of wool growth and liveweight gain in Merino sheep fed transgenic lupin seed containing sunflower albumin. *J. Sci. Food. Agric.*, **81**: 147–154.

PART II
FORAGE LEGUMES

GENETIC TRANSFORMATION OF *MEDICAGO* SPECIES

RAY J. ROSE*, KIM E. NOLAN AND CHEN NIU
School of Biological and Chemical Sciences, The University of Newcastle NSW 2308, Australia
**e-mail: birjr@alinga.newcastle.edu.au*

Abstract

The *Medicago* genus includes the world's most widely grown forage legume, the perennial *Medicago sativa*, as well as annual pasture legumes such as the model legume *Medicago truncatula*. Transformation of all *Medicago* species, until quite recently, was dependent on a high frequency of regeneration by somatic embryogenesis. Obtaining high rates of regenerability and the use of *Agrobacterium tumefaciens* have been the keys to transformation. High rates of regenerability have been obtained through identification of highly regenerable genotypes, recurrent selection and cycling through tissue culture. High regenerability can be transferred to suitable agronomic lines by traditional breeding techniques. Selection of a suitably virulent *A. tumefaciens* will aid transformation, and the selected strain needs to be used with the most suitable antibiotic to prevent overgowth. Kanamycin, hygromycin and phosphinothricin are all suitable as selection agents. A new generation of selectable markers or a strategy to eliminate the marker transgene are desirable for future field releases to overcome negative perceptions in the public domain of antibiotic resistance and herbicide resistance markers. More recent *in planta* transformation techniques in *M. truncatula* have facilitated the production of large numbers of transgenics (e.g. insertional mutants) and the robustness and general applicability to the *Medicago* genus needs to be assessed. Transgenic strategies in *Medicago* have developed to the stage that defined genes can most likely be introduced into the cultivar of choice. The assessment of transgenes for agronomic performance is well advanced in *Medicago*. Genetic modification using cell fusion has also been successful in the *Medicago* genus. Asymmetric and highly asymmetric hybridisations are strategies that are technically feasible. Future developments in regenerability and transformation are likely to lead to simpler and more rapid strategies for the production of transgenics that can also be applied to other legumes. Research in the model legume *M. truncatula* will also contribute new genetic knowledge that can be applied directly to the *Medicago* genus.

1. Introduction

The genus *Medicago* includes diploid, tetraploid and polyploid, wild and cultivated species. Tetraploid *Medicago sativa* L. (alfalfa or lucerne) is one of the most widely cultivated

P. K. Jaiwal and R. P. Singh (eds.), Applied Genetics of Leguminosae Biotechnology, 223–237.

forage legumes in the world with an estimated worldwide area of 32 million ha (Michaud *et al.*, 1988). *M. sativa* (2n = 4x = 32) is a herbaceous perennial plant that is allogamous and insect pollinated. Annual *Medicagos* are also important agriculturally being grown as pasture species and as a source of nitrogen in Mediterranean areas, South Africa and Australia. In Australia, annual medics are found in more than 50 million ha of agricultural land (Crawford *et al.*, 1989). *Medicago truncatula* Gaertn. (barrel medic) is one of the major annual medic species. *M. truncatula* (2n = 2x = 16) is diploid, autogamous with a relatively small genome (9×10^8 bp per haploid genome) and has gained popularity as a model legume for molecular genetic analysis (Barker *et al.*, 1990; Cook, 1999).

The perennial *Medicagos* have historically been easier to transform than the annuals because of their better regenerability. This may mean that perenniality is associated with an easier propensity to regenerate (Scarpa *et al.*, 1993). Such statements have to be tempered with the knowledge that an understanding of the fundamental mechanisms of regenerability are lacking. However, given the current knowledge of media, screens across cultivars will find regenerability amongst a higher proportion of cultivars in perennials. The perennials *M. varia*, *M. coerulea* (Deak *et al.*, 1986) and *M. sativa* (Deak *et al.*, 1986; Shahin *et al.*, 1986) were first transformed in 1986 using *Agrobacterium tumefaciens*. In 1987, *M. sativa* was also transformed using *Agrobacterium rhizogenes* (Spano *et al.*, 1987; Sukhapinda *et al.*, 1987). The annual *M. truncatula* was first transformed in 1992 using *A. tumefaciens* and *A. rhizogenes* (Thomas *et al.*, 1992). Transformation of all *Medicago* species, until quite recently, was dependent on the high frequency of regeneration by somatic embryogenesis. Obtaining high rates of regenerability has been the key, together with the use of *A. tumefaciens*.

In reviewing and discussing the genetic transformation of *Medicago* species we will consider regenerability via somatic embryogenesis (Fig. 1A and 1B) and the utilisation of *A. tumefaciens* in tranformation, other transformation strategies, and some of the current applications of transformation. A brief consideration will also be given to genetic modification via cell fusion and the use of *M. truncatula* as a model legume.

2. Regenerability

2.1. PERENNIALS

In consideration of regeneration of perennials the main focus will be on *Medicago sativa* where most of the work has been done. *M. sativa* was first regenerated from tissue culture by Saunders and Bingham in 1972 and *M. sativa* tissue culture has been reviewed by Mgroginski and Karther (1984), Bingham *et al.* (1988) and McKersie and Brown (1996). Our emphasis will be on two aspects – the role of genotype and the hormonal control of somatic embryogenesis (Fig. 1B). Direct somatic embryogenesis is possible in *M. sativa* (Maheswaran and Williams, 1984) and *M. falcata* (Denchev *et al.*, 1993) but has not been used in transformation systems.

Somatic embryogenesis and plant regeneration have been reported in many other perennials – *M. coerulea* Less. and *M. glutinosa* Mars. (Arcioni *et al.*, 1982), *M. falcata* L. (Teoule, 1983), *M. arborea* L. (Mariotti *et al.*, 1984), *M. difalcata* Sinsk., *M. hemicycla*

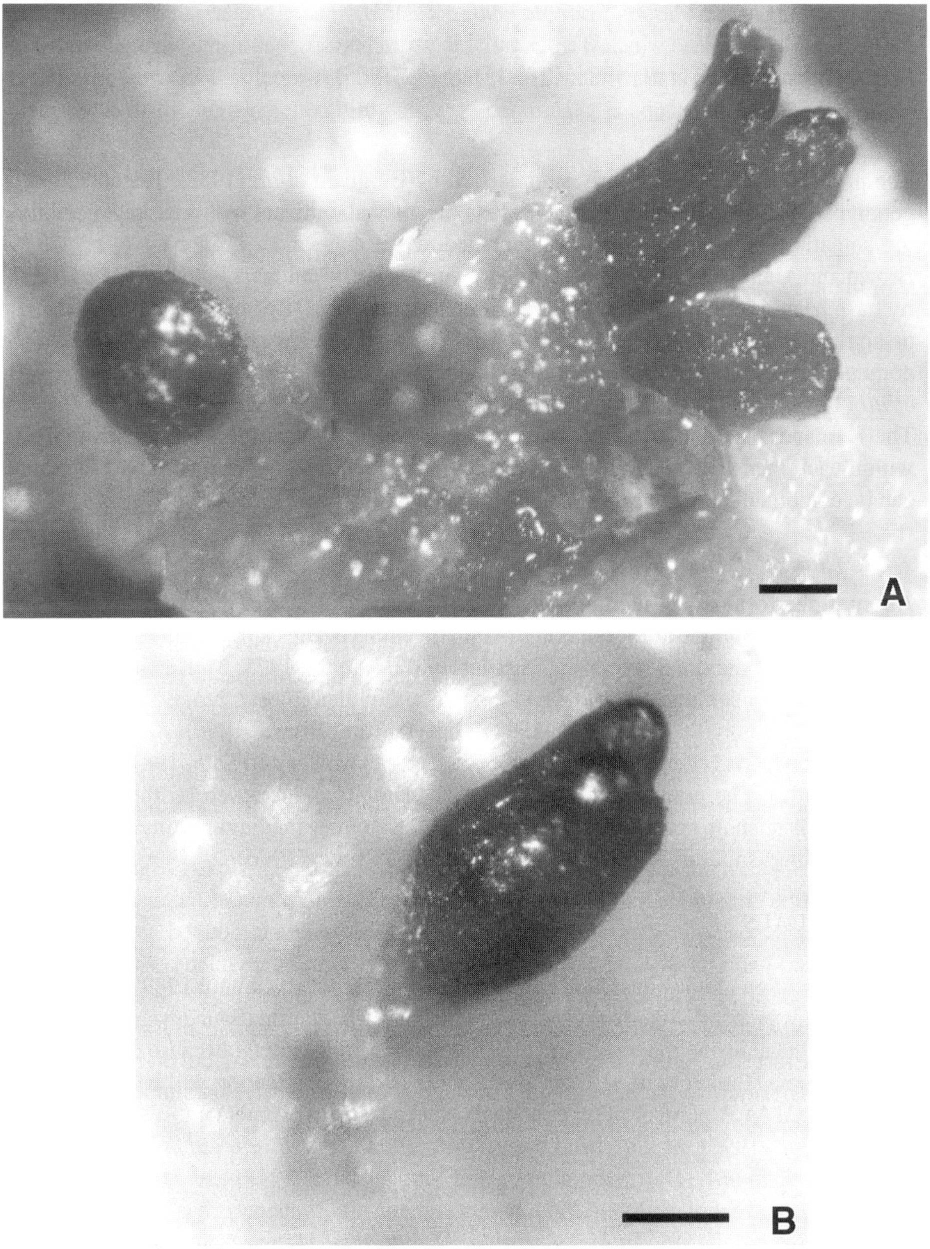

Figure 1. (A) Somatic embryos from M. truncatula *Jemalong 2HA tissue culture at different developmental stages, Bar = 1 mm (colour print of Fig. 1, Nolan et al., 1989). (B) Somatic embryo from* M. sativa *Regen S, Bar = 1 mm (J. T. Fitter and R. J. Rose, unpublished data).*

Table 1. Selected examples of Medicago sativa *and* Medicago truncatula *transformation methods*

Plant species	Transformation method	Explant	Marker	*Agrobacterium* killing	Reference
M. varia A2	*At* A281	Stem	NPTII	300 μg/mL Cb	Deak *et al.*, 1986
M. sativa	*Ar* A4	Stem	T-DNA and nopaline synthase	250 μg/mL Cf	Sukhapinda *et al.*, 1987
M. varia A2	*At* A281 LBA4404	Leaf, petiole	NPTII	300 μg/mL Cb	Chabaud *et al.*, 1988
M. sativa	*At* GV3101	Stem, leaf	NPTII	500 μg/mL Cb +500 μg/mL Cf	Kuchuk *et al.*, 1990
M. sativa	*At* LBA 4404	Stem	NPTII	1000 μg/mL Cb	Pezzotti *et al.*, 1991
M. sativa	*At* A281	Petiole	NPTII	500 μg/mL Cb	Du *et al.*, 1994
M. sativa	*At* LBA 4404	Leaf	NPTII and GUS	100 μg/mL Timentin	Oommen *et al.*, 1994
M. sativa	PB	Petiole, and stem calli	NPTII	Not applicable	Pereira and Erickson, 1995
M. sativa	PB	Pollen	NPTII and GUS	Not applicable	Ramaiah and Skinner, 1997
M. sativa	*At* C58C1 RifpMP90	Petiole	NPTII	500 μg/mL Claforan	McKersie *et al.*, 1999
M. sativa	*At* GV 3101	Petiole, leaf	NPTII and GUS	400 μg/mL Cb	Stotz and Long, 1999
M. sativa ssp. falcata and *M. truncatula*	*At* EHA 105, GV3101	Leaf	NPTII or HPT and GUS	400 μg/mL Aug or 500 μg/mL Ticarpin	Trinh *et al.*, 1998
M. truncatula	*At* EHA101	Leaf	NPTII	250 μg/mL Cb +750 μg/mL Va	Thomas *et al.*, 1992
	Ar R1022	Leaf	NPTII	500 μg/mL Cb	
M. truncatula	*At* LBA 4404	Leaf, petiole	NPTII and GUS	500 μg/mL Cb	Chabaud *et al.*, 1996
M. truncatula	*At* A281, GV 2260	Leaf, petiole, hypocotyl, root	NPTII or HPT	400 μg/mL Aug	Hoffman *et al.*, 1997
M. truncatula	*At* LBA 4404	Cotyledon + 1–2 mm embryonic axis	PAT	500 μg/mL Cb	Trieu and Harrison, 1996
M. truncatula	*At* LBA 4404	Leaf	NPTII and GUS	300 μg/mL Cb	Wang *et al.*, 1996
M. truncatula	*At* ASE1, EHA105, GV3101	Seedling, flowers *In planta* infiltration	PAT	Not applicable (transferred to pots)	Trieu *et al.*, 2000

Abbreviations: *Agrobacterium tumefaciens* (*At*), *Agrobacterium rhizogenes* (*Ar*), Augmentin (Aug), Carbenicillin (Cb), Cefotaxime (Cf), Vancomycin (Va), β-glucuronidase (GUS), Neomycin phosphotransferase (NPT II), Phosphinothricin acetyl transferase (PAT), Hygromycin phosphotransferase (HPT), Electroporation (EP), Particle bombardment (PB), *M. varia* A2 = *M. sativa* × *M. falcata*.

Agrobacterium strains (Desgagnes *et al.*, 1995; Samac, 1995). Other parameters such as co-cultivation period and the antibiotic type and concentration used for *Agrobacterium* killing need to be optimised. The time to obtain transformation, from inoculation to transfer to soil, can be as short as 3–4 months (Trinh *et al.*, 1998). Du *et al.* (1994) reported that those plants regenerated after a prolonged selection phase *in vitro* generally had increased copy number and integration sites of inserts.

3.2. ANNUALS

The first transformation of annuals (Thomas *et al.*, 1992) with *Agrobacterium tumefaciens* used the *Agrobacterium* strain EHA101 and the binary vector pMT2. Neomycin phosphotransferase was the selectable marker and *A. tumefaciens* was killed with carbenicillin. In subsequent studies we have found timentin and augmentin (Jayasena *et al.*, 2001) to be more effective than carbenicillin (Wang *et al.*, 1996) in preventing *Agrobacterium* overgrowth. Subsequent transformation studies have used a range of disarmed *A. tumefaciens* strains – EHA105, LBA4404, GV2260, GV3101 (Table I). GUS (Chabaud *et al.*, 1996; Wang *et al.*, 1996) and the green fluorescent protein (Kamate *et al.*, 2000) have been used as reporter genes. While somatic embryogenesis has worked well as a transformation procedure and is quite robust (Thomas *et al.*, 1992; Wang *et al.*, 1996; Chabaud *et al.*, 1996; Hoffman *et al.*, 1997; Trinh *et al.*, 1998). Trieu and Harrison (1996) have successfully used organogenesis from cotyledon explants. The key to the transformation of *M. truncatula* via somatic embryogenesis has been the development of high regenerators (Nolan *et al.*, 1989; Chabaud *et al.*, 1996; Hoffman *et al.*, 1997) and the principles outlined above for *M. sativa* transformation also apply to *M. truncatula*. A range of explants from the vegetative part of the plant have been used for somatic embryogenesis and transformation (Table I) and more recently Kamate *et al.* (2000) have obtained high rates of embryogenesis from all parts of flowers. As for *M. sativa* the time to obtain transformation, from inoculation to transfer to soil, can be as short as 3–4 months (Trinh *et al.*, 1998).

The Trieu and Harrison procedure (1996) relies on organogenetic explants and strong selection pressure using phosphinothricin acetyl transferase. The explants are prepared from 3 day old seedlings, with each explant consisting of one cotyledon and 1–2 mm of split embryonic axis. Shoots developed from the epidermal cell layer adjacent to the primary apical meristem and embryonic axis. The transformation procedure used LBA4404 and the binary plasmid pSLJ525. Transformed plants suitable for transfer into soil are produced in approximately 2.5 months. This method has the potential to be cultivar independent. In our unpublished studies, high regenerators such as Jemalong 2HA work better in producing more transgenic shoots.

The *Agrobacterium tumefaciens* strategies in *Medicago* are basically the universal ones and developments in the binary vector systems and promoter strategies will continue. There is a public perception that herbicide resistance has the potential for negative environmental consequences and that antibiotic resistance has the potential for negative health effects (Puchta, 2000). To eliminate the use of these markers in the field there are strategies to either use markers that relate to plant metabolism, co-transformation and segregation or excision of the selectable marker genes (Puchta, 2000). These strategies need to be developed in *Medicago*.

4. Other transformation procedures

4.1. *AGROBACTERIUM RHIZOGENES*

A. rhizogenes has been used to transform *Medicago sativa* L. (Spano *et al.*, 1987; Sukhapinda *et al.*, 1987), *Medicago arborea* L. (Damiani and Arcioni, 1991) and also *M. truncatula* (Thomas *et al.*, 1992) but its use has not been seriously developed for transformation in *Medicago* species. Though the Ri plasmid can be used as a helper plasmid (Sukhapinda *et al.*, 1987), there has been more of an interest in the effect of the *rol* genes on morphology, where it caused an increase in stem number per plant in *M. sativa* cv. Rangelander (Frugis *et al.*, 1995) and an extensive shallow root system.

4.2. ELECTROPORATION

Kuchuk *et al.* (1990) have carried out electroporation of *Medicago borealis* protoplasts using the plasmid pGA472. Kanamycin resistant plants were regenerated. Electroporation has not been widely used in *Medicago* given that protoplast formation is required and *Agrobacteriun* methods have been successful. Both *M. sativa* (Kao and Michayluck, 1980; Johnson *et al.*, 1981) and *M. truncatula* (Rose and Nolan, 1995) can be regenerated from protoplasts.

4.3. PARTICLE BOMBARDMENT

Stable transformation of *M. sativa* has been obtained by particle bombardment of calli derived from petiole and stem sections (Pereira and Erickson, 1995). This method has however not become widely used and *A. tumefaciens*-mediated transformation has become the method of choice in most laboratories. An interesting application of particle bombardment has been its use in the bombardment of pollen of *M. sativa* by Ramaiah and Skinner (1997) with the pBI121 plasmid. This pollen was successfully used to produce transgenic plants by pollination. However there was loss of integrated DNA after several vegetative generations. This is a promising avenue for future research given its application to all cultivars.

4.4. *IN PLANTA* TRANSFORMATION METHODS USING VACUUM INFILTRATION

There are two recent *in planta* transformation methods that have been published for *Medicago truncatula* which show much promise (Trieu *et al.*, 2000). The methods involve vacuum infiltration of flowering plants and vacuum infiltration of seedlings. Technically the methods are very simple and can produce large numbers of transformants. Even though the percentage of transformation is mainly less than 10%, the selection system based on phosphinothricin resistance can produce large numbers of resistant seedlings. The selection system makes use of the *bar* gene, kanamycin resistance does not work using this technique.

The flowering method (Trieu *et al.*, 2000) makes use of Chabaud and Journet's vernalisation procedure (Trieu *et al.*, 2000), where seeds are imbibed and incubated at 4°C for two weeks, during which time they germinate. The resulting plants then flower in 3–4 weeks while they are still relatively small. When plants have flower buds and a few open flowers they are submerged in a suspension of *Agrobacterium* under a vacuum of 25 inches of mercury. Seeds are ultimately collected from the infiltrated plants and seedlings selected with phosphinothricin (PPT).

The initial stages of the seedling method (Trieu *et al.*, 2000) is similar to the flower transformation method. Seeds are imbibed at 4°C for two weeks and then vacuum infiltrated with *Agrobacterium*. After infiltration under a vacuum of 25 inches of mercury the seedlings are incubated on a co-cultivation medium, planted and allowed to set seed in 3–4 weeks. Seeds are selected with PPT.

Trieu *et al.* (2000) point out that both *in planta* transformation procedures work with a range of binary vetors, as long as the T-DNA contains the *bar* gene. *Agrobacterium* strains EHA105, ASE1 and GV3101 were used successfully, but surprisingly LBA4404, a strain used widely in *Medicago* transformation was unsuccessful. LBA4404 is recognised as having relatively low virulence (Trieu and Harrison, 1996).

5. Applications of transformation in *Medicago*

5.1. AGRONOMIC CHARACTERS

Table 2 gives examples of the transformation of *M. sativa* for genes connected to important agronomic characters. In all the listed examples there were useful responses to the introduced transgenes. These studies indicate that transgenesis in *M. sativa* can be readily applied to plant improvement utilising the basic transformation strategies considered in this chapter. The main limiting factor is that the genotype of choice is not always regenerable. This can be overcome by backcrossing, as transgenic alfalfa has been successfully backcrossed to cultivars of choice (Micaleff *et al.*, 1995). In this latter study three backcrosses were considered optimal, and it was indicated that backcrossing using a dominant marker required only as much time as the original transformation experiment. Alfalfa mosaic virus resistance has also been obtained with transgenic *M. truncatula* transformed with the AMV coat protein (Jayasena *et al.*, 2001).

Table 2. Examples of successful transformations of Medicago sativa *for agronomic traits*

Characteristic	Transgene	Reference
Winter-hardiness	Superoxide dismutase	McKersie *et al.*, 1993, 1999
Resistance to phytophthora disease	β-1,3-glucanase	Masoud *et al.*, 1996
Resistance to thrip predation	Proteinase inhibitor	Thomas *et al.*, 1994
Resistance to AMV infection	AMV coat protein	Halk *et al.*, 1989
High sulphur containing protein	Chicken ovalbumin gene	Schroeder *et al.*, 1991

5.2. STUDY OF THE RHIZOBIUM–LEGUME SYMBIOSIS

Transgenic plants have been produced to aid in the understanding of the *Rhizobium*–plant interaction. Chabaud *et al.* (1996) introduced into *M. truncatula* the early nodulin MtENOD12 promoter and the GUS reporter gene. Charon *et al.* (1999) have overexpressed *enod40* in *M. truncatula*. Bauchrowitz *et al.* (1996) introduced promoter – *gusA* fusions for each of three lectin genes into *Medicago varia*. Large scale insertional mutagenesis is under way in *M. truncatula* (Trieu *et al.*, 2000). Transformation is now part of the technical strategies that will enable progress in understanding fundamental legume biology.

6. Genetic modifications using cell fusion

In the *Medicago* genus, successful somatic hybridisation has been possible (Pupilli *et al.*, 1992), including hybridisation between sexually incompatible *Medicago* species (Table 3) such as *M. sativa* + *M. arborea* (Nenz *et al.*, 1996). Asymmetric hybridisation has been

Table 3. Somatic hybrid plants obtained in the genus Medicago *using either* Medicago sativa *or* Medicago truncatula *as fusion partners*

Recipient species	Donor species	Hybridity (chromosome number)	Fertility	Reference
M. sativa (2n = 4x = 32)	M. *falcata* (2n = 4x = 32)	SH ASH (54–63)	Fertile[a]	Teoule, 1983, Mendis *et al.*, 1991
M. sativa (2n = 4x = 32)	*M. coerulea* (2n = 2x = 16)	SH ASH (45–50)	Fertile	Pupilli *et al.*, 1992
M. sativa (2n = 4x = 32)	*Onobrychis viciifolia* cv 'Othello' (2n = 48)	SH ASH HASH (32, 30–78)	Fertile	Li *et al.*, 1993
M. sativa (2n = 4x = 32)	*M. arborea* (2n = 4x = 32)	ASH (56–57)	Fertile[a]	Nenz *et al.*, 1996
M. sativa (2n = 4x = 32)	*M. falcata* (2n = 2x = 16)	ASH HASH (2n = 33)	Fertile[a]	Crea *et al.*, 1997
M. sativa (2n = 4x = 32)	*Onobrychis viciaefolia* Scop (2n = 28)	ASH HASH (30–60)	—	Xu and Jia, 1997
M. truncatula (2n = 2x = 16)	*M. scutellata* (2n = 30)	ASH HASH (16–26)	Fertile	Tian and Rose, 1999

[a]Hybrid flowered, no information on seed production, – No fertility information, ASH – Asymmetric somatic hybrid; HASH – Highly asymmetric somatic hybrid; SH – Symmetric somatic hybrid.

possible between *M. sativa* and *Onobrychis viciaefolia* (Li *et al.*, 1993) and the annuals *M. truncatula* and *M. scutellata* (Tian and Rose, 1999). Asymmetric somatic hybridisation involves transfer of donor chromosomes, or in the case of highly asymmetric somatic hybridisation only a few donor chromosomes (Hinnisdaels *et al.*, 1994) or pieces of donor DNA without chromosome transfer where the introduced DNA can be detected by molecular techniques (Tian and Rose, 1999). Highly asymmetric somatic hybridisation without chromosome transfer (Tian and Rose, 1999) has analogies with transformation except there are larger amounts of DNA transferred and the function of the introduced DNA can only be deduced by the use of selection agents, e.g. sensitivity to pests or pathogens. Thomas *et al.* (1990) produced infertile plantlets from a somatic hybridisation between the annual *M. intertexta* and *M. sativa* in an attempt to transfer insect resistance from an annual to a perennial.

7. The model legume *M. truncatula*

In 1990 *M. truncatula* was proposed as a model legume (Barker *et al.*, 1990) to study the molecular genetics of legumes, because of it being diploid, autogamous and having a relatively small genome. At that time a transformation system was a limiting factor, but since that time transformation systems have developed rapidly (Thomas *et al.*, 1992; Chabaud *et al.*, 1996; Wang *et al.*, 1996; Trieu *et al.*, 1996; Hoffman *et al.*, 1997; Trinh *et al.*, 1998; Trieu *et al.*, 2000) and mutants are becoming available (Cook, 1999). Further, with the development of large scale insertional mutagenesis (Trieu *et al.*, 2000) and tools for functional genomics such as microarrays and proteomics rapid progress can be expected in *Medicago* and legume biology.

8. Conclusions and future developments

As discussed in this chapter, transformation is now readily achievable in the *Medicago* genus and is utilised extensively to produce transgenics for the introduction of genes of agronomic importance and to study legume biology. The most successful systems are *Agrobacterium tumefaciens*-based. Genetic manipulation using cell fusion is also possible in this genus.

An important question is how do you produce transgenics in cultivars that have yet to be transformed? A proven way is to transform a cultivar that is amenable to transformation and to backcross the gene into the cultivar of interest. *In planta* transformation methods need to be explored further, particularly with low regenerating strains and species. The *Medicago* genus also has useful systems to explore the molecular basis of regenerability, in both *M. sativa* (Gyorgyey *et al.*, 1991) and our favoured *M. truncatula* system using Jemalong and Jemalong 2HA (Rose *et al.*, 1999). Ultimately a given transformed cell may be able to be induced to regenerate.

Future developments in regenerability and transformation are likely to lead to simpler and more rapid strategies for the production of transgenics, that can also be applied to other legumes. Research in the model legume *M. truncatula* will also contribute new genetic knowledge that can be applied directly to *Medicago*.

References

Arcioni S, Davey M R, Dos Santos A V P and Cocking E C (1982) Somatic embryogenesis in tissues from mesophyll and cell suspension protoplasts of *Medicago coerula* and *M. glutinosa*. *Z. Pflanzenphysiol.*, **106**: 105–110.

Barker D G, Bianchi S, Blondon F, Dattee Y, Duc G, Essad S, Flament P, Gallusci P, Genier G, Guy P, Muel X, Tourneur J, Denarie J and Huguet T (1990) *Medicago truncatula*, a model plant for studying the molecular genetics of the *Rhizobium*-legume symbiosis. *Plant Mol. Biol. Rep.*, **8**: 40–49.

Bauchrowitz M A, Barker G and Truchet G (1996) Lectin genes are expressed throughout root nodule development and during nitrogen fixation in the *Rhizobium–Medicago* symbiosis. *Plant J.*, **9**: 31–43.

Bingham E T, Hurley L V, Kaatz D M and Saunders J W (1975) Breeding alfalfa which regenerates from callus tissue in culture. *Crop Sci.*, **15**: 719–721.

Bingham E T, McCoy T J and Walker K A (1988) Alfalfa tissue culture. In: *Alfalfa and Alfalfa Improvement*, Vol. 29, American Society of Agronomy, 903–929.

Bowley S R, Kielly G A, Anandarajah K, Mckersie B D and Senaratna T (1993) Field evaluation following two cycles of backcross transfer of somatic embryogenesis to commercial alfalfa germplasm. *Can. J. Plant Sci.*, **73**: 131–137.

Brown D C W and Atanassov A (1985) Role of genetic background in somatic embryogenesis in *Medicago. Plant Cell Tiss. Org. Cult.*, **4**: 111–132.

Chabaud M, Passiatore J E, Cannon F and Buchanan-Wollaston V (1988) Parameters affecting the frequency of kanamycin resistant alfalfa obtained by *Agrobacterium tumefaciens* mediated transformation. *Plant Cell Rep.*, **7**: 512–516.

Chabaud M, Larsonneau C, Marmouget C and Huguet T (1996) Transformation of barrel medic (*Medicago truncatula* Gaetrn.) by *Agrobacterium tumefaciens* and regeneration *via* somatic embryogenesis of transgenic plants with the *MtENOD12* nodulin promoter fused to the *gus* reporter gene. *Plant Cell Rep.*, **15**: 305–310.

Charon C, Sousa C, Crespi M and Kondorosi A (1999) Alteration of *enod40* expression modifies *Medicago truncatula* root nodule development induced by *Sinorhizobium meliloti*. *Plant Cell*, **11**: 1953–1965.

Christou P (1994) The biotechnology of crop legumes. *Euphytica*, **74**: 165–185.

Cook D R (1999) *Medicago truncatula* – a model in the making! *Curr. Opin. Plant Biol.*, **2**: 301–304.

Crawford E J, Lake A W H and Boyce K G (1989) Breeding annual *Medicago* species for semi-arid conditions in Southern Australia. *Adv. Agron.*, **42**: 399–437.

Crea F, Calderini O, Nenz E, Cluster P D, Damiani F and Arcioni S (1997) Chromosomal and molecular rearrangements in somatic hybrids between tetraploid *Medicago sativa* and diploid *Medicago falcata. Theor. Appl. Genet.*, **95**: 1112–1118.

Damiani F and Arcioni S (1991) Transformation of *Medicago arborea* L. with an *Agrobacterium rhizogenes* binary vector carrying the hygromycin resistance gene. *Plant Cell Rep.*, **10**: 300–303.

Deak M, Kiss G B, Koncz C, Dudits D (1986) Transformation of *Medicago* by *Agrobacterium*-mediated gene transfer. *Plant Cell Rep.*, **5**: 97–100.

Denchev P D, Kuklin A I, Atanassov A I and Scragg A H (1993) Kinetic studies of embryo development and nutrient utilization in an alfalfa direct somatic embryogenic system. *Plant Cell Tiss. Org. Cult.*, **33**: 67–73.

Desgagnes R, Laberge S, Allard G, Khoudi H, Castonguay Y, Lapointe J, Michaud R and Vezina L P (1995) Genetic transformation of commercial breeding lines of alfalfa (*Medicago sativa* L.). *Plant Cell Tiss. Org. Cult.*, **42**: 129–140.

Du S, Erickson L and Bowley S (1994) Effect of plant genotype on the transformation of cultivated alfalfa (*Medicago sativa*) by *Agrobacterium tumefaciens. Plant Cell Rep.*, **13**: 330–334.

Dudits D, Bogre L and Gyorgyey J (1991) Molecular and cellular approaches to the analysis of plant embryo development from somatic cells *in vitro. J. Cell Sci.*, **99**: 475–484.

Frugis G, Caretto S, Santini L and Mariotti D (1995) *Agrobacterium rhizogenes rol* genes induce productivity-related modifications in 'creeping-rooted' alfalfa types. *Plant Cell Rep.*, **14**: 488–492.

Gilmour D M, Davey M R and Cocking E C (1987) Plant regeneration from cotyledon protoplasts of wild *Medicago* species *Plant Sci.*, **48**: 107–112.

Gyorgyey J, Gartner A, Nemeth K, Magyar Z, Hirt H, Heberle-Bors E and Dudits D (1991) Alfalfa heat shock genes are differentially expressed during somatic embryogenesis. *Plant Mol. Biol.*, **16**: 999–1007.

Halk E L, Merlo D J, Liao L W, Jarvis N P, Nelson S E, Krahn K J, Hill K K, Rashka, K E and Loesch-Fries L S (1989) Resistance to alfalfa mosaic virus in transgenic tobacco and alfalfa. In: *Molecular Biology of Plant–Pathogen Interactions*, Vol. 101 (Eds Staskawicz B, Ahlquist O and Yoder P), Alan R Liss Inc., New York, 283–296.

Hernandez-Fernandez M M and Christie B R (1989) Inheritance of somatic embryogenesis in alfalfa (*Medicago sativa* L.). *Genome*, **32**: 318–321.

Hinnisdaels S, Jacobs M and Negrutiu I (1994) Asymmetric somatic hybrids. In: *Biotechnology in Agriculture and Forestry*. Vol. 27, *Somatic Hybridization in Crop Improvement I* (Ed. Bajaj Y P S), Springer-Verlag, Berlin, 57–71.

Hoffmann B, Trinh T H, Leung J, Kondorosi A and Kondorosi E (1997) A new *Medicago truncatula* line with superior *in vitro* regeneration, transformation, and symbiotic properties isolated through cell culture selection. *Mol. Plant Microbe Interact.*, **10**: 307–315.

Iantcheva A, Vlahova M, Bakalova E, Kondorosi E, Elliott M C and Atanassov A (1999) Regeneration of diploid annual medics via direct somatic embryogenesis promoted by thidiazuron and benzylaminopurine. *Plant Cell Rep.*, **18**: 904–910.

Ibragimova S S and Smolenskaya S E (1997) Plant regeneration from seedling apex in annual medics. *Acta Agronomica Hungarica*, **45**: 109–116.

Jayasena K W, Hajimorad M R, Law E G, Rehman A-U, Nolan K E, Zanker T, Rose R J and Randles J W (2001) Resistance to Alfalfa mosaic virus in transgenic barrel medic lines containing the virus coat protein gene. *Aust. J. Agri. Res.*, **52**: 67–72.

Johnson L B, Stuteville D L, Higgins R K and Skinner D Z (1981) Regeneration of alfalfa plants from protoplasts of selected Regen S clones. *Plant Sci. Lett.*, **20**: 297–304.

Kamate K, Rodriguez-Llorente I D, Scholte M, Durand P, Ratet P, Kondorosi E, Kondorosi, A and Trinh T H (2000) Transformation of floral organs with GFP in *Medicago truncatula. Plant Cell Rep.*, **19**: 647–653.

Kao K N and Michayluck M R (1980) Plant regeneration from mesophyll protoplasts of alfalfa. *Z. Planzenphysiol.*, **96**: 135–141.

Kielly G A and Bowley S R (1992) Genetic control of somatic embryogenesis in alfalfa. *Genome*, **35**: 474–477.

Kuchuk N, Komarnitski I, Shakhovsky A and Gleba Y (1990) Genetic transformation of *Medicago* species by *Agrobacterium tumefaciens* and electroporation of protoplasts. *Plant Cell Rep.*, **8**: 660–663.

Li X-Q and Demarly Y (1995) Characterization of factors affecting regeneration frequency of *Medicago lupulina* L. *Euphytica*, **86**: 143–148.

Li, X-Q and Demarly Y (1996) Somatic embryogenesis and plant regeneration in *Medicago suffruticosa. Plant Cell Tiss. Org. Cult.*, **44**: 79–81.

Li X-Q, Teoule E, Dattee Y, and Demarly Y (1986) Plant regeneration by *in vitro* culture of inflorescence axis of *Medicago lupulina* L. *C. R. Acad. Sci. Paris*, **303**: 601–606.

Li Y-G, Tanner G J, Delves A C and Larkin P J (1993) Asymmetric somatic hybrid plants between *Medicago sativa* L. (alfalfa, lucerne) and *Onobrychis viciifolia* Scop (sainfoin). *Theor. Appl. Genet.*, **87**: 455–463.

Maheswaran G and Williams E G (1984) Direct somatic embryoid formation on immature embryos of *Trifolium repens*, *T. pratense* and *Medicago sativa*, and rapid clonal propagation of *T. repens. Ann. Bot.*, **54**: 201–211.

Mariotti D, Arcioni S and Pezzotti M (1984) Regeneration of *Medicago arborea* L. plants from tissue and protoplast cultures of different organ origin. *Plant Sci. Lett.*, **37**: 149–156.

Masoud S A, Zhu Q, Lamb C and Dixon R A (1996) Constitutive expression of an inducible beta-1,3-glucanase in alfalfa reduces disease severity caused by the oomycete pathogen *Phytophthora megasperma* f sp *medicaginis*, but does not reduce disease severity of chitin-containing fungi. *Transgenic Res.*, **5**: 313–323.

McKersie B D and Brown D C W (1996) Somatic embryogenesis and artificial seeds in forage legumes. *Seed Science Res.*, **6**: 109–126.

McKersie B D, Bowley S R and Jones K S (1999) Winter survival of transgenic alfalfa overexpressing superoxide dismutase. *Plant Physiol.*, **119**: 839–847.

McKersie B D, Chen Y R, Debeus M, Bowley S R, Bowler C, Inze D, Dhalluin K and Botterman J (1993) Superoxide dismutase enhances tolerance of freezing stress in transgenic alfalfa (*Medicago sativa* L.). *Plant Physiol.*, **103**: 1155–1163.

Meijer E G M and Brown D C W (1985) Screening of diploid *Medicago sativa* germplasm for somatic embryogenesis. *Plant Cell Rep.*, **4**: 285–288.

Meijer E G M and Brown D C W (1987) A novel system for rapid high frequency somatic embryogenesis in *Medicago sativa. Physiol. Plant.*, **69**: 591–596.

Mendis M H, Power J B and Davey M R (1991) Somatic hybrids of the forage legumes *Medicago sativa* L. and *Medicago falcata* L. *J. Exp. Bot.*, **42**: 1565–1573.

Mgroginski L A and Karther K K (1984) Tissue culture of legumes for crop improvement. *Plant Breeding Rev.*, **2**: 215–263.

Micaleff M C, Austin S and Bingham E T (1995) Improvement of transgenic alfalfa by backcrossing. *In Vitro Cell. Dev. Biol. Plant.*, **31**: 187–192.

Michaud E, Leman W F and Rumbaugh M D (1988) World distribution and historical development. In: *Alfalfa and Alfalfa Improvement*, Vol. 29, American Society of Agronomy, 25–91.

Nagarajan P, McKenzie J S and Walton P D (1986) Embryogenesis and plant regeneration of *Medicago* species in tissue culture. *Plant Cell Rep.*, **5**: 77–80.

Nenz E, Pupilli F, Damiani F, and Arcionim S (1996) Somatic hybrid plants between the forage legumes *Medicago sativa* L. and *Medicago arborea* L. *Theor. Appl. Genet.*, **93**: 183–189.

Ninkovic S, Miljusdjuka J and Neskovic M (1995) Genetic transformation of alfalfa somatic embryos and their clonal propagation through repetitive somatic embryogenesis. *Plant Cell Tiss. Org. Cult.*, **42**: 255–260.

Nolan K E, Rose R J and Gorst J G (1989) Regeneration of *Medicago truncatula* from tissue culture : increased somatic embryogenesis using explants from regenerated plants. *Plant Cell Rep.*, **8**: 278–281.

Nolan K E and Rose R J (1998) Plant regeneration from cultured *Medicago truncatula* with particular reference to abscisic acid and light treatments. *Aust. J. Bot.*, **46**: 151–160.

Oommen A, Dixon R A and Paiva N L (1994) The elicitor-inducible alfalfa isoflavone reductase promoter confers different patterns of developmental expression in homologous and heterologous transgenic plants. *Plant Cell*, **6**: 1789–1803.

Pereira L F and Erickson L (1995) Stable transformation of alfalfa (*Medicago sativa* L.) by particle bombardment. *Plant Cell Rep.*, **14**: 290–293.

Pezzotti M, Pupilli F, Damiani F and Arcioni S (1991) Transformation of *Medicago sativa* L. using a Ti plasmid derived vector. *Plant Breed.*, **106**: 39–46.

Puchta H (2000) Removing selectable marker genes: taking the shortcut. *Trends Plant Sci.*, **5**: 273–274.

Pupilli F, Scarpa G M, Damiani F and Arcioni S (1992) Production of interspecific somatic hybrid plants in the genus *Medicago* through protoplast fusion. *Theor. Appl. Genet.*, **84**: 792–797.

Ramaiah, S M and Skinner D Z (1997) Particle bombardment – a simple and efficient method of alfalfa (*Medicago sativa* L.) pollen transformation. *Curr. Sci.*, **73**: 674–682.

Reisch B and Bingham E T (1980) The genetic control of bud formation from callus cultures of diploid alfalfa. *Plant Sci. Lett.*, **20**: 71–77.

Rose R J and Nolan K E (1995) Regeneration of *Medicago truncatula* from protoplasts isolated from kanamycin-sensitive and kanamycin-resistant plants. *Plant Cell Rep.*, **14**: 349–354.

Rose R J, Nolan K E and Bicego L (1999) The development of the highly regenerable seed line Jemalong 2HA for transformation of *Medicago truncatula* – implications for regenerability via somatic embryogenesis. *J. Plant Physiol.*, **155**: 788–791.

Samac D A (1995) Strain specificity in transformation of alfalfa by *Agrobacterium tumefaciens. Plant Cell Tiss. Org. Cult.*, **43**: 271–277.

Saunders J and Bingham E T (1972) Production of alfalfa plants from callus tissue. *Crop Sci.*, **12**: 804–808.

Saunders J W and Bingham E T (1975) Growth regulator effects on bud initiation in callus cultures of *Medicago sativa. Amer. J. Bot.*, **62**: 850–855.

Scarpa G M, Pupilli F, Damiani F and Arcioni S (1993) Plant regeneration from callus and protoplasts in *Medicago polymorpha. Plant Cell Tiss. Org. Cult.*, **35**: 49–57.

Schroeder H E, Khan M R I, Knibb W R, Spencer D and Higgins T J V (1991) Expression of a chicken ovalbumin gene in three lucerne cultivars. *Aust. J. Plant Physiol.*, **18**: 495–505.

Shabnam S, Zafar Y and Malik K A (1996) Transformation of alfalfa (*Medicago sativa* L.) plants with GUS marker containing intron. *Pakistan J. Bot.*, **28**: 167–172.

Shahin E A, Spielmann A, Sukhapinda K, Simpson R B and Yashar M (1986) Transformation of cultivated alfalfa using disarmed *Agrobacterium tumefaciens. Crop Sci.*, **26**: 1235–1239.

Spano L, Mariotti D, Pezzotti M, Damiani F and Arcioni F (1987) Hairy root transformation in alfalfa (*Medicago sativa* L.). *Theor. Appl. Genet.*, **73**: 523–530.

Stotz H U and Long S R (1999) Expression of the pea (*Pisum sativum* L.) alpha-tubulin gene *TubA1* is correlated with cell division activity. *Plant Mol. Biol.*, **41**: 601–614.

Sukhapinda K, Spivey R and Shahin E A (1987) Ri-plasmid as a helper for introducing vector DNA into alfalfa plants. *Plant Mol. Biol.*, **8**: 209–216.

Teoule E (1983) Hybridization somatique entre *Medicago sativa* L. et *Medicago falcata* L. *C. R. Acad. Sci. Paris Serie III*, **297**: 13–16.

Thomas J C, Wasmann C C, Echt C, Dunn R L, Bohnert H J and McCoy T J (1994) Introduction and expression of an insect proteinase inhibitor in alfalfa (*Medicago sativa* L.). *Plant Cell Rep.*, **14**: 31–36.

Thomas M R, Johnson L B and White F F (1990) Selection of interspecific somatic hybrids of *Medicago* by using *Agrobacterium* – transformed tissues. *Plant Sci.*, **69**: 189–198.

Thomas M R, Rose R J and Nolan K E (1992) Genetic transformation of *Medicago truncatula* using *Agrobacterium* with genetically modified Ri and disarmed Ti plasmids. *Plant Cell Rep.*, **11**: 113–117.

Tian D and Rose R J (1999) Asymmetric somatic hybridisation between the annual legumes *Medicago truncatula* and *Medicago scutellata. Plant Cell Rep.*, **18**: 989–996.

Trieu A T and Harrison M J (1996) Rapid transformation of *Medicago truncatula. Plant Cell Rep.*, **16**: 6–11.

Trieu A T, Burleigh S H, Kardailsky I V, Maldonado-Mendoza I E, Versaw W K, Blaylock L A, Shin H, Chiou T-J, Kateg H, Katagi H, Dewbre G R, Weigel D and Harrison M J (2000) Transformation of flowering plants with *Agrobacterium. Plant J.*, **22**: 531–541.

Trinh T H, Ratet P, Kondorosi E, Durand P, Kamate K, Bauer P and Kondorosi A (1998) Rapid and efficient transformation of diploid *Medicago truncatula* and *Medicago sativa* ssp. *falcata* lines improved in somatic embryogenesis. *Plant Cell Rep.*, **17**: 345–355.

Walker K A and Sato S J (1981) Morphogenesis in callus tissue of *Medicago sativa*: the role of ammonium ion in somatic embryogenesis. *Plant Cell Tiss. Org. Cult.*, **1**: 109–121.

Wan Y, Sorensen E L and Liang G H (1988) Genetic control of *in vitro* regeneration in alfalfa (*Medicago sativa* L.). *Euphytica*, **39**: 3–9.

Wang J H, Rose R J and Donaldson B I (1996) *Agrobacterium*-mediated transformation and expression of foreign genes in *Medicago truncatula. Aust. J. Plant Physiol.*, **23**: 265–270.

Xu Z Q and Jia J F (1997) Regeneration of intergeneric somatic hybrids by protoplast fusion between *Onobrychis viciaefolia* and *Medicago sativa. Science in China, Series C Life Sciences*, **40**: 363–370.

Zafar Y, Nenz E, Damiani F, Pupilli F and Arcioni S (1995) Plant regeneration from explant and protoplast derived calluses of *Medicago littoralis. Plant Cell Tiss. Org. Cult.*, **41**: 41–48.

MOLECULAR GENETICS OF WHITE CLOVER

K. JUDITH WEBB*, MICHAEL T. ABBERTON AND STEPHEN R. YOUNG
*Institute of Grassland and Environmental Research,
Aberystwyth, Ceredigion, SY23 3EB, UK
e-mail: judith.webb@bbsrc.ac.uk

Abstract

Molecular genetics offers a major new impetus to breeding programmes of white clover throughout the world. Researchers have identified key agronomic traits governing reliability and persistence in the field. These characters reflect both tolerance of a variety of biotic and abiotic stresses and root architecture. Current access to a wealth of newly characterised plant material and an arsenal of new molecular technologies will facilitate development of genetic markers for use in breeding programmes. The advent of transgenic approaches has brought opportunities to identify new genes and to understand the role of specific genes in complex interactions between plants and the environment. These transgenic techniques have also brought public perception issues into sharp focus, since such transgenic plants could provide a platform for the development of new varieties. Future goals for white clover breeding include improved forage quality and more efficient use of resources to support environmental sustainability.

1. Introduction

Legumes play an important role in global agriculture by providing sources of high quality protein and reducing reliance on inorganic fertilizers. The forage species, white clover (*Trifolium repens* L.) is a major component of temperate grasslands, growing alongside species such as perennial ryegrass (*Lolium perenne* L.), Italian ryegrass (*Lolium multiflorum* Lam.), *Festuca* species or cocksfoot (*Dactylis glomerata* L.). It is used both for grazing, particularly by sheep and cattle, and for conservation as silage.

White clover traditionally plays a major role in sheep and beef production, and its importance is increasing in the dairy sector and in organic agriculture in the UK. Its value lies in two key areas: its nutritional quality and its ability to form two important symbiotic relationships. Its symbionts, *Rhizobium* bacteria and arbuscular mycorrhiza fungi, improve plant nutrition and plant health by fixing nitrogen and improving phosphate

P. K. Jaiwal and R. P. Singh (eds.), Applied Genetics of Leguminosae Biotechnology, 239–253.

uptake respectively. Fixed nitrogen becomes available to clover and its companion grasses, helping to create nutritious forage for livestock. White clover contains high protein levels and its use can reduce the need for concentrates, thereby lowering farm costs (Thomson, 1984).

Because of these factors, clover has a key role to play in sustainable farming systems. In organic systems, where legumes are needed to add nitrogen to the soil, generally as part of a crop rotation, its use is even more important. In the UK, white clover is the most widely used forage legume species. In the year ending June 1998, a total of 248 tonnes of seed of white clover was delivered for use in the UK, as compared to 56 tonnes of red clover (MAFF, 1998). Worldwide, the seed market for white clover is 7,422 tonnes per year.

Many aspects of the growth, agronomy and breeding of white clover have been previously reviewed (Gibson and Cope, 1985; Frame and Newbould, 1986; Baker and Williams, 1987; Woodfield and Caradus, 1994). In this chapter, we will discuss the current objectives of white clover breeding programmes, the available techniques and plant materials and the implications of these new breeding strategies.

Historically, white clover breeders have successfully created varieties suited to a range of uses and climates by exploiting the natural variation found in white clover grown under different conditions (Rhodes and Ortega, 1996; Rhodes *et al.*, 1998). By combining classical and molecular approaches, breeders can now create a new generation of varieties with novel characteristics.

1.1. OBJECTIVES FOR GERMPLASM IMPROVEMENT

The benefits of white clover can only be realised when it persists in the sward at a level where it can make a significant impact. This is generally taken to be around 30–40% of the sward by dry weight (Rhodes and Ortega, 1996), although this proportion varies throughout the year. It has been the overriding breeding objective to produce white clover varieties that are persistent and reliably yielding from one year to the next. To this end there has been considerable focus on those factors that limit persistence (Caradus and Mackay, 1991; Rhodes and Ortega, 1996).

In many parts of Europe, a major consideration has been tolerance of cold winters (Collins *et al.*, 1997). This is not only a question of direct tolerance of low temperatures but also of desiccation and depletion of reserves. Corre and co-workers (1996) have studied the role of vegetative storage proteins in immobilising nitrogen prior to growth or regrowth following defoliation. Reserves are also stored as carbohydrates and, again, the level of these may be an important factor in winter survival (Boller and Nosberger, 1983). Allied to this are questions of resource uptake and utilisation, particularly response to nitrogen, which is generally considered to be faster in grass. To overcome some of these problems, new varieties have been produced that have greater cold tolerance and respond more quickly to the changes in daylength and temperature at the onset of spring (Rhodes and Webb, 1993). Efforts have also been made to improve the tolerance of white clover to a range of edaphic stresses, including salt (Rogers *et al.*, 1997), drought (Brink and Pederson, 1998) and low pH (Voigt *et al.*, 1997).

Grass-clover compatibility has also attracted much interest, both from an ecological and a physiological perspective (Frame and Newbould, 1984; Hill, 1996). Factors important in determining the outcome of inter-specific interactions have been identified, although more work is needed in this area. Similarly, the transfer of fixed nitrogen from clover to grass has been recognised as an important factor in the total performance of the sward but little is known of the underlying mechanisms or the genetic variation for this phenomenon (Ledgard, 1991).

White clover is susceptible to a range of pests and diseases. Breeding effort has focused on stem nematode (*Ditylenchus dipsaci*), cyst nematode (*Heterodera trifolii*), clover rot (*Sclerotinia trifolii*) and viral diseases, particularly white clover mosaic virus (WCMV) (Baker and Williams, 1987). Sitona weevil has also been identified as a major pest of white clover and slugs may be particularly important during the establishment phase (Baker and Williams, 1987). Insect damage, due to pests such as porina caterpillars, may be a serious problem in New Zealand pastures (Voisey *et al.*, 1994b).

Although white clover is grown for utilisation of its vegetative parts, the production and commercial impact of new varieties is dependent on economically viable seed yield (Williams *et al.*, 1998a). Under some climatic conditions (e.g. in the UK) this can be difficult to achieve. The new variety AberDale addresses such problems by having strong, erect peduncles that are both attractive to pollinating insects and capable of high seed set without bending back into the canopy (Rhodes and Webb, 1993; Marshall, 1995). However, there is scope for further improvement by increasing allocation of resources to reproductive parts without detriment to agronomic yield.

The development of a white clover plant from seed involves initially the growth of a thick taproot that breaks down after approximately 18 months and is replaced by a large number of roots from stolon nodes (Baker and Williams, 1987). This nodal rooting is important for adhesion and tolerance of grazing through resistance to up-rooting, as well as nutrient uptake. Where plants are grazed heavily and continuously, they may break down into smaller units following taproot senescence and be more susceptible to stress-related mortality. Thus, root architecture is an important feature of plant growth and survival as well as a factor in grass/clover compatibility. The recent delineation of novel mutants defective in nodal root development points towards one way of unravelling the physiology and genetics of important traits (White *et al.*, 1998).

1.2. BREEDING STRATEGIES

Classical approaches to germplasm improvement in white clover have been typical of those for outbreeding species. A combination of mass selection and progeny testing has been used to incorporate useful material from germplasm accessions and collections into new varieties of white clover (Rhodes and Ortega, 1996). The result is varieties in which improvements can be traced back to traits introduced from germplasm that evolved under particular ecological conditions. For example, varieties with greatly improved cold tolerance and early spring growth have arisen from material collected in Switzerland. Similarly, material selected from within ecotypes from USA or New Zealand has proved well-adapted to conditions in those countries.

However, the rate of genetic improvement in white clover has been modest for most traits and low for traits contributing to reproductive potential (Woodfield and Caradus, 1994). This may be due to several factors. Although white clover is highly heterozygous and populations typically genetically heterogeneous, relatively little use is made of heterosis. Selection for agronomic traits, is often hampered by relatively low heritabilities and confounded by negative phenotypic and genetic correlations between traits (Caradus and Chapman, 1996; Annicchiarico *et al.*, 1999). The species also shows a high degree of phenotypic plasticity and genotype/environment interactions that complicate selection regimes and may reduce genetic gains (Annicchiarico and Piano, 2000). The white clover gene pool, although diverse, may contain relatively little variation for some desirable traits including drought tolerance and seed yield (Woodfield and Caradus, 1994; Brink and Pederson, 1998). Finally, the length of the breeding cycle, with trials of more than three years in plots with companion grasses required to evaluate persistency, increases further the difficulties in achieving rapid improvements.

2. Molecular approaches to breeding

The problems described above have led to the development of new strategies to increase the capability to make significant alterations to agronomic potential on a shorter time-scale. Some of these biotechnological approaches to white clover breeding have been reviewed previously (Humphreys *et al.*, 1997; Webb and Abberton, 1999).

2.1. INTERSPECIFIC HYBRIDIZATION AND INTROGRESSION

Although a large range of interspecific crosses involving white clover have been carried out (Trimble and Hovin, 1960; Hovin, 1962b; Chen and Gibson, 1970), present research and breeding studies focus largely on two: with *Trifolium nigrescens* Viv. and *T. ambiguum* M. Bieb (Meredith *et al.*, 1995; Abberton *et al.*, 1998a). *T. nigrescens* is a diploid ($2n = 2x = 16$) annual species that is a putative source of one of the two white clover genomes (Hovin, 1963; Ansari *et al.*, 1999). It is profusely flowering with a high seed yield. Hybrids between *T. nigrescens* and white clover have been produced by hand-crossing and a backcrossing programme developed with white clover as recurrent parent with a view to increasing seed yield (Brewbaker and Keim, 1953; Hovin, 1962a, b; Marshall *et al.*, 1995; Hussain *et al.*, 1997a, b; Abberton *et al.*, 1998a; Marshall *et al.*, 1998). Studies have shown that nitrogen fixation in the hybrids is not impaired and that they exhibit considerable agronomic potential for traits, such as yield and persistency (Abberton *et al.*, 1999). Hussain and co-workers (1997a, b) have produced hybrids in which clover cyst nematode resistance was transferred from *T. nigrescens* to white clover. Research on *T. repens* × *T. nigrescens* hybrids has also led to the first evidence for the occurrence of functional unreduced gametes in white clover (Hussain and Williams, 1997a).

Trifolium ambiguum is a long-lived drought tolerant species with a rhizomatous habit that is extremely persistent although not as vigorous or competitive as white clover (Taylor and Smith, 1998). Hybrids between these two species have been produced by ovule culture (Meredith *et al.*, 1995). Breeding objectives have focused on the introduction of virus resistance (Pederson and McLaughlin, 1989) and on increasing stress tolerance and persistency, in particular by the introduction of the rhizomatous habit into white clover (Abberton *et al.*, 1998b, c; Marshall *et al.*, 2000).

Hybrids between *T. ambiguum* and white clover have been analysed by a range of techniques. Classical cytological approaches have been used to study and select hexaploid hybrids that are easily crossed with hexaploid *T. ambiguum*. Hussain and Williams (1997b) describe this as a 'fertile genetic bridge' allowing traits to be introduced without the need for ovule culture. Molecular marker techniques have been developed for use in marker assisted introgression of the rhizomatous trait. In particular, bulk segregant amplified fragment length polymorphisms (AFLPs) are being used to detect markers linked to the gene(s) controlling the rhizomatous trait in segregating families of the third generation backcross families (M. T. Abberton, unpublished data). It seems likely that within the next five years markers will be used to increase the speed and precision of several introgression programmes in white clover. This will allow a reduction in problems of linkage drag and cut down on the number of cycles of selection required.

Additionally, molecular cytogenetic approaches have been used to explore species relationships in the genus *Trifolium* by the use of FISH mapping of 5S and 18S-26S rDNA loci (Ansari *et al.*, 1999). This study supports the proposition that *T. nigrescens* ssp. *petrisavii* is a present day relation of one of the diploid ancestors of white clover.

2.2. GENOME MAPPING

The development of molecular marker maps offers a means of improving the efficiency of selection and widening the scope of breeding programmes. Progress in white clover improvement has been hampered by the allotetraploid, highly heterozygous nature of the crop (Williams *et al.*, 1998b). To overcome this, a strategy based on the use of the rare Sf allele, conferring self-compatibility, has been adopted (Yamada *et al.*, 1989; Michaelson-Yeates *et al.*, 1996; Minchin *et al.*, 1997; Eason *et al.*, 1998; Webb and Abberton, 1999; Mizen *et al.*, 1998). Inbred lines of white clover have been produced from eight generations of selfing and single seed descent. Within-line variation for phenotypic traits is low and the uncovering of recessive alleles usually hidden in heterozygotes has given rise to a range of plants with characteristics not normally seen in white clover (Michaelson-Yeates *et al.*, 1998). The greater homozygosity of inbred lines means that they can be used in ways facilitating genome mapping.

Two approaches have been adopted. Randomly amplified polymorphic DNA (RAPDs) were used by Joyce and colleagues (1999) to determine genetic distances between a range of inbred lines of different origins. These data informed decisions concerning which lines were used to produce an F_1 that was selfed to give an F_2 mapping family. The homozygosity of the parents allows greater clarity in assignations of loci and simplifies the process of building up linkage groups. Molecular marker analysis is being carried out on this population using both AFLP and microsatellite (simple sequence repeat, SSR) markers. In another approach, lines inbred for several generations have been crossed to elite material to produce an F_1 mapping family (Woodfield, personal communication). In both cases the objective is to build a map that can be used as a basis for identifying quantitative trait loci (QTL) and for subsequent marker assisted selection (MAS) of qualitative and quantitative traits. MAS has the capability of increasing both the speed and precision of breeding for agronomically important traits such as yield and cold hardiness. However, it seems likely that the greatest effect of such methods will be on breeding directed towards improvements in quality and environmental impact.

Classical breeding for traits that can be assessed only by animal studies, or by measurement of environmental impacts at the field level, is prohibited by logistical and financial constraints for all but a very limited range of germplasm. Selection for such traits becomes possible on a much larger scale when markers are used, particularly those that can be identified via polymerase chain reaction (PCR) techniques, such as AFLPs and SSRs, which require very little DNA. Additionally, the use of markers offers a route to overcoming genetic and phenotypic correlations that have hindered progress in germplasm improvement (Collins *et al.*, 1998a, b).

The use of markers in programmes of introgression and selection within the white clover gene pool meshes with prospects for molecular characterisation of germplasm accessions in gene pools round the world. Possibilities exist not only for tailoring white clover varieties to particular environments or management types (intensive, extensive, or organic) but also widening the adaptive range of the species itself – to drier, warmer areas, for example.

2.3. GENETIC DIVERSITY

There is considerable interest in delineating the extent of genetic variation within white clover at a range of levels. Restriction fragment length polymorphisms (RFLPs) have been used to study variation at the *Li* locus controlling synthesis of the enzyme linamarase responsible for the release of hydrogen cyanide by hydrolysis following tissue damage (Hughes, 1991). Nuclear ribosomal DNA has been used to examine the extent of clonal growth (Capossela *et al.*, 1992) and such studies in evolutionary ecology are being complemented by molecular approaches to taxonomy (Ansari *et al.*, 1999) and the characterisation of diversity in gene banks and wild populations.

Randomly amplified polymorphic DNA (RAPD) markers have been used to investigate population level variation in permanent pastures (Gustine and Huff, 1999). Molecular characterisation of genetic resources by AFLPs and SSRs has clear relevance to germplasm improvement programmes in terms of both breeding strategies and incorporation of particular traits found only outside elite gene pools.

2.4. GENE IDENTIFICATION AND ISOLATION

Inbred lines of white clover developed at IGER are an invaluable resource for understanding the genetic bases for efficient functioning of arbuscular mycorrhiza (AM) fungi, which are important for plant health and nutrition, and for resistance/susceptibility to stem nematode (*Ditylenchus dipsaci*), an important pest of white clover. Comparisons between genetically similar inbred lines of white clover have revealed differences in their performance with arbuscular mycorrhiza (Eason *et al.*, 1998) and in response to infection with stem nematode (M. T. Abberton, unpublished data), with concomitant differences in gene expression detected by differential display. Such approaches will play a key role in attempts to isolate plant genes involved in both interactions.

3. Transgenic approaches

Genetic modification or transformation has two potential roles in white clover breeding. The first is as a research tool. Techniques such as gene tagging allow researchers to identify genes that are involved in key plant processes, providing information that can be fed into conventional plant breeding programmes. Transformation may also be used as a tool to demonstrate that particular breeding strategies, such as adding certain genes, or changing their expression, will have the desired agronomic effect. Such strategies can then be pursued by conventional techniques. Genetic modification has also been considered in a second major role in variety development, where it allows breeders to add certain traits, such as virus or insect resistance, in a relatively precise fashion.

3.1. PLANT REGENERATION AND GENETIC MODIFICATION

Plants of white clover can be regenerated from meristematic tissues, such as cotyledons of immature (Maheswaran and Williams, 1984; Parrott, 1991) or mature embryos (Webb *et al.*, 1987; Larkin *et al.*, 1996) and from seedling shoots, including cotyledons (Voisey *et al.*, 1994a). In addition, stolon tips from mature plants and early generation inbred lines of certain genotypes regenerate shoots (Bond and Webb, 1989; Webb and Michaelson-Yeates, 1991). As in many species, there is a marked effect of the genotype of the donor plant that can only be overcome by selecting responsive plants. The use of juvenile tissues has eliminated many of the problems posed by the genetic make-up of the plant, by targeting the most actively growing cells, resulting in successful procedures for plant regeneration.

The process of genetic modification involves several distinct stages, insertion, integration expression and inheritance of the new DNA. The bacterial vectors, *Agrobacterium tumefaciens* and *A. rhizogenes*, are routinely used to genetically modify or transform material. These bacteria naturally infect a range of forage legumes, including clover plants (Webb, 1986). During infection, they deliver some of their own genes to the cells of the host plant, causing crown gall and hairy root diseases, respectively. The ability of the bacteria to transfer genes directly into the genome of the host plant is harnessed in research programmes in which new genes either replace the disease-causing genes (*A. tumefaciens*) or are added alongside them (*A. rhizogenes*).

Embryos or young seedlings are routinely used for genetic modification by either species of *Agrobacterium*. Hairy roots can be rapidly produced by inoculating seedlings with *A. rhizogenes* (Diaz *et al.*, 1989, 1995a, b) and can be maintained as axenic root cultures on hormone-free media (Webb *et al.*, 1990). By contrast, modification with *A. tumefaciens* requires stimulation of shoot regeneration by growing tissues on a medium containing auxin and cytokinin (Voisey *et al.*, 1994a; Larkin *et al.*, 1996).

Selection of transformed cells after co-cultivation with *A. tumefaciens* is essential for rescuing transformed shoots. Clover cells, containing the gene for neomycin phosphotransferase II from *E. coli*, grow on the antibiotics, kanamycin or G418 (Voisey *et al.*, 1994a, b; Sharma *et al.*, 1998) and cells with the gene for phosphinothricin N-acetyl transferase from *Streptomyces hygroscopicus* grow on phosphinothricin or on the herbicides, bialophos or Basta (Larkin *et al.*, 1996). The use of selectable markers is surrounded with controversy, and while they are appropriate in plants used as research tools, they would not be appropriate for use in future clover varieties. In the development of varieties, for most markets, other options, such as co-transformation, must be considered. In co-transformation, cells are transformed with two constructs, one of which contains a selectable marker gene and the other the gene required in the plant. If the two separate constructs insert into different regions of the plant's genome, then the marker gene could be removed by conventional breeding.

3.2. GENETIC MODIFICATION AS A RESEARCH TOOL

Recent developments in molecular genetics of model legumes, *Lotus japonicus* (Regel) Larsen and *Medicago truncatula* Gaertn (Cook *et al.*, 1997) offer many opportunities for identifying genes that control agronomically important characters, inaccessible in the model plant, *Arabidopsis*. Such genes include those involved in secondary metabolism and symbiotic interactions.

One approach to identifying genes involved in symbiosis is gene tagging. This involves creating genetically modified plants that incorporate a marker (a sequence of known DNA) so that any disrupted or mutant gene can be easily identified. Knowing the sequence of the inserted DNA allows the isolation of DNA on either side of the inserted transgene – so yielding the gene of interest.

Two symbiotic genes have been identified from genetically tagged plants of *L. japonicus*. One causes a mutation in nodulation inception (*nin*; Schauser *et al.*, 1999), whilst the other has homology to a calcium binding protein and is expressed specifically in response to challenge by both *Rhizobium* and arbuscular mycorrhiza (*cbp*1; Webb *et al.*, 2000). Others will no doubt follow and will be incorporated as markers into developing genetic maps of legumes, including white clover.

Modifying gene expression in white clover has also offered insights into genetic control of lateral root development, effects of gravity and plant-rhizobial symbiosis. White clover plants expressing an auxin-responsive promoter, GH3, fused to β-glucuronidase (GUS; *uidA*) reporter gene were used to investigate plant responses to auxin – such as gravitropism and lateral root development (Larkin *et al.*, 1996). Experiments using these same plants have suggested the involvement of auxin and a role for flavonoids secreted by roots in nodule organogenesis (Mathesius *et al.*, 1998a, b).

Using genetically modified hairy roots, expression of pea lectins in white clover broadened the normal host range of *Rhizobium*, enabling the pea symbiont, *Rhizobium leguminosarum* biovar *viciae* to infect white clover (Diaz *et al.*, 1995a, b). Genetic modification with a mutant pea lectin clearly showed that sugar-binding activity of pea lectin is needed for successful infection in this system (van Eijsden *et al.*, 1995).

3.3. DEVELOPMENT AND FIELD TRIALS OF GENETICALLY MODIFIED WHITE CLOVER

Researchers around the world are investigating the possibility of using genetic modification to address a range of breeding targets in forage legumes (McKersie, 1997). Suitable targets include forage quality (with emphasis on lignin and tannin levels), virus resistance, cold tolerance and production of industrial chemicals. Although alfalfa is at the centre of several programmes, there is also significant interest in white clover (Young *et al.*, 1999).

One team has produced white clover plants containing a gene designed to protect against white clover mosaic virus, which has a serious impact on yield. The gene in question is of viral origin and codes for the coat protein of the virus (Dudas *et al.*, 1998). In Australia, researchers have equipped white clover with a coat protein gene from alfalfa mosaic virus. The virus occurs widely in the state of Victoria, and has been linked to loss of clover production of 60% and consequent losses in milk production. Transformed plants are resistant to attack by at least two strains of the virus (GMAC, 1999).

New Zealand grasslands can suffer significant damage from insects, such as porina caterpillars (larvae of *Wiseana* moths), leading to economic losses of millions of dollars (Burgess and Gatehouse, 1997; Voisey *et al.*, 1994b). Attempts have been made to produce transgenic insect resistant clovers by inserting genes targeted against attacking insects. Candidates include toxin genes from the bacterium *Bacillus thuringiensis* (Voisey *et al.*, 1994b).

Other research has focused on equipping white clover with genes that make nutritious sulphur-rich proteins able to resist breakdown in the rumen of sheep and cattle. When consumed by sheep, such proteins are associated with enhanced growth of wool. One candidate gene, the δ-zein gene from maize, has been introduced into white clover, and the resulting plants expressed the protein in leaves, roots, seeds and other tissues (Sharma *et al.*, 1998).

Genetically modified white clovers have yet to be commercialised, but field trials on transgenic plants have been staged in Australia and New Zealand. A GM white clover designed to resist white clover mosaic virus was the subject of a field trial in New Zealand in 1995–6 (New Zealand Gazette, 1998). In Australia, a field trial on white clover resistant to alfalfa mosaic virus began in 1996 (GMAC, 1999).

3.4. TRANSGENE BEHAVIOUR IN WHITE CLOVER

The controversy surrounding genetic modification has highlighted the need for accurate, detailed characterisation of transgenic plants and the inheritance of their novel genes. Genetic modification involves introducing a small number of genes into a plant cell, instead of the much larger numbers that are inserted by traditional breeding methods. At the moment it is not possible to predict how many copies of the transgenes will be transferred, or their sites of insertion (Dale, 1999). Molecular methods can be used to identify transgenic plants with relatively uncomplicated insertions.

Plants transformed with *Agrobacterium* may have single or multiple copies of T-DNA, which may be in the same orientation, or in different orientations (Gheysen *et al.*, 1998). Experiments on white clover (using *Agrobacterium*) have resulted in a likely copy number of one (Voisey *et al.*, 1994a) or in copy numbers ranging from one to two or more (Sharma *et al.*, 1998). White clover plants transformed with a β-glucuronidase (GUS; *uidA*) reporter gene contained 1–4 copies of the transgene (Larkin *et al.*, 1996)

Inheritance of genes introduced by genetic modification has become another important focus of research. Work on alfalfa, for example, in which transgenic plants were backcrossed to plants from a range of cultivars, showed that a *uidA* gene was stably inherited over four generations (Micallef *et al.*, 1995).

Scott and co-workers (1998) studied expression and inheritance of the *uidA* reporter gene in white clover. They crossed a transgenic plant, containing a single T-DNA insert, with a non-transgenic plant and then forwarded the progeny (the BC1 generation) into other crosses. The results yielded evidence of stable inheritance and expression of the transgene, but there were some signs that genetic background influenced level of GUS expression. One cross between two BC1 plants revealed a departure from Mendelian ratios, in that the ratio of plants expressing GUS to plants not expressing GUS was lower than anticipated. This was not because the *uidA* gene had been silenced.

Experiments on *Lotus corniculatus* L., in which researchers examined the progeny of crosses between transformed plants and non-transformed control plants, have also revealed complex patterns of transgene inheritance, which could have been due to gene silencing (Webb *et al.*, 1996). Later work has followed inheritance of transgenes through two subsequent generations (Webb *et al.*, 1999). The expression and inheritance pattern of transgenes varied considerably between transformed lines. The careful monitoring of transgene expression and not just molecular analysis for presence of the transgene is essential to select for stable transmission and expression.

3.5. RELEASE OF GENETICALLY MODIFIED WHITE CLOVER

The out-breeding nature of white clover means that breeders cannot produce a uniform variety by self-pollination of parent plants. Instead, they arrange a number of rounds of random mating between groups of carefully chosen progenitors. The upshot is a 'synthetic' variety that contains a degree of genetic diversity. This diversity enables the variety to survive and flourish in the range of microenvironments of a pasture (Voisey *et al.*, 1994b). Incorporating a genetic modification into such a programme presents a challenge. If a small number of transgenic parents were to be included in a conventional breeding programme, for example, many progeny would not carry the transgene and the resulting variety might not be sufficiently uniform. A method for creating a transgenic white clover variety has been suggested by Voisey *et al.* (1994b). The method involves crossing transformed plants with elite individuals and then allowing transgenic progeny from those crosses to interbreed. F_2 individuals homozygous for the transgene would then be selected and interbred for five or six generations.

Genetic modification in agriculture is currently very controversial in some countries, with a major debate surrounding matters such as food safety and labelling, environmental impacts and ethical, social and political issues. As a forage crop, white clover occupies a special niche in this debate. One reason is that agricultural grassland is a major component in the environment of certain countries, such as the UK, where it accounts for about half of the land surface. Another is that white clover is often cultivated in proximity to wild populations of the same species, creating opportunities for gene escape via pollen. However, as white clover is generally grown for herbage, not seed, there could be scope for blocking gene escape by methods such as prevention of flowering or pollen production. From some perspectives, it might seem unlikely that white clover would arouse food safety concerns, but feeding of livestock on GM crops has become a trigger for anxiety in some countries. If GM white clover was widely grown, there might also be concerns over the presence of clover pollen in honey.

4. Future prospects

Quality components are becoming increasingly important in forage breeding programmes as a result of changes in policy and consumer preferences. The move towards organic and less intensive farming systems and the need to optimise home grown protein sources mean that white clover is likely to play an increasingly important role in the livestock sector. Future targets may well include changes to proteins, water-soluble carbohydrates and digestibility. Allied to this is the need to increase the efficiency of protein utilisation in the rumen. Here, protein protection strategies, including changes in secondary metabolites such as tannins, may decrease the rate of protein digestion and increase efficient utilisation as well as reducing the incidence of bloat in cattle (Waghorn and Caradus, 1994). This is a complex area as tannins may have both beneficial and harmful effects depending on the amount in the diet. They also exist in a wide range of different structural forms and in different locations within the plant. However, the alteration of secondary metabolite status remains a long term objective of breeding programmes (Morris and Robbins, 1997; Robbins and Morris, 1999; Webb and Abberton, 1999).

The need to increase quality and efficiently utilise resources (such as nitrogen and phosphorus) overlaps with the need to develop varieties more in tune with environmental sustainability. Although white clover is regarded as an environmentally friendly crop, there remains a need to optimise its environmental impact. Future objectives may include breeding for reduced leaching (Macduff *et al*., 1990; Wu and McGechan, 1999), increased uptake and utilisation of resources by improving nitrogen fixation by rhizobium and phosphate uptake via arbuscular mycorrhiza fungi. Similarly, levels of proteins and carbohydrates in the crop should be better synchronised with the companion grass to give a balanced diet with reduced pollution.

Acknowledgements

IGER is grant aided by the Biotechnology and Biological Sciences Research Council. The UK Ministry of Agriculture, Fisheries and Food (MAFF) funds research on the germplasm development of white clover at IGER.

References

Abberton M T, Macduff J H, Marshall A H and Michaelson-Yeates T P T (1999) Nitrogen fixation by hybrids of white clover (*Trifolium repens* L.) and *Trifolium nigrescens*. *J. Agron. Crop. Sci.*, **183**: 27–33.

Abberton M T, Michaelson-Yeates T P T and Macduff J H (1998a) Characterisation of novel inbred lines of white clover (*Trifolium repens* L.). I. Dynamics of plant growth and nodule development in flowing solution culture. *Euphytica*, **103**: 35–43.

Abberton M T, Michaelson-Yeates T P T, Macduff J H, Marshall A H and Rhodes I (1998b) New approaches to legume improvement for sustainable agriculture. In: *Breeding for a Multi-Functional Agriculture* (Eds Boller B and Stadlemann F J), Proc. of the 21st Meeting of the Fodder Crops and Amenity Grasses Section of EUCARPIA, Switzerland 1997, FAL Reckenholz, Zurich, 91–93.

Abberton M T, Michaelson-Yeates T P T, Marshall A H, Holdbrook-Smith K and Rhodes I (1998c) Morphological characteristics of hybrids between white clover, *Trifolium repen*s L., and Caucasian clover, *Trifolium ambiguum* M. Bieb. *Plant Breed.*, **117**: 491–493.

Annicchiarico P and Piano E (2000) Response of white clover genotypes to evaluation environments of dense and spaced planting, and implications for selection. *Euphytica*, **111**: 111–120.

Annicchiarico P, Piano E and Rhodes I (1999) Heritability of, and genetic correlations among, forage and seed yield traits in Ladino white clover. *Plant Breed.*, **118**: 341–346.

Ansari H A, Ellison N W, Reader S M, Badaevas E D, Friebe B, Miller T E and Williams W M (1999) Molecular cytogenetic organisation of 5S and 18S-26S rDNA loci in white clover (*Trifolium repens* L.) and related species. *Ann. Bot.*, **83**: 199–206.

Baker M J and Williams W M (1987) *White Clover*. CAB International, Wallingford, UK.

Boller B C and Nosberger J (1983) Effects of temperature and photoperiod on stolon characteristics, dry matter partitioning and nonstructural carbohydrate of two white clover ecotypes. *Crop Sci.*, **23**: 1057–1062.

Bond J E and Webb K J (1989) Regeneration and analysis of plants from stolon segments of *Trifolium repens* (white clover). *Plant Sci.*, **61**: 119–125.

Brewbaker J L and Keim W F (1953) A fertile interspecific hybrid in *Trifolium* (4n *T. repens* L × 4n *T. nigrescens* Viv.). *Am. Nat.*, **87**: 323–326.

Brink G E and Pederson G A (1998) White clover response to a water-application gradient. *Crop Sci.*, **38**: 771–775.

Burgess E P J and Gatehouse A M R (1997) Engineering for insect pest resistance. In: *Biotechnology and the Improvement of Forage Legumes* (Eds McKersie B D and Brown D C W). CAB International, Wallingford, 229–258.

Capossela A, Silander J A, Jansen R K, Bergen B and Talbot D R (1992) Nuclear ribosomal DNA variation among ramets and genets of white clover. *Evolution*, **46**: 1240–1247.

Caradus J R and Chapman D F (1996) Selection for and heritability of stolon characteristics in two cultivars of white clover. *Crop Sci.*, **36**: 900–904.

Caradus J R and Mackay A C (1991) Performance of white clover cultivars and breeding lines in a mixed species sward. 2. Plant characters contributing to differences in clover proportion in swards. *N. Z. J. Ag. Res.*, **34**: 155–160.

Chen C and Gibson P B (1970) Chromosome pairing in two interspecific hybrids of *Trifolium. Can. J. Genet. Cytol.*, **12**: 790–794.

Collins R P, Abberton M T, Michaelson-Yeates T P T, Marshall A H and Rhodes I (1998a) Effects of divergent selection on correlations between morphological traits in white clover (*Trifolium repens* L.) *Euphytica*, **101**: 301–305.

Collins R P, Abberton M T, Michaelson-Yeates T P T and Rhodes I (1998b) Response to divergent selection for stolon characters in white clover (*Trifolium repens*). *J. Agric. Sci.*, **129**: 279–285.

Collins R P, Michaelson-Yeates T P T, Fothergill M, Abberton M T and Rhodes I (1997) Comparison of the freezing tolerance of populations of white clover grown in a range of locations in Europe. In: *Proceedings of the Cost 814 Meeting, Ireland, August 1997.*

Cook D R, VadenBosch K, de Bruijn F J and Huguet T (1997) Model legumes get the nod. *Plant Cell*, **9**: 275–281.

Corre N, Bouchart V, Ourry A and Boucaud J (1996) Mobilisation of nitrogen reserves during regrowth of defoliated *Trifolium repens* L. and identification of potential vegetative storage proteins. *J. Exp. Bot.*, **301**: 1111–1118.

Dale P J (1999) Public reactions and scientific responses to transgenic crops. *Curr. Opin. Biotech.*, **10**: 203–208.

Diaz C L, Logman T J J, Stam H C and Kijne J W (1995a) Sugar-binding activity of pea lectin expressed in white clover hairy roots. *Plant Physiol.*, **109**: 1167–1177.

Diaz C L, Melchers L S, Hooykaas P J J, Lugtenberg B J J and Kijne J W (1989) Root lectin as a determinant of host-plant specificity in the *Rhizobium*-legume symbiosis. *Nature*, **338**: 579–581.

Diaz C L, Spaink H, Wijffelman C A and Kijne J W (1995b) Genomic requirements of rhizobium nodulation of white clover hairy roots transformed with the pea lectin gene. *Mol. Plant-Microbe Interact.*, **8**: 348–356.

Dudas B, Woodfield D R, Tong P M, Nicholls M F, Cousins G R, Burgess R, White D W R, Beck D L, Lough T J and Forster R L S (1998) Estimating the agronomic impact of white clover mosaic virus on white clover performance in the North Island of New Zealand. *N. Z. J. Ag. Res.*, **41**: 171–178.

Eason W, Michaelson-Yeates T P T, Abberton M T, Culshaw C and Griffiths G (1998) Response of near isogenic lines of *Trifolium repens* to *Glomus mosseae* and their potential use in molecular studies of the arbuscular mycorrhizal (AM) symbiosis. *Proceedings of 2nd International Conference on Mycorrhiza*, Uppsala, 5–10 July 1998.

Frame J and Newbould P (1984) Herbage production from grass/white clover swards. In: *Forage Legumes* (Ed. Thomson D J), *Proceedings of the British Grassland Society Occasional Symposium No. 16*, British Grassland Society, 15–35.

Frame J and Newbould P (1986) Agronomy of white clover. *Adv. Agron.*, **40**: 1–87.

Gheysen G, Angenon G and Van Montagu M (1998) *Agrobacterium*-mediated plant transformation: a scientifically intriguing story with significant applications. In: *Transgenic Plant Research* (Ed. Lindsey K), Harwood, Amsterdam, 1–33.

Gibson PB and Cope W A (1985) White Clover. In: *Clover science and Technology* (Ed. Taylor N L), ASA-CSSA-SSSA, Madison, Wisconsin, 471–490.

GMAC (1999) http://www.health.gov.au/tga/gene/gmac/piscont.htm

Gustine D L and Huff D R (1999) Genetic variation within and among white clover populations from managed permanent pastures of the Northeastern USA. *Crop Sci.*, **39**: 524–530.

Hill J (1996) Breeding components for mixture components. *Euphytica*, **92**: 135–138.

Hovin A W (1962a) Interspecific hybridisation between *Trifolium repens* L. and *T. nigrescens Viv*. and the analysis of hybrid meiosis. *Crop Sci.*, **2**: 251–254.

Hovin A W (1962b) Species compatibility in Subsection Euamoria of *Trifolium. Crop Sci.*, **2**: 527–530.

Hovin A W (1963) Compatibilty reactions of interspecific hybrids between *Trifolium repens* L. and *T. nigrescens* Viv. *Crop Sci.*, **3**: 487–489.

Hughes M A (1991) The cyanogenic polymorphism in *Trifolium repens* L. (white clover). *Heredity*, **66**: 105–115.

Humphreys M O, Hayward M D, Morris P and Webb K J (1997) Applying biotechnology to grass and forage legume breeding; new solutions, new horizons and new challenges. In: *Ecological Aspects of Breeding Fodder Crops and Amenity Grasses. Proceedings of Eucarpia 20th Fodder Crops and Amenity Grass Section Meeting* (Eds Staszewski Z, Waleria Mlyniec and Osinski R), Plant Breeding and Acclimatisation Institute, Radzikow, Poland, 7–10 October 1996, 283–292.

Hussain S W and Williams W M (1997a) Evidence of functional unreduced gametes in *Trifolium repens* L. *Euphytica*, **97**: 21–24.

Hussain S W and Williams W M (1997b) Development of a fertile genetic bridge between *Trifolium ambiguum* M. Bieb and *T. repens* L. *Theor. Appl. Genet.*, **95**: 678–690.

Hussain S W, Williams W M, Mercer C F and White D W R (1997a) Transfer of clover cyst nematode resistance from *Trifolium nigrescens* Viv. to T. *repens* L. by interspecific hybridisation. *Theor. Appl. Genet.*, **95**: 1274–1281.

Hussain S W, Williams W M, Woodfield D R and Hampton J G (1997b) Development of a ploidy series from a single interspecific *Trifolium repens* L. × *T. nigrescens* Viv. F_1 hybrid. *Theor. Appl. Genet.*, **94**: 821–831.

Joyce T A, Abberton M T, Michaelson-Yeates T P T and Forster J W (1999) Relationships between genetic distance measured by RAPD-PCR and heterosis in inbred lines of white clover (*Trifolium repens). Euphytica*, **107**: 159–165.

Larkin P J, Gibson J M, Mathesius U, Weinman J J, Gartner E, Hall E, Tanner G J, Rolfe B G and Djordjevic M A (1996) Transgenic white clover. Studies with the auxin-responsive promoter, GH3, in root gravitropism and lateral root development. *Transgenic Res.*, **5**: 325–335.

Ledgard S F (1991) Transfer of fixed nitrogen from white clover to associated grasses in swards grazed by dairy cows, estimated using 15N methods. *Plant Soil*, **131**: 215–223.

Macduff J H, Jarvis S C and Roberts D H (1990) Nitrates: leaching from grazed grassland systems. In: *Nitrates – Agriculture – Eau.*, (Ed Calvet R), INRA, Paris, France, 405–410.

MAFF (1998) *Seed Traders' Annual Return Summary*. MAFF Seeds Branch, Cambridge.

Maheswaran G and Williams E G (1984) Direct somatic embryoid formation on immature embryos of *Trifolium repens*, *T. pratense* and *Medicago sativa* and rapid clonal propagation of *T. repens. Ann. Bot.*, **54**: 201–211.

Marshall A H (1995) Peduncle characteristics, inflorescence survival and reproductive growth of white clover (*Trifolium repens* L.). *Grass Forage Sci.*, **50**: 324–330.

Marshall A H, Abberton M T, Michaelson-Yeates T P T, Rhodes I and Williams T A (2000) Drought tolerance of interspecific hybrids between *Trifolium repens* and *Trifolium ambiguum*. XIX International Grassland Congress, Universidade de Sao Paolo, 2001.

Marshall A H, Holdbrook-Smith K, Michaelson-Yeates T P T, Abberton M T and Rhodes I (1998) Growth and reproductive characteristics in backcross hybrids derived from *Trifolium repens* L. × *T. nigrescens*. Viv. Interspecific crosses. *Euphytica*, **104**: 61–66.

Marshall A H, Michaelson-Yeates T P T, Aluka P and Meredith M (1995) Reproductive characters of interspecific hybrids between *Trifolium repens* L. and *T. nigrescens* Viv. *Heredity*, **74**: 136–145.

Mathesius U, Bayliss C, Weinman J J, Schlaman H R M, Spaink H P, Rolfe B G, McCully M E and Djordjevic M A (1998a) Flavonoids synthesised in cortical cells during nodule intiation are early developmental makers in white clover. *Mol. Plant-Microbe Interact.*, **11**: 1223–1232.

Mathesius U, Schlaman H R M, Spaink H P, Sautter C, Rolfe B G and Djordjevic M A (1998b) Auxin transport inhibition precedes root nodule formation in white clover roots and is regulated by flavonoids and derivatives of chitin oligosaccharides. *The Plant J.*, **14**: 23–24.

McKersie B D (1997) Summary and future prospects for the improvement of forage legumes using biotechnology. In: *Biotechnology and the Improvement of Forage Legumes* (Eds McKersie B D, Brown D C W), CAB International, Wallingford, UK, 427–435.

Meredith M R, Michaelson-Yeates T P T, Ougham H J and Thomas H (1995) *Trifolium ambiguum* as a source of variation in the breeding of white clover. *Euphytica*, **82**: 185–191.

Micallef M C, Austin S and Bingham E T (1995) Improvement of transgenic alfalfa by backcrossing. *In Vitro Cell. Dev. Biol. Plant*, **31**: 187–192.

Michaelson-Yeates T P T, Marshall A H, Abberton M T and Rhodes I (1996) Self-fertility and heterosis in white clover (*Trifolium repens* L.). *Euphytica*, **94**: 341–348.

Michaelson-Yeates T P T, Macduff J H, Abberton M T and Raistrick N (1998) Characterisation of novel inbred lines of white clover (*Trifolium repens* L.). II. Variation in N_2 fixation, NO_3^- uptake and their interactions. *Euphytica*, **103**: 45–54.

Minchin F R, Macduff J H, Michaelson-Yeates T P T and Abberton M T (1997) Reduced nitrate sensitivity in white clover. In: *Proceedings of the 11th International Congress on N_2 Fixation*, Paris, June.

Mizen K A, Abberton M T and Cook R (1998) Stem nematode resistance in white clover and variation in nematode virulence. *Association of Applied Biologists Offered Papers in Nematology*, London, 18 December (Abstract).

Morris P and Robbins M P (1997) Manipulating condensed tannins in forage legumes. In: *Biotechnology and the Improvement of Forage Legumes* (Eds McKersie B D, Brown D C W), CAB International, Wallingford, 147–173.

New Zealand Gazette (1998) Hazardous Substances and New Organisms (Genetically Modified Organisms Approvals) Order 1998. Supplement to New Zealand Gazette of Thursday 30 July 1998. Wellington: Friday 31 July 1998, Issue 101.

Parrott W A (1991) Auxin-stimulated somatic embryogenesis from immature cotyledons of white clover. *Plant Cell Rep.*, **10**: 17–21.

Pederson G A and McLaughlin M R (1989) Resistance to viruses in *Trifolium* interspecific hybrids related to white clover. *Plant Dis.*, **73**: 997–999.

Rhodes I, Collins R P, Williams T A, Michaelson-Yeates T P T and Marshall A (1998) Breeding white clover for cool wet climates. *COST 814 II*. Workshop on Crop Development for Cool and Wet Climate of Europe. Pamplona, Spain, October 19–21, 64–78.

Rhodes I and Ortega F (1996) Progress in Forage Legume Breeding. In: *Legumes in Sustainable Farming Systems* (Ed. Younie D), BGS Occasional Symposium, SAC, Craibstone, Aberdeen, 62–71.

Rhodes I and Ortega F (1996) Progress in Forage Legume Breeding. In: *Legumes in Sustainable Farming Systems* (Ed. Younie D), Occasional Symposium No. 30, British Grassland Society, 62–71.

Rhodes I and Webb K J (1993) Improvement of white clover. *Outlook Agric.*, **22**: 181–194.

Robbins M P and Morris P (1999) Metabolic engineering of condensed tannins and other phenolic pathways in forage and fodder crops. In: *Metabolic Engineering of Plant Secondary Metabolism* (Eds Verporte R and Alfermann A W), Kluwer Academic Publishers, Netherlands, 165–177.

Rogers M E, Noble C L, Halloran G M and Nicolas M E (1997) Selecting for salt tolerance in white clover (*Trifolium repens* L.): chloride ion exclusion and its heritability. *New Phytol.*, **135**: 645–654.

Schauser L, Roussis A, Stiller J and Stougaard J (1999) A plant regulator controlling development of symbiotic root nodules. *Nature*, **402**: 191–195.

Scott A, Woodfield D and White D W R (1998) Allelic composition and genetic background effects on transgene expression and inheritance in white clover. *Mol. Breed.*, **4**: 479–490.

Sharma S B, Hancock K R, Ealing P M and White D W R (1998) Expression of a sulphur-rich maize seed storage protein, delta zein, in white clover (*Trifolium repens*) to improve forage quality. *Mol. Breed.*, **4**: 435–448.

Taylor N L and Smith R R (1998) Kura clover (*Trifolium ambiguum* M.B.). Breeding, culture, and utilisation. *Adv. Agron.*, **63**: 154–179.

Thomson D J (1984) The nutritive value of white clover. In: *Forage Legumes* (Ed. Thomson D J), Proceedings of the British Grassland Society Occasional Symposium No.16, 15–35.

Trimble J P and Hovin A W (1960) Interspecific hybridisation of certain *Trifolium* species. *Agron. J.*, **52**: 485.

van Eijsden R R, Diaz C L, de Pater B S and Kijne J W (1995) Sugar-binding activity of pea (*Pisum sativum*) lectin is essential for heterologous infection of transgenic white clover hairy roots by *Rhizobium leguminosarum* biovar *viciae*. *Plant Mol. Biol.*, **29**: 431–439.

Voigt P W, Morris D R and Godwin H W (1997) A soil-on-agar method to evaluate acid-soil resistance in white clover. *Crop Sci.*, **37**: 1493–1496.

Voisey C R, White D W R, Dudas B, Appleby R D, Ealing P M and Scott A G (1994a) *Agrobacterium*-mediated transformation of white clover using direct shoot organogenesis. *Plant Cell Rep.*, **13**: 309–314.

Voisey C R, White D W R, Wigley P J, Chilcott C N, McGregor P G and Woodfield D R (1994b) Release of transgenic white clover plants expressing *Bacillus thuringiensis* genes: an ecological perspective. *Biocontrol Sci. Technol.*, **4**: 475–481.

Waghorn G C and Caradus J R (1994) Screening white clover cultivars for improved nutritive value-development of a method. *Proceedings of the New Zealand Grassland Society*, **56**: 49–53.

Webb K J (1986) Transformation of forage legumes using *Agrobacterium tumefaciens*. *Theor. Appl. Genet.*, **72**: 53–58.

Webb K J and Abberton M T (1999) Molecular genetics of white clover. In: *COST 814* – II Crop development for the cool and wet climate of Europe. Pamplona, Spain, October 19–21, 53–63.

Webb K J, Fay M F and Dale P J (1987) An investigation of morphogenesis within the genus *Trifolium*. *Plant Cell. Tiss. Org. Cult.*, **11**: 37–46.

Webb K J, Gibbs M J, Mizen S, Skøt L and Gatehouse J A (1996) Genetic transformation of *Lotus corniculatus* with *Agrobacterium tumefaciens* and the analysis of the inheritance of transgenes in the T_1 generation. *Transgenic Res.*, **5**: 303–312.

Webb K J, Humphreys M O, Skøt L, Gibbs M and Gatehouse J (1999) Inheritance and expression of transgenes in T_2 and T_3 generations of *Lotus corniculatus* transformed using *Agrobacterium tumefaciens*. *Euphytica*, **108**: 169–179.

Webb K J, Jones S, Robbins M P and Minchin F R (1990) Characterization of transgenic root cultures of *Trifolium repens*, *T. pratense* and *Lotus corniculatus* and transgenic plants of *L. corniculatus*. *Plant Sci.*, **70**: 243–254.

Webb K J and Michaelson-Yeates T P T (1991) Selection of self-fertile lines of white clover (*Trifolium repens* L.) for shoot regeneration in tissue culture. *University College of Wales, Agricultural Society Journal*, **71**: 142–152.

Webb K J, Skøt L, Nicholson M N, Jørgensen B and Mizen S (2000) *Mesorhizobium loti* increases root-specific expression of a calcium-binding homologue identified by promoter tagging in *Lotus japonicus*. *Mol. Plant-Microbe Interact.*, **13**: 606–616.

White D W R, Woodfield D R and Caradus J R (1998) *Mortal:* A mutant of white clover defective in nodal root development. *Plant Physiol.*, **116**: 913–921.

Williams T A, Abberton, M T, Thornley W, Evans D R and Rhodes I (1998a) Evaluation of seed production potential in white clover (*Trifolium repens* L.) varietal improvement programmes. *Grass Forage Sci.*, **53**: 197–207.

Williams W M, Mason K M and Williamson M L (1998b) Genetic analysis of shikimate dehydrogenase allozymes in *Trifolium. repens* L. *Theor. Appl. Genet.*, **96**: 859–868.

Woodfield D R and Caradus J R (1994) Genetic improvement in white clover representing six decades of plant breeding. *Crop Sci.*, **34**: 1205–1213.

Wu L and McGechan M B (1999) Simulation of nitrogen uptake, fixation and leaching in a grass/white clover mixture. *Grass Forage Sci.*, **54**: 30–41.

Yamada T, Fukuoka H and Wakamatsu T (1989) Recurrent selection programs for white clover (*Trifolium repens* L.) using self-compatible plants. I. Selection of self-compatible plants and inheritance of a self-compatibility factor. *Euphytica*, **44**: 167–172.

Young S R, Humphreys M O, Abberton M T, Robbins M P and Webb K J (1999) The risks associated with the introduction of genetically modified forage grasses and forage legumes. Report to MAFF, UK.

AGROBACTERIUM-MEDIATED TRANSFORMATION OF *LOTUS* SPECIES

PHIL OGER[1,2] AND YVES DESSAUX[1,*]
[1,]Institut des Sciences Végétales, Bâtiment 23, CNRS, Avenue de la terrasse, F-91198 Gif sur Yvette, France*
**e-mail: dessaux@isv.cnrs-gif.fr*
[2]Present address: Laboratoire des Sciences de la Terre, ENS-Lyon, 46 allée d'Italie, 69364 Lyon, Cedex of France

Abstract

Lotus species, and especially *Lotus corniculatus*, are becoming important alternate forage crops. First, *Lotus* can be easily grown under a wide range of soil conditions; second, it can be heavily grazed; and third, it does not cause bloat. The development of the culture of *Lotus corniculatus* is, however, impaired by specificities of that species such as seed shattering, which increase seed cost production, or ploidy number and self-incompatibility, which makes genetic improvement tedious. Genetic engineering of *Lotus* can help to solve some of the above problems, by allowing the rapid and efficient introduction of agronomically important traits into that species. Several species of *Lotus* have been transformed using *Agrobacterium* as a gene transfer system. These include *L. corniculatus*, *L. tenuis*, *L. angustissimus* and *L. uliginosus*. *Lotus* species are easily maintained *in vivo* and *in vitro*. The efficiency of the regeneration process differs from one species to the other, but overall *Lotus* species are easily regenerated for calli, explants or cell cultures. In the present chapter, we present a survey of the available protocols for *Lotus* transformation, with an emphasis on laboratory and procedure tips to allow the people interested in genetic transformation of a novel *Lotus* species to adapt the existing protocols to the particularities of their favorite species.

1. Introduction

The genus *Lotus* (*Leguminosae*) consists of over 200 species, including xerophytic, desert and alpine perennials and salt tolerant annuals (Larsen, 1958; Urbanska, 1984). The regional centre for the origin of *Lotus* is probably the Mediterranean Basin, where the greatest diversity of the species can be observed (Swanson *et al.*, 1990). Most of the species of *Lotus* can be classified as weeds, although three species have attained agronomical importance (Grant and Marten, 1985). These are *Lotus corniculatus* (birdsfoot trefoil),

P. K. Jaiwal and R. P. Singh (eds.), Applied Genetics of Leguminosae Biotechnology, 255–281.

which is the most widely grown of the three species, *Lotus tenuis*, and *Lotus uliginosus*. Interestingly, *L. corniculatus* is postulated to be a hybrid of the two other cultivated species *L. uliginosus* and *L. tenuis* (Ross and Jones, 1985). This point, however, remains controversial (Campos *et al.*, 1994; Grant and Small, 1996; Gauthier *et al.*, 1997).

Lotus corniculatus is grown over large regions in North and South America, much of Europe and parts of Asia. *L. corniculatus* has the same feed value as alfalfa. Furthermore, it can be easily grown under a wide range of soil conditions (pH, fertility and moisture), can resist heavy grazing and does not cause bloat (Seaney and Henson, 1970; Smith, 1975; Tesar, 1977). There are some drawbacks with *L. corniculatus*, such as plant lodging and seed shattering (Grant, 1996), which reduce yield and severely increase seed production costs (McGraw and Beuselink, 1983). Besides, breeding of *L. corniculatus* is hampered by its ploidy number (tetraploid, Angulo and Real, 1977), self-incompatibility and indeterminate growth habit. However, the development of new techniques in biotechnology could help to solve some of the problems related to *Lotus* breeding and/or bring new species to significant agronomical status. In this respect, one such species, *Lotus japonicus*, although not of agronomical importance is yet emerging as an important model system (Handberg and Stougaard, 1992; Jiang and Gresshoff, 1997). This species has been chosen as an alternate plant model for the studies on nitrogen fixation by legumes, and more precisely on the establishment of determinate root nodule during the *Rhizobium*/ Legume symbiosis. The model can also be used to study another plant/microbe interaction, the symbiosis between the legume and mycorrhizae and its influence on nitrogen fixation by symbiotic *Rhizobium*. The research efforts on *L. japonicus* have already given new insights on these topics which shall be applicable to other *Lotus* or legume species (Thykjaer *et al.*, 1995; Szczyglowski *et al.*, 1997; Schauser *et al.*, 1999; Szczyglowski *et al.*, 1998; Wopereis *et al.*, 2000).

2. Genetic engineering of *Lotus*

The genetic transformation of different *Lotus* species has been reported. These include *Lotus corniculatus* (Petit *et al.*, 1987; Webb *et al.*, 1996; Akashi *et al.*, 1998). *Lotus japonicus* (Handberg and Stougaard, 1992; Oger *et al.*, 1996; Stiller *et al.*, 1997), *Lotus tenuis* (Damiani *et al.*, 1993), and *Lotus angustissimus* (Nenz *et al.*, 1996). All protocols developed for *Lotus* transformation are based on the sensitivity of most *Lotus* species to *Agrobacterium*-mediated gene transfer to plants using either *A. tumefaciens* (Oger *et al.*, 1996; Webb *et al.*, 1996; Akashi *et al.*, 1998) or *A. rhizogenes* (Petit *et al.*, 1987; Handberg and Stougaard, 1992; Daminai *et al.*, 1993; Nenz *et al.*, 1996; Oger *et al.*, 1996; Stiller *et al.*, 1997). This does not necessarily imply that other methods such as particle bombardment may not work efficiently. However, *Agrobacterium* represents a versatile, cheap and easy way to transform plants and does not require expensive equipments. The biotechnological approaches in this area of research include the identification, transfer and expression of genes responsible for low seed shattering, herbicide resistance and disease tolerance (Swanson *et al.*, 1990).

Over the past years, significant progress in *Lotus* biotechnology have been achieved on callus isolation (Swanson and Tomes, 1980; Pupilli *et al.*, 1990; Piccirili *et al.*, 1998;

Nenz *et al.*, 1996), suspension cultures (Swanson and Tomes, 1980; Vessabutr and Grant, 1995), and *in vitro* culture (Petit *et al.*, 1987; Picciriili *et al.*, 1988; Swanson *et al.*, 1990; Handberg and Stougaard, 1992; Pupilli *et al.*, 1990; Vessabutr and Grant, 1995; Nenz *et al.*, 1996; Stiller *et al.*, 1997), etc. The aim of the present chapter is not to review *Lotus* species and their usefulness as an alternate forage crop. Rather, we aim at presenting protocols that are applicable to the transformation of most species of the *Lotus* genus, and therefore may be useful to anyone interested in the genetic engineering of *Lotus*. Therefore, we have focused on our contribution and procedure tips to help users to adapt the existing protocols to the transformation of their favorite *Lotus* species. The two protocols presented here (Petit *et al.*, 1987; Oger *et al.*, 1996) approach genetic transformation and regeneration in a different way, a feature which should broaden the spectrum of *Lotus* species potentially amenable to the genetic transformation.

2.1. BACTERIAL STRAINS AND PLASMIDS IN *LOTUS* TRANSFORMATION

Bacterial strains used for *Lotus* transformation are described in Table 1. The *Agrobacterium* strains can be divided into two groups: wild-type strains, which harbor wild-type Ti plasmids (for a review on *Agrobacterium* Ti plasmids, see Zambryski, 1988), and therefore are parthogenic on *Lotus*, and *vir*-helper strains (Hoekema *et al.*, 1983, see details below). The pros and cons of each type of strains are discussed at the end of this review in the General comments section. All the *Lotus* species tested so far are sensitive to *Agrobacterium* infection (Petit *et al.*, 1987; Piccirilli *et al.*, 1988; Nenz *et al.*, 1996; Oger *et al.*, 1996; Webb *et al.*, 1996). Therefore, almost every strain of *Agrobacterium* can be used to transform *Lotus*. There is however some differences in terms of transformation efficiency as reported by Stiller *et al.* (1997) for the pathogenicity of *A. rhizogenes* strains on *Lotus japonicus*.

As a first approach, we recommend the use of wild-type strains to evaluate the sensitivity of the *Lotus* species or cultivar to *Agrobacterium* infection. Their use allows to quickly and reliably evaluate the sensitivity of a given species of *Lotus* to *Agrobacterium* transformation just by scoring for the apparition of the disease symptoms (for more details see section ‘Important considerations’). *Agrobacterium tumefaciens* induces the formation of tumors (de Cleene and De Ley, 1976), while *Agrobacterium rhizogenes* induces the formation of roots at the wound site (de Cleene and De Ley, 1981). Though imperfect, the weight of tumor tissues (for *A. tumefaciens*) or the number of roots (for *A. rhizogenes*) emerging at the infection site is an indication of the infectivity of the *Agrobacterium* strain.

Wild-type strains have been used for genetic engineering, since they allowed a high efficiency of the transformation/regeneration process (Oger *et al.*, 1996; Stiller *et al.*, 1997; Tepfer, 1984). Preference should be given to strains inducing hairy-root disease, because each root emerging at the wound site results from the transformation of a single cell. It is therefore, extremely easy to obtain and determine independent transformants. On the contrary, wild-type *Agrobacterium tumefaciens* strains induce the formation of tumors which are a chimerical cluster of cells. The regeneration can, therefore, lead to chimerical plants. Furthermore, in some cases, the metabolic perturbation generated by the insertion of the pTi T-DNA is incompatible with the regeneration process. However,

Table 1. Bacterial strains and characteristics

Strain name		Reference
Wild-type	**Tumor inducing**	
Ach5	pTiAch5; octopine-type; wild-type	Hamilton and Fall, 1971
Chry5	pTiChry5; chrysopine-type; wild-type; Supervirulent on soybean	Miller, 1975
A281	pTiBo542; agropine-type; C58 background; Supervirulent	Montoya *et al.*, 1977
C58	pTiC58; nopaline-type; wild-type	Sciaky *et al.*, 1978
T37	pTiT37; nopaline-type; wild-type	Sciaky *et al.*, 1978
Bo542	pTiBo542; agropine-type; wild-type	Sciaky *et al.*, 1978
C58C1 (pTiChry5)	pTiChry5; chrysopine-type; C58 background; Supervirulent on soybean	Oger, unpublished
Wild-type	**Hairy root inducing**	
C58C1 (pRiA4)	pRiA4	Petit *et al.*, 1983
8196	pRi8196; mannopine-type; wild-type	Petit *et al.*, 1987
C58C1 (pRi8196)	pRi8196; mannopine-type; C58 background	Petit *et al.*, 1987
15834	pRi15834; agropine-type; wild-type	Petit *et al.*, 1987
C58C1 (pRi15834)	pRi15834; agropine-type; C58 background	Petit *et al.*, 1987
A4	pRiA4; wild-type	Petit *et al.*, 1987
1855	pRi1855; wild-type	Hamill *et al.*, 1987
K599	pRiK599; cucumopine-type	Savka *et al.*, 1990
LBA9402	pRi1855; 1855 background	Damiani *et al.*, 1993
***vir* helper**		
LBA4404	disarmed pTiAch5 derivative; Ach5 background	Ooms *et al.*, 1981
C58C1 (pGV3850)	disarmed pTiAch5 derivative; C58 background	Zambryski *et al.*, 1983
ASE-1	disarmed pTiT37 derivative	Fraley *et al.*, 1985
C58Z707	disarmed pTiC58 derivative	
EHA101	disarmed pTiBo542 derivative;	Hood *et al.*, 1986;
EHA105	Supervirulent	Hood *et al.*, 1993
GV3101	disarmed pTiC58 derivative	Baker *et al.*, 1986
Mini-Ri	disarmed pRi derivative	Vilaine and Casse-Delbart, 1987
KYRT1	disarmed pTiChry5 derivative; Supervirulent on soybean	Torisky *et al.*, 1997
KPSF2	disarmed pTiChry5 derivative; Supervirulent on soybean	Palanichelvam *et al.*, 2000

derivatives of these strains might be useful as they can be a source of genes promoting disarmed T-DNA transfer. These constructs are renowned as the *vir*-helper plasmids.

Vir-helper plasmids are Ti or Ri plasmid derivatives, engineered to delete the wild-type T-DNA (Hoekema *et al.*, 1983). As a result, they cannot transfer any part of the helper plasmid itself to the plant cell during the infectious process. However, they are still capable of promoting T-DNA transfer, provided both T-DNA and helper plasmids are in the same cell (hence the name helper plasmid; Hoekema *et al.*, 1983). *Vir*-helper strains have been obtained from several broad host range *Agrobacterium tumefaciens* strains (Table 1) such as the nopaline-type C58 (C58Z707) and T37 (ASE-1, Fraley *et al.*, 1985), the octopine-type 15955 (LBA4404, Ooms *et al.*, 1981), the agropine-type Bo542 (EHA101, EHA105, Hood *et al.*, 1986, 1993) or the chrysopine-type Chy5 (KYRT1, KPSF2, Torisky *et al.*, 1997; Palanichelvam *et al.*, 2000). One *vir*-helper plasmid has also been derived from an *Agrobacterium rhizogenes* strain (Vilaine and Casse-Dellart, 1987). *Vir*-helper strains harbor the same host spectrum as the wild-type strains they originate from.

Many plasmids have been designed to provide only an engineered T-DNA *in trans* into a strain harboring a *vir*-helper plasmid (Table 2). These are commonly known as binary vectors. The *vir*-helper/binary plasmids are extremely valuable tools since they normally promote the transfer of only the engineered T-DNA to the plant cell. However, parts of the binary vector backbone are seldom transferred (Kononov *et al.*, 1997; Wenck *et al.*, 1997; Porsch *et al.*, 1998). As of today, the T-DNA of most of the binary vectors is transferred efficiently by all the *vir*-helper strains, as well as by the wild-type hairy-root or tumor-inducing *Agrobacterium* strains. Most binary vectors harbor the following features: a bacterial origin of replication, a bacterial selectable marker, a plant selectable marker and multiple cloning site for the insertion of the gene of interest into the T-DNA. Additional features include: the *lacZ* gene for gene for blue/white selection of cloned fragments, plant promoters (35S, *nos*, *mas*, etc.), plant reporter genes (*gus*, *gfp*), plant gene regulatory sequences (poly A signal, transcription terminator), etc. The choice of binary plasmids should therefore be driven by convenience of use, presence of adequate restriction sites for cloning, and antibiotic markers for selection of transformed tissues. A short list of binaries is given in Table 2.

2.2. GROWTH CONDITIONS

2.2.1. Bacteria

Agrobacterium strains can be indifferently grown on one of the following media: Rich media, low salt LB (5 g NaCl, Vaudequin-Dransart *et al.*, 1995), YEB (Ausubel *et al.*, 1989). Semi rich medium: YEM (0.5 g K_2HPO_4, 0.2 g $MgSO_4$, 0.1 g NaCl, 1 g yeast extract, 10 g mannitol). Minimal media: AT (Petit *et al.*, 1978) or AB (Chilton *et al.*, 1974) based medium, supplemented with 0.2% of a carbon source (glucose or mannitol) and 4 μg/mL D-biotin. Media are solidified with 16 g/L Bacto agar. Common antibiotics are used at the following concentration (in μg/mL): kanamycin, 50; streptomycin, 250; rifampicin, 100; tetracycline, 5; carbenicillin, 100 (to be used in place of ampicillin, to which *Agrobacterium* is resistant); gentamycin, 100. Strains are grown at 28°C with shaking to ensure sufficient aeration of liquid cultures. Growth temperatures higher than 29°C

Table 2. Binary vectors for use in genetic engineering

Plasmid name	Selectable marker: Bacterial[a]	Selectable marker: Plant[b]	MCS/BW[c]	Promoter[d]	Reporter gene	Remarks	Reference
pBin19	Kn	*nptII*	Y/Y	—	—	Complete sequence available. A large number of other binaries derived from pBin19	Bevan, 1984
pUCD series	Kn	*hpt*	(Y)/N	35S	—	—	Zyprian and Kado, 1990
pGPTV series	Kn	*nptII* *hpt* *dhfr* *bar* *ble*	(Y)/N	—	Promoterless GUS	A large variety of plant selectable marker is available in this series which makes multiple transformation or crossing possible	Becker *et al.*, 1992
pART 7/27	St	*nptII*	N/Y	35S	Convenient 35S cloning shuttle	—	Gleave, 1992
pSB series	Tc/ Sp	*nptII* *hpt* no marker	Y/N	—	GUS	A complete series of vectors for the creation of marker free transgenic lines	Komari *et al.*, 1996
pTi18/ pTi20	Kn	*nptII* + *dhfr*	N/N	—	GUS	*dhfr* confers resistance to methotrexate	Irdani *et al.*, 1998
pART 54/ pCre-1	St	*nptII*, *hpt*	N/N	35S	GUS	Derived from the pART7/27 vectors. Contains the clean gene locus *cre*/*lox*	Gleave *et al.*, 1999
pCB series	Kn	*bar*	Y/N	35S	Promoterless GUS or *gfp*	The series comprises vectors with one or more of the available features. pCB301 is a binary vector without selectable marker which makes it possible to get marker free transgenic lines	Xiang *et al.*, 1999
pBECKS pBECKS 2000 series	Sp	*hpt* + *nptII*, *bar* + *nptII*	Y/Y	—	GUS	A complete series of binary vector which includes the clean gene facility *cre*/*lox*	McCormac *et al.*, 1997; McCormac *et al.*, 1999

Table 2. (Continued)

Plasmid name	Selectable marker: Bacterial[a]	Selectable marker: Plant[b]	MCS/BW[c]	Promoter[d]	Reporter gene	Remarks	Reference
pCAM-BIA series	Kn, Cm	*hpt* *nptII* *bar*	Y/Y	—	GUS *gfp*	A very complete set of binary vector with modular selectable markers	Roberts *et al.*, 2000
pGreen series	Tc/Kn	*nptII* *hpt* *bar* *sul*	Y/Y	35S *nos*	GUS *luc* *gfp*	The most complete series available; each combination of the elements is available upon request. Marker free T-DNA are also available to create marker free transgenic lines following the protocol of Komari *et al.*, 1996	Hellens *et al.*, 2000
pMAT series	Kn	*nptII* *ipt*	Y/Y	—	GUS	A series of vectors designed to obtain marker free transgenic lines without sexual crossing, making it possible to obtain marker free transgenic lines of vegetatively propagated species or polyploid species	Ebunima *et al.*, 1997a, b; Ebunima *et al.*, 1997b; Sugita *et al.*, 1999; Sugita *et al.*, 2000

a: Kn, kanamycin resistance; St, streptomycin resistance; Sp, spectinomycin resistance; Tc, tetracycline resistance; Cm, chloramphenicol resistance. b: *nptII*, neomycin phosphotransferase; *hpt*, hygromycin phosphotransferase; *dhfr*, dihydrofolate reductase; *bar*, phosphinothricin acetyl transferase; *ble*, bleomyin phosphotransferase; *sul*, dihydropteroate synthase; *codA*, cytosine deaminase; *ipt*, isopentenyl transferase. c: presence of multiple cloning site (MCS) or blue white selection screening (BW). d: plant expressed promoter for the expression of cloned fragments.

lead to a decrease of transformation efficiency. Acetosyringone (100 μM) and low pH (5.2) can be used to induce the virulence system of *Agrobacterium* prior to inoculation.

2.2.2. Plants

Plants are grown *in vitro* on MSM/2 medium, containing half-concentrated Murashige and Skoog salts (Murashige and Skoog, 1962, M-0153, Sigma Chemical Co., St. Louis, Mo.) and 20 g/L sucrose. MSM/2 is supplemented with Morel and Wetmore vitamins

(Morel and Wetmore, 1951) but does not contain growth regulators unless otherwise stated. Alternatively, plants can be grown on Gamborg's B_5 medium (Gamborg, 1970). Media are solidified using 1 g/L gelrite (Gellam Gum K9 A40, Serva Feinbiochemica GmBH, Heidelberg) and 0.85 g/L phytagel (Sigma Chemical, St. Louis, Mo.). We have found that gelrite and phytagel increase the rooting ability of some *Lotus* cultivars, especially that of *L. japonicus* compared to the standard agar containing medium (A. Petit, unpublished observations). However if unavailable, gelrite and phytagel can be substituted with 12 g/L high quality agar. Plants and plant fragments are grown *in vitro* at 23°C under long daylight conditions (16 h, neon light, ca. 10 W m^{-2}). In the greenhouse, they are grown under long daylight conditions (16 h, sodium light, minimum 45 W m^{-2}) in nonsterile soil substrate composed of a mixture of soil, sand and clay (1/1/1) watered with tap water. Temperatures were 24°C (day) and 17°C (night).

2.3. SEED GERMINATION

Lotus seeds are first processed for 10 s with liquid nitrogen to scarify the seed coat. Scarified seeds are treated 15 s with 70% ethanol, then sterilized for 15 to 45 min in a mixture of calcium hypochlorite (70 g/L) and Teepol (1 ml/L) and washed extensively in sterile distilled water. Seeds are germinated on MSM/2 or B_5 medium *in vitro* as described above.

2.4. DETERMINATION OF THE TRANSFORMED STATUS OF PLANT MATERIAL

Agrobacterium-mediated plant transformation leads to stable integration of the T-DNA into the plant nuclear genome. The number and location of integrated copies vary from one regenerant to another. To verify the transformed status of the plant material, several procedures, such as PCR amplification or southern blot hybridization of total genomic DNA have been developed. Catabolic markers, such as the plant-expressed opine synthesis genes, can be used as transformation markers (Petit and Dessaux, 1994). One should know that the only method proving the stable integration of the transgenes into the plant genome is the study of the inheritance of the introduced traits in the progeny of the transgenic lines, e.g. by Southern blot hybridization.

2.4.1. PCR amplification of transgenes

DNA from individual regenerants is prepared by a modified Dellaporta protocol (Dellaporta *et al.*, 1983; Oger *et al.*, 1996). Two sets of PCR reactions are carried out systematically: the first one to detect the possible presence of contaminating agrobacterial cells, and the second one to amplify the target, transferred gene. Primer sets for the amplification of target genes are given in Table 3. The PCR reactions are performed in a final reaction volume of 50 μL, consisting of: 1X reaction buffer (50 mM Tris-HCl, pH–9.0, 50 mM NaCl, 2.5 mM $MgCl_2$ and 170 μg/mL BSA), 100 μM of each of the dNTP and circa 100 ng template DNA. Additionally, the mix contained 2.5 μM of each of the appropriate primers. Reactions are hot started (4 min at 94°C) by adding 2 units of Taq DNA

Table 3. Primer pairs for the amplification of common target transferred and non-transferred genes

Locus	Primer name	Primer sequence (5′–3′)	Reference
T-DNA specific			
nptII	—	GAACAAGATGGATTGCACGC GAAGAACTCGTCAAGAAGGC	Trieu *et al.*, 2000
aphIV	—	GCTGGGGCGTCGGTTTCCACTATCGG CGCATAACAGCGGTCATTGACTGGAGC	Wang *et al.*, 1992
rol	*rolB*1 *rolB*2	GCAATCTATAGCCGTGAC GGTTCCTCCGTGCGGCTG	Stiller *et al.*, 1997
nos	FGP*nos*14 FGP*nos*1236′	GGCAATTACCTTATCCGCAA CACCATCTCGTCCTTATTGA	Ponsonnet and Nesme, 1994
gus	*gus*5 *gus*6	CCATCGCAGCGTAATGCTCT GCCGACAGCAGCAGTTTCAT	Stiller *et al.*, 1997
—	—	TAGCGGGACTTTGCAAGTG GTTTTTGCAGCAGAAAAGCC	Webb *et al.*, 1996
bar	—	GAGCCCAGAACGACGCCCG TATCCGAGCCTCGTGC	Trieu *et al.*, 2000
tms	FGP$tms_2$194′ FGPtms_1, 884	CCTACTCCGGCGTTTCCATG CGGATCCCCCCCATTTC	Ponsonnet and Nesme, 1994
Non-T-DNA genes			
virG	*vir*G1 *vir*G2	CGATGACGATGTCGCTATGC CAGCACCTCTTGCAGTCTTG	Stiller *et al.*, 1997
virAG	FGP*virA*2275 FGP$virG_2$164	TCAAAAGGCAAGCAAGCAGATCTGG TCAGTGCCGCCACCTGCAGATTG	Ponsonnet and Nesme, 1994
virBG	FGP*virG*15′ FGP$virB_{11}$21	GAACGTGTTTCAACGGTTCA TGCCGCATGGCGCGTTGTAG	Ponsonnet and Nesme, 1994

polymerase (Promega, Madison, WI). The amplification was performed as follows. 1 min at 93°C; 1 min at 55°C; 1 min at 72°C (35 cycles). The last extension phase was prolonged to 10 min to allow chain termination. PCR fragments are detected by standard gel electrophoresis followed by ethydium bromide staining.

2.4.2. Southern analysis

Stable integration is confirmed by Southern analysis of restricted total genomic DNA from putative transformed lines. Southern blots are performed on plant DNA preparations obtained by the methods described above. Circa 20 μg of DNA are restricted using appropriate endonucleases, transferred onto a nylon membrane and probed with cold-labeled DNA probe. All steps of the analysis are performed according to the instructions of the manufacturer (non-radioactive DNA labeling and detection kit, Roche Biochemicals) and as reviewed by Dessaux *et al.* (1995). To prove that the DNA is integrated in the genome of the plant, it is essential to prove it with border fragments of the T-DNA. It is also essential to use restriction enzymes that do not release the probed fragment from the T-DNA, but on the contrary release T-DNA/plant DNA junction fragments, and therefore will reveal the heterogeneity of the T-DNA-border fragment length polymorphism. Since the right border of the T-DNA is more efficiently processed, it should be preferred, over the left border, to probe restricted plant DNA.

3. *Agrobacterium rhizogenes*-mediated transformation of *L. corniculatus*

This protocol was originally designed to generate transgenic roots of *L. corniculatus* for the study of *Rhizobium*-induced nodules on this species (Petit *et al.*, 1987). It is based on the ability of hairy-root strains to induce the formation of transformed roots at the infection site. Roots induced by *A. rhizogenes* arise from a single transformation event and therefore are clonal in nature (Binns and Thomashow, 1988). Consequently, separate regenerants from the same root are identical, but regenerants from separate roots are independent since they arise from independent transformation events. The major drawback of the procedure is that the regenerated T_0 lines harbor the T-DNA from the wild-type *A. rhizogenes* strain used for the transformation. For many applications, especially for research purpose, the presence of this extra piece of DNA material into the plant genome does not impair the value of the transgenic lines. However, then 'clean' transgenic plants are required (e.g. for *Lotus* breeding), the wild-type T-DNA can be eliminated from the transformed lines by successive back-crosses with the parent line (Komari *et al.*, 1996). Overall, this technique is valuable in terms of transformation efficiency, time spent and labor, and can be used as first approach for the transformation of new species or cultivars of *Lotus*.

3.1. STEP BY STEP PROTOCOL

3.1.1. Wound site infection

Lotus corniculatus seeds are surface-sterilized as described in materials and methods, germinated on MSM/2 plates and grown *in vitro* for 12 days (Fig. 1). Plantlets are wounded on the hypocotyl using a needle or scalpel dipped in *A. rhizogenes*. Bacteria can be taken directly from fresh colonies grown on selective LBm agar plates at 30°C for 2 days. Plants are left 2 weeks in sealed petri dishes for roots to develop, at 23°C continuous light. Alternatively, hypocotyls can be excised, cut longitudinally and dipped into a suspension of *Agrobacterium* (Handberg and Stougaard, 1992; Stiller *et al.*, 1997).

3.1.2. Root culture

Roots were excised from infected hypocotyls harboring hairy-roots, and transferred onto MSM/2 plates supplemented with 500 μg/mL cefotaxime. Root cultures were established 8–10 days later on new MSM/2 cefotaxime plates from individual fast growing roots. Growth conditions were as above, but plant organs are kept in the dark.

3.1.3. Regeneration and root induction

Five-cm long root pieces from well-developed axenic root cultures were sub-cultured on MSM/2 and plates were transferred to light conditions. In the light, hairy roots of

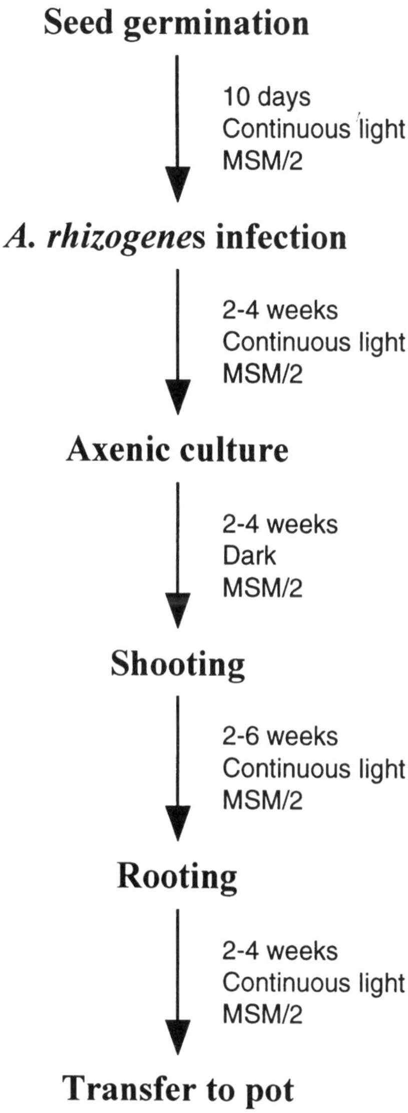

Figure 1. The flow chart for hairy-root transformation of Lotus corniculatus *using hormone-free media, according to Petit* et al. *(1987).*

L. corniculatus spontaneously differentiate, a feature which leads the formation of green shoots (Petit *et al.*, 1987). These shoots are excised when 3 to 4 cm long and transferred to fresh, hormone-free MSM/2 to stimulate root formation. Roots normally appear in approximately 1 or 2 weeks. Rooted plantlets can be maintained *in vitro* or transferred to the greenhouse.

3.1.4. Transfer to the greenhouse

Rooted plantlets are transferred to the greenhouse when their roots are approximately 1 to 2 cm long. Pots are covered with a plastic bag and grown as indicated above (long daylight, 23°C day, 17°C night). After the third day, the bags are gradually removed from the pots for a few hours per day to harden the plants. The sensitivity of the plants to greenhouse transfer differ from one species to the other. Thus, *L. corniculatus* plants withstand greenhouse transfer very easily, while *L. japonicus* plants are more fragile. However, in both cases, ca. 100% of the transfer events are successful.

3.1.5. Elimination of the wild-type T-DNA

Plants obtained by this strategy harbor both the T-DNA of interest and the wild-type, T-DNA from the *A. rhizogenes* strains used for the transformation. To obtain wild-type T-DNA-free transgenic plants, it is necessary to cross the transformed and regenerated lines with the wild-type parent line, screening the progeny for segregation of the T-DNA (Komar *et al.*, 1996). Segregating lines can be easily scored by the use of PCR amplification.

3.2. ALTERNATE REGENERATION PROCEDURE

The above protocol takes advantage of the ability of some *Lotus* species, such as *L. corniculatus*, to spontaneously regenerate shoots from root explants. In case spontaneous regeneration is infrequent or does not occur, one can use the regeneration procedure developed by Handberg and Stougaard (1992), modified by Stiller *et al.* (1997) to induce shoot formation in hairy-root cultures of recalcitrant species (Fig. 2).

3.2.1. Initiation of callus formation

One-cm-long root explants are transferred onto Gamborg's B_5 medium supplemented with 3 μg/mL of 2,4-D (2,4-dichlorophenoxyacetic acid) and kinetin. Roots are grown for 4 weeks to allow callus formation.

3.2.2. Shoot induction and elongation

After 4 weeks, calli are large enough to be transferred on shoot induction medium, which consists of B_5 medium supplemented with 0.2 μg/mL BAP (benzylaminopurine) for 5 weeks. A slightly higher frequency of shoot induction is observed when the B_5 + BAP medium is supplemented with 10 mM $(NH_4)_2SO_4$. After 8 weeks, 58% of the calli grown on ammonium containing B_5 medium harbor shoots, whereas only 50% of those on B_5 without ammonium do. However, ammonium interferes with shoot elongation, and therefore the use of B_5 + BAP + NH_4 medium implies that shooting calli be transferred into ammonium-free B_5 after 2 weeks. Frequencies of shooting calli does not increase dramatically (from 50% to 58% but the time needed for shoots to appear is significantly reduced, and the number of shoots per callus increased. During the process calli are transferred weekly to fresh medium.

3.2.3. Root induction and elongation

Shoots that are more than 1 cm long are excised from the calli and transferred onto half-strength B_5 medium supplemented with 0.5 μg/mL of the auxin analog NAA (α-naphthalene

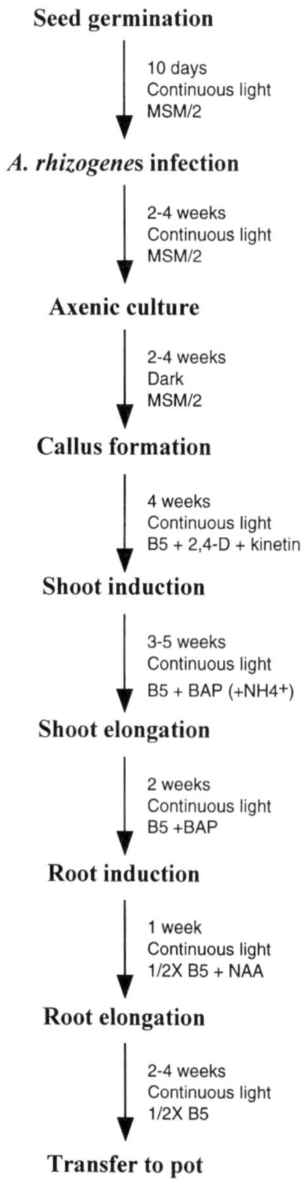

Figure 2. Alternate regeneration procedure on 2,4-D (Handberg and Stougaard, 1992).

acetic acid) to induce cell differentiation. Shoot explants are transferred into hormone-free B_5 medium after 1 week and root growth is continued until roots are approximately 1 to 2 cm long to allow transfer to the greenhouse. Rooted plants are adapted to greenhouse conditions as described above.

3.3. IMPORTANT CONSIDERATIONS

3.3.1. Antibiotic selection

We voluntarily omitted to mention antibiotic selection in the protocol for several reasons. First, when using the *Agrobacterium* induced rhizogenesis, there is no actual need for antibiotic selection, since each root emerging at the wound/inoculation site results from a single integration even of the wild-type T-DNA. Furthermore, co-transfer of the binary T-DNA with the wild-type T-DNA is frequent (50% and 75%, Tepfer, 1984; Hamill *et al.*, 1987; Stiller *et al.*, 1997; Cho *et al.*, 2000). Therefore, most hairy roots arising at the inoculation site are co-transformed with the wild-type T-DNA and the T-DNA of interest. Co-transformed roots can easily be screened by PCR amplification targeting transferred genes. Second, we and others have observed that antibiotics interfere with the regeneration and root induction processes in some *Lotus* species (for further details see general comments at the end of this chapter). This is especially true for kanamycin, even used at low concentration in *L. corniculatus* and *L. japonicus* (Handberg and Stougaard, 1992; Oger *et al.*, 1996). If antibiotic markers are used, selection for antibiotic resistant tissues can be done at either the root level (immediately after excision from the hypocotyl), or after regeneration of shoots (at the rooting step). For use with *Lotus* species, one might prefer the use of hygromycin to that of kanamycin.

3.3.2. Optimizing transformation efficiency

The above procedure can be optimized in term of transformation efficiency. Since infection by hairy-root strains of *Agrobacterium* leads to the formation of roots at the infection site, and because each root originates from a single transformation event, the number of roots appearing at the inoculation site can be regarded as an estimation of the transformation efficiency of the strain on a given *Lotus* species. For *Lotus corniculatus* and *Lotus japonicus*, the hypocotyl or cotyledonary node are the organs most sensitive to *Agrobacterium* infection (Swanson *et al.*, 1990; Handberg and Stougaard, 1992; Stiller *et al.*, 1997). However, this parameter might differ from one species to the other. Therefore, it is advisable to test root, stem and leaf section as targets for *Agrobacterium* infection.

3.3.3. Bacterial strains and plasmids

The protocol is based on the formation of roots at the inoculation site. Therefore, only hairy-root strains of *Agrobacterium* can be used (see Table 1 for details). Two such strains are routinely used in our laboratory for *Lotus corniculatus* transformation experiments. These are strains 8196 and C58C1RS (pRi15834) (see Table 1 for details). These strains have a broad host range and similar efficiency on *Lotus corniculatus*.

4. Direct transformation and regeneration of *L. japonicus* using disarmed *Agrobacterium* strains

The following protocol was developed to transform the model legume *Lotus japonicus* (Handberg and Stougaard, 1992; Jiang and Gresshoff, 1997). All experiments have been performed on the 'Gifu' ecotype of this species (Stougaard and Beuselinck, 1996). Seeds were collected originally in Japan on a river bank near Gifu by Pr. I. Hirayoshi, Kyoto University. Seeds were obtained from Dr. J. Stougaard, Laboratory of Gene Expression, University of Denmark. The transformation/regeneration protocol (Oger *et al.*, 1996) is based on the ability of some plant species, such as *L. japonicus*, to spontaneously regenerate shoots from various organs. Indeed, Dr. Petit (Institut des Sciences Végétales, CNRS, Gif sur Yvette, France) observed that section of the main stem, seedlings of *L. japonicus* would regenerate shoots near or at the cotyledon attachment site (unpublished observations). Furthermore, this regeneration process can be maintained on a long term provided the arising shoots are excised regularly (Oger *et al.*, 1996). This approach takes advantage of the regeneration potential of *L. japonicus* and the ability of *Agrobacterium* to transform meristematic cells to generate transformed shoots. Consequently, no hormone-containing media are required, limiting the generation of somaclonal variants.

4.1. STEP BY STEP PROTOCOL

4.1.1. Germination of seeds

Lotus japonicus seeds are surface-sterilized as described in materials and methods, germinated onto MSM/2 plates and grown *in vitro* for 7 to 15 days. Approximately 10 sterile seeds are placed on hormone-free MSM/2 medium with a spacing between seeds of about 2–3 cm to facilitate access to the seedlings and reduce further manipulations (Fig. 3).

4.1.2. Wound site inoculation

Seven to 15 day-old seedlings were decapitated at the cotyledon attachment site (see Figs 3 and 4), removing the main stem and both cotyledons from the plants to leave only the root system and hypocotyl. Wounded seedlings should not be transferred onto a fresh medium in this state since this often results in the death of the decapitated plants. Bacteria taken directly from fresh colonies grown on selective LB plates at 28°C for 2 days were applied at the wound site. Inoculated, decapitated seedlings are grown *in vitro* at 23°C in continuous light.

4.1.3. Regeneration

Within a few days post inoculation, shoots appear at the inoculation site. These are carefully excised, placed onto fresh MSM/2 medium supplemented with 400 μg/mL cefotaxime and allowed to root *in vitro* under long daylight conditions. After excision of the shoots, *Agrobacterium*-inoculated seedlings are returned to the growth chamber. New shoots appear at the inoculation site weekly. These new shoots are processed as described above.

4.1.4. Determination of the transformed nature of the plantlets

Transformed shoots can be screened by PCR amplification of an internal fragment of the T-DNA, or by the use of other screenable markers (*gus* activity, opine production, etc.).

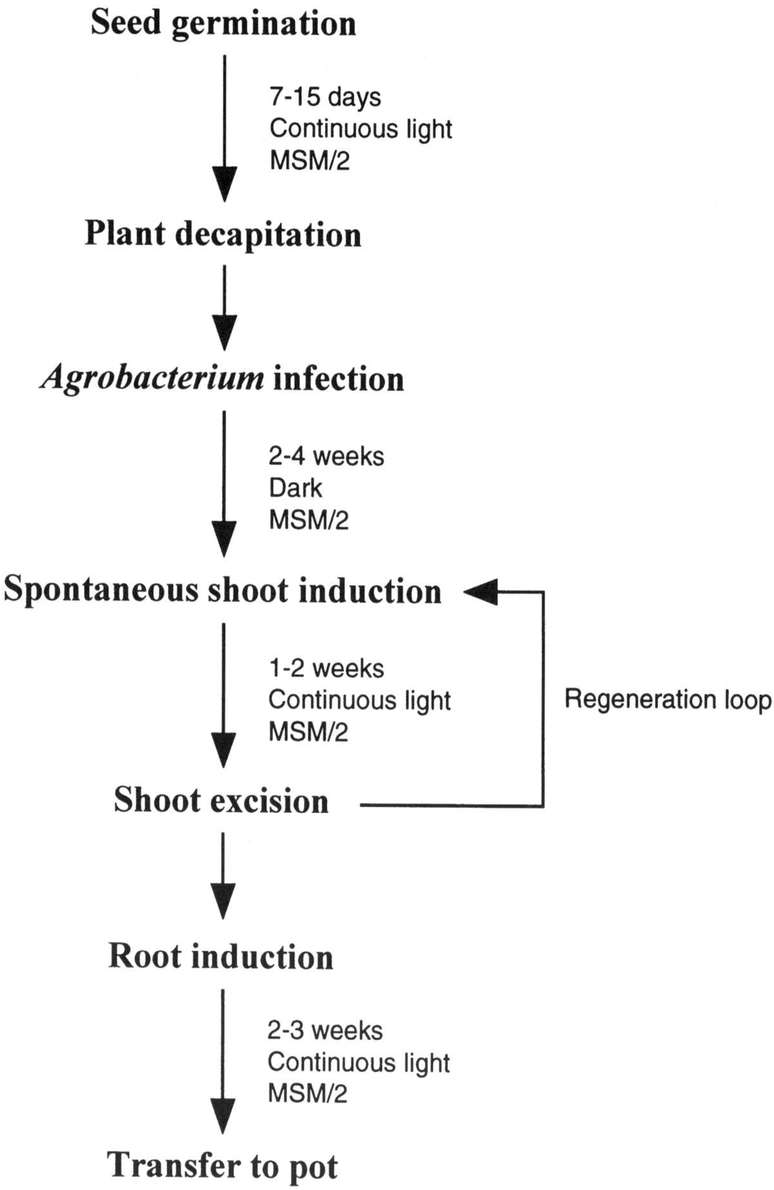

Figure 3. Schematic representation of the direct transformation/regeneration of Lotus japonicus.

PCR-positive shoots are maintained *in vitro*, and then adapted to the greenhouse as described above. Mature plants are selfed in plastic bags to set seeds. Seeds are germinated *in vitro* to generate plants whose phenotype is further investigated by PCR amplification and DNA hybridization.

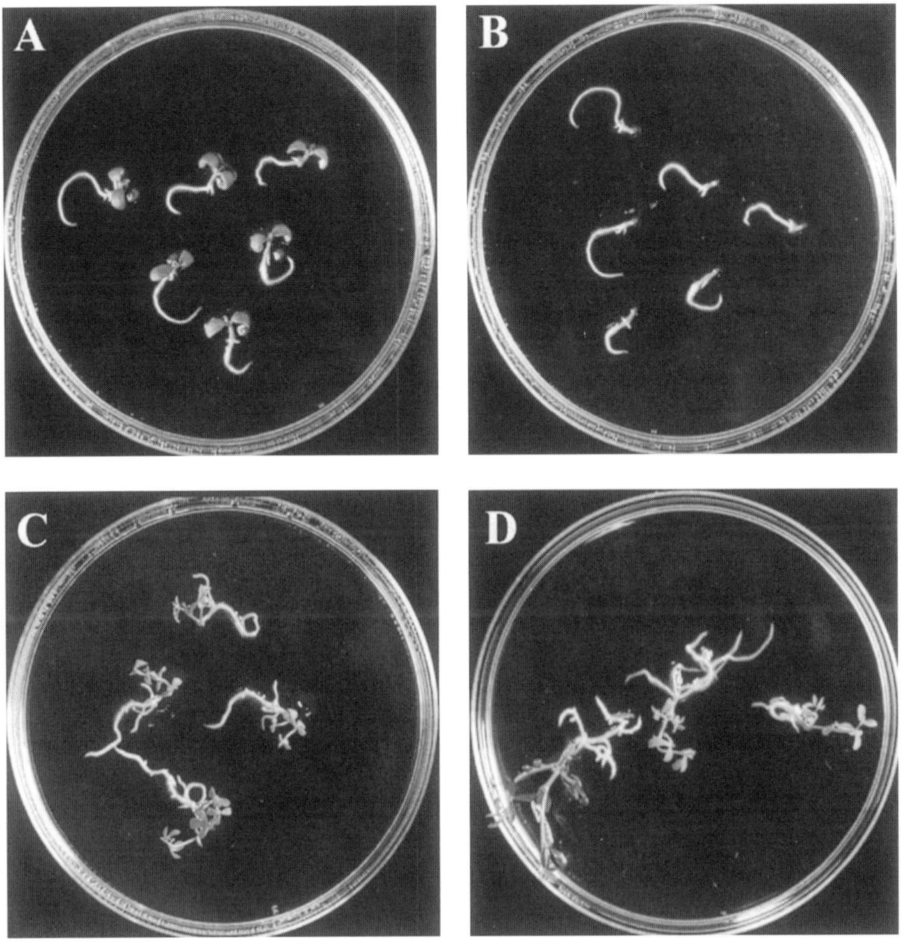

Figure 4. Inoculation of Lotus japonicus *seedlings.* Lotus *seeds are germinated* in vitro. *Panel A, seven-day seedlings. Panel B, the seedlings are cut at the cotyledon attachment site just prior to inoculation with a disarmed* Agrobacterium *strain. Panel C, a few day later, shoots appear at the wound site. They can be excised carefully and transferred to a cefotaxime-containing medium, new shoots will keep on appearing at the inoculation site for over 10 weeks (Panel D).*

4.2. IMPORTANT CONSIDERATION

4.2.1. Trouble shooting

The above protocol is labor-friendly, and gives ca. 5% of transformed shoots from which stably transformed seeds can be recovered (Oger *et al.*, 1996). However, the efficiency of this protocol mainly depends on the ability of the plant to repeatedly regenerate shoots at the inoculation site. In our many attempts to transform *Lotus japonicus*, we have observed that *Lotus japonicus*, and other legume and non-legume species, do not regenerate new shoots when the seedlings are wounded below the cotyledon attachment site. When seedlings are decapitated above the cotyledon attachment site, usually one set of two shoots originating from the cotyledonary nodes arise, but no other shoot appear after excision of this first set. Only seedling which were wounded at the cotyledon attachment site, and which wound included part of the cotyledonary node meristems, although we have not confirmed it yet, were able to regenerate spontaneously past the first set of shoots. This step is therefore crucial to ensure optimal efficiency of the transformation protocol.

4.2.2. Transformed status of the plants

Approximately 5% to 10% of the shoots appearing at the inoculation site will contain transformed cells. In contrast to what is observed with hairy-root regenerants, most shoots are chimeras of transformed and wild-type cells (Oger *et al.*, 1996). Furthermore, the proportion of wild-type vs transformed cell can be drastically different from one shoot to another. It is therefore important to be able to screen the population of shoots with a method sensitive enough to detect the transformed cells in the chimeras. In our laboratory, we routinely use opine as markers of transformation (Petit and Dessaux, 1994). These molecules can be detected in a chimerical environment, even at trace amounts (Dessaux *et al.*, 1992). Amongst other available methods, the most efficient and versatile is undoubtfully the PCR amplification of transferred genes. However, this amplification can only be performed on genes or gene fragments internal to the transferred DNA. Furthermore, the PCR amplification will be able to amplify the target gene in the T-DNA which is present in the original *Agrobacterium* strain used to inoculate the plants. Consequently, the PCR screening can only be performed on *Agrobacterium*-free material, hence a couple of weeks after, the shoots are excised and transferred on cefotaxime containing MSM/2. One way to circumvent the PCR amplification of T-DNA genes from the inciting strain is to amplify the target messenger RNA by RT-PCR on DNA-free, mRNA preparation from chimerical shoots. Results from the amplifications are only indicative of the transformed nature of the shoots if intron containing genes are targeted. In any case, this phenotype must be confirmed by Southern hybridization on the progeny of the regenerants.

4.2.3. Increasing the efficiency of transformation

The use of hairy-root or tumor-inducing wild-type strains of *Agrobacterium rhizogenes* significantly increases the transformation and regeneration efficiency of *Lotus japonicus* in our protocol. To explain this, one can propose that the T-DNA transformation leads to a hormone balance favoring the division rate of transformed cells at this site, and increasing ratio of transformed vs wild-type cells. Consequently, transformed shoots can be more readily obtained when using a combination of wild-type and engineered T-DNAs.

4.2.4. Getting seeds from chimerical shoots

Common sense tells us that the higher the ratio between transformed cells and wild-type cells in the chimera, the greater the probability that the transformed cells will be present in the germ line, and therefore in the seeds. This is especially true since maintaining selection for transformed cells is usually not possible in the greenhouse (unless resistance to a herbicide is used as a marker). This ratio can be estimated by different techniques depending on the transgenes transferred into the plant genome (opine production, *gus* expression, etc.). The yield of the PCR reactions can give an estimate of it.

Finally, we observed that the ratio of transformed cells to wild-type cells tended to decrease over time in chimeras (Oger *et al.*, 1996). This is especially true when the starting ratio is low. On the contrary, when we used wild-type hairy-root strains (rather than *vir*-helper strains) to transform *L. japonicus*, we observed that the ratio of transformed to wild-type cells did not decrease over time, and even increased in some cases (Oger *et al.*, 1996), leading to an increase in the number of lines from which transformed seeds could be obtained.

5. General comments

5.1. ON BACTERIAL STRAINS AND PLASMIDS

As mentioned above, one might choose indifferently between wild-type hairy-root or tumor-inducing strains, or non-pathogenic *vir*-helper *Agrobacterium* strains to transform *Lotus*. There is no intrinsic difference in terms of T-DNA transfer to the plant cell. However, there is one in terms of ability to obtain transformed lines from calli, which, for the end user, is the apparent transformation efficiency. Therefore, it is our understanding that wild-type strains should be used to estimate the value of a transformation protocol for a new species or cultivar. It is easier and more convenient to use and generates more transformed lines per calli. A list of wild-type hairy-root and tumor-inducing strains are listed in Table 1. They can be obtained from our laboratory, and/or their respective creators. When protocols are established, the choice to use either wild-type strains or *vir*-helper strains (see Table 1) to transfer the engineered T-DNA to the plant cell should be mostly driven by convenience of use. As a rule of the thumb, one should favour *vir*-helper strains when the species to be transformed is tetraploid or when it is self-incompatible, or when the expression of wild-type T-DNA genes are incompatible with the regeneration process. On the other hand, one should favour wild-type strains when the transformation/regeneration efficiency is limited, or when 'clean' transgenic lines are not needed, or when the selection of transformed vs wild-type cells is inefficient.

There has been an large number of binary plasmids developed over the last two decades. These plasmids come with different plant and bacterium selectable markers, such as antibiotic resistance (kanamycin, hygromycin, Table 4). From the first binary vectors developed at the beginning of the '80s (Hoekema *et al.*, 1983; Zamkryski *et al.*, 1983; Bevan, 1984) to the last generation of binaries (McCormac *et al.*, 1999; Hellens *et al.*, 2000; Roberts *et al.*, 2000; Sugita *et al.*, 2000), there has been an important effort to create versatile, powerful and convenient binary vectors. Most of the vectors developed recently include diverse plant selectable markers, reporter genes, plant promoters, which make selection of multiple transformation events possible on the same line. More recently, vectors and protocols have been developed that allow the generation of marker free transgenic plants (Abuin and Bradley, 1996; Komari *et al.*, 1996; Ebunima *et al.*,

Table 4. Plant selectable marker

Gene	Gene product	Selection	Binary	Reference
nptII	neomycin phosphotransferase	kanamycin, G418, neomycin	Yes	Bevan *et al.*, 1983; Fraley *et al.*, 1983; Herrera-Estrella *et al.*, 1983
ble	bleomycin phosphotransferase	bleomycin phleomycin	Yes	Hille *et al.*, 1986; Perez *et al.*, 1989
dhfr	dihydrofolate reductase	methotrexate	Yes	Herrera-Estrella *et al.*, 1983
cat	chloramphenicol acetyl transferase	chloramphenicol		Herrera-Estrella *et al.*, 1983
aphIV, *hpt*	hygromycin phosphotransferase	hygromycin B	Yes	van den Elzen *et al.*, 1985; Waldron *et al.*, 1985
SPT	streptomycin acetyl transferase	streptomycin	—	Jones *et al.*, 1987
accC3	gentamycin acetyl transferase	gentamycin	—	Hayford *et al.*, 1988
bar	phosphinothricin acetyl transferase	phosphinothricin, bialophos	Yes	de Block *et al.*, 1987; Thompson *et al.*, 1987
EPSP	5-enolpyruvyl-shikimate-3-phosphate synthase	glyphosate	Yes	Shah *et al.*, 1986
bxn	bromoxynil specific nitrilase	bromoxynil	—	Stalker *et al.*, 1988
psbA	QB protein 2,4-D	atrazine	—	Cheung *et al.*, 1988
tfdA	monooxygenase	2,4-D	—	Lyon *et al.*, 1989
DHPS	dihydropi-colinate synthase	S-aminoethyl L-cysteine	—	Perl *et al.*, 1993
AK	aspartate kinase	lysine, threonine	—	Perl *et al.*, 1993
sul	dihydropteroate synthase	sulfonamide	Yes	Guerineau *et al.*, 1990
csr1-1	acetolactate synthase	sulphonylurea herbicides	—	Naughn *et al.*, 1988
Tdc	tryptophane decarboxylase	4-methyl-tryptophan	—	Goddjin *et al.*,. 1993
pmi	phosphomannose isomerase	mannose	Yes	Goldsworthy and Street, 1965; Hernandez and de la Fuente, 1989; Joersbo *et al.*, 1998
codA	cytosine deaminase	5-fluorouracyl	Yes	Stougaard, 1993
sat3	streptothricin acetyl transferase	streptothricin	—	Tietze and Brevet, 1995
ipt	isopentenyl transferase	—	Yes	Ebunima *et al.*, 1997

1997; Gleave *et al.*, 1999; Hansen and Wright, 1999; McCormac *et al.*, 1999; Sugita *et al.*, 1999) (Table 2).

5.2. ON CALLUS FORMATION/SHOOT REGENERATION

In vitro methods have been applied successfully to several species and cultivars of *Lotus* (Swanson and Tomes, 1980; Grant and Marten, 1985; Piccirilli *et al.*, 1988; Pupilli *et al.*, 1990; Swanson *et al.*, 1990; Vessabutr and Grant, 1995; Nenz *et al.*, 1996). *Lotus* species appear to be relatively easy to grow and maintain *in vitro* (Swanson *et al.*, 1990; Handberg and Stougaard, 1992; Nenz *et al.*, 1996; Petit *et al.*, 1987). However, the *in vitro* culture and selection on hormone-containing media of *Lotus* cells and calli leads to the selection of somaclonal variants (Swanson, 1983; Damiani *et al.*, 1985; Nikolic *et al.*, 1997) similarly to what has been observed for other plant species. Differences in terms of regeneration capabilities can be observed between *Lotus* species or within a given species. The most critical variable in determining callus regeneration ability is the culture genotype. For *L. corniculatus*, Swanson (1983) reports regeneration frequencies ranging from 1 to 3 depending on the genotype, while the origin of the organ used to generate the callus has little effect on regeneration frequencies. Plant regeneration capabilities seem to be governed by a few genes, and improved regeneration capabilities are accessible to phenotypic selection (Swanson and Tomes, 1980). The same phenomenon has been observed for other legume species, including non-*Lotus* species, such as *Medicago truncatula* (Hoffmann *et al.*, 1997; Rose *et al.*, 1999). For species other than *L. corniculatus*, differences in regeneration efficiency have been reported for calli originating from different plant organs (leaf, root or hypocotyls). For example, Handberg and Stougaard (1992) report that in *Lotus japonicus* cotyledon-derived calli are more prone to regenerate shoots (100% of calli regenerate shoots) than root- or hypocotyl-derived calli (33% calli regenerate shoots). The ability to regenerate shoot from calli is also affected in that species by the salt composition of the medium. Two media were used: Murashige and Skoog Medium (MSM) and Gamborg's B_5 medium. Cotyledon-derived calli of *L. japonicus* did not give rise to any shoots on MS medium, while 100% did on B_5. It is therefore, difficult to predict the regeneration capabilities of a given *Lotus* species, and those of defined ecotypes or cultivars, without experimentation. However, most studies converge to point out that the cotyledons and cotyledonary nodes may be the more suitable parts of *Lotus* for the genetic transformation/regeneration process. That they can be obtained from young sterile seedlings makes them a very convenient plant material.

5.3. ON TRANSFORMATION TISSUE SELECTION

Several studies report problems with the selection of *Lotus* plants on kanamycin containing media (Handberg and Stougaard, 1992; Oger *et al.*, 1996). Handberg and Stougaard (1992) noted that untransformed calli of *L. japonicus* were not efficiently counterselected on kanamycin containing medium. We observed that kanamycin strongly reduced the rooting efficiency of kanamycin resistant *L. japonicus* shoots, which makes it inapplicable for the selection of transformed vs wild-type shoots. The problem with the use of kanamycin as a resistance marker is not restricted to *Lotus*. Several authors now use geneticin (G418) as a replacement for kanamycin in selection medium. This antibiotic gives promising results for the selection of transformed calli of *Lotus japonicus* (Stiller *et al*., 1997), although no comparative studies have been conducted so far. Several other antibiotic markers are available for the selection of transformed tissues (Table 4). One of them, namely hygromycin resistance, has been tested successfully in *L. japonicus* (Handberg and Stougaard, 1992) and *L. corniculatus* (Damiani *et al*., 1993) and *Lotus tenuis* (Damiani *et al*., 1993). It permits the direct, and efficient selection of transformed tissues on hygromycin containing media. Additional research is needed to determine which, amongst the additional markers available for plant transformation, can be useful for *Lotus* transformation.

More recently a new generation of non-antibiotic markers has been developed by different laboratories for their use in obtaining antibiotic-marker-free transgenic plant (Ebunima *et al*., 1997; Joersbo *et al*., 1998). One of these is based on the toxicity of a simple sugar to the plant cell, e.g. mannose (Stenlid, 1954; Goldsworthy and Street, 1965; Hernandez and de la Fuente, 1989; Stein and Hansen, 1999), which is detoxified by the phosphate-6-mannose-isomerase (*pmi*) gene of *E. coli* (Joersbo *et al*., 1998). This gene catabolizes the transformation of mannose into glucose, a harmless sugar to the plant cell and, importantly, harmless for cattle and/or human consumption. In the presence of mannose in the selection plates, untransformed calli do not grow or form shoots, while transformed cells prosper. Practically, the system is a positive selection screen very similar to the antibiotic markers that have been generalized in molecular biology.

6. Conclusions and future prospects

From the published and unpublished observations of the various laboratories which have tried to transform and regenerate *Lotus* plants, we believe that most of the *Lotus* species should be amenable to genetic transformation with little difficulties following the published protocols (Tepfer, 1984; Petit *et al*., 1987; Pupilli *et al*., 1990; Handberg and Stougaard, 1992; Damiani *et al*., 1993; Nenz *et al*., 1996; Oger *et al*., 1996; Stiller *et al*., 1997). However, it is possible if not probable that the efficiency of transformation will vary mostly due to specificities of the cultivar chosen for transformation. It has been known for some times now that certain cultivars of *Medicago* do not regenerate easily, while others regenerate proficiently (Hoffmann *et al*., 1997). However, as we mentioned above, the ability to regenerate seems to be coded for by a few genes only, while this phenotype can be selected effectively. The two protocols described in this chapter should allow the transformation of most *Lotus* species. The transformation/regeneration efficiency of the methods will mainly depend on three parameters:

1. the sensitivity of the *Lotus* species to be transformed to infection by *Agrobacterium* species.
2. the efficiency of the selection of transformed cells.
3. the ability to regenerate plants from transformed tissues.

Each separate critical step can be tested and optimized experimentally and individually by the use of standard plant molecular techniques that have been applied to other species.

The two protocols described here have been successfully applied to other plant species, in and outside the legume family. For example, the plant legume *L. corniculatus* (A. Petit, personal communication), soybean (Chee and Slightom, 1995) and the non-legumes *Solanum nigrum* (A. Mansouri and coworkers, unpublished data) and *Arabidopsis thaliana* (Chang *et al*., 1994) can be transformed efficiently by a direct transformation/regeneration protocol similar to the one developed originally for *Lotus japonicus*.

References

Abuin A and Bradley A (1996) Recycling selectable markers in mouse embryonic stem cells. *Mol. Cell Biol.*, **16**: 1851–1856.

Akashi R, Uchiyama T, Sakamoto A, Kawamura O and Hoffmann F (1998) High-frequency embryogenesis from cotyledons of bird's-foot treefoil (*Lotus corniculatus*) and its effective utilization in *Agrobacterium*-medium transformation. *J. Plant Physiol.*, **152**: 84–91.

Angulo M D and Real M C (1977) A new basic chromosome number in the genus *Lotus*. *Can. J. Bot.*, **55**: 1848–1850.

Ausubel F M, Brent R, Kingston R E, Morre D D, Seidman J G, Smith J A and Struhl K (1989) *Current Protocols in Molecular Biology*. Greene Publishing Associates and Wiley-Interscience, John Wiley, New York.

Baker B, Schell J, Borz H and Federoff N (1986) Transposition of the maize controlling element 'Activator' in tobacco. *Proc. Natl. Acad. Sci. USA*, **83**: 4844–4848.

Banns A B and Thomshow M F (1988) Cell biology of *Agrobacterium* infection and transformation of plants. *Annu. Rev. Microbiol.*, **42**: 575–606.

Becker D, Kemper E, Schell J and Masterson R (1992) New plant binary vectors with selectable markers located proximal to the left T-DNA border. *Plant Mol. Biol.*, **20**: 1195–1197.

Bevan M (1984) *Agrobacterium* vectors for plant transformation. *Nucleic Acid Res.*, **12**: 8711–8721.

Bevan M W, Flavell R B and Chilton M D (1983) A chimeric antibiotic resistance gene as a selectable marker for plant cell transformation. *Nature*, **304**: 184–187.

Campos L P, Raelson J V and Grant W F (1994) Genome relationships among *Lotus* species based on random amplified polymorphic DNA (RAPD). *Theor. Appl. Genet.*, **88**: 417–422.

Chang S S, Park S K, Kim B C, Kang B J, Kim D U and Nam H G (1994) Stable genetic transformation of *Arabidopsis thaliana* by *Agrobacterium* inoculation *in planta*. *Plant J.*, **5**: 551–558.

Chee P P and Slightom J L (1995) Transformation of soybean (*Glycine max*) via *Agrobacterium tumefaciens* and analysis of transformed plants. *Methods Mol. Biol.*, **44**: 101–119.

Cheung A L, Bogorad L, van Montagu M and Schell J (1988) Relocating a gene for herbicide tolerance: A chloroplast gene is converted into a nuclear gene. *Proc. Natl. Acad. Sci. USA*, **85**: 391–395.

Chilton M D, Currier T C, Farrand S K, Bendich A J, Gordon M P and Nester E W (1974) *Agrobacterium tumefaciens* and PS8 bacteriophage DNA not detected in grown gall tumors. *Proc. Natl. Acad. Sci. USA*, **71**: 3672–3676.

Cho H J, Farrand S K, Noel G R and Widhokm J M (2000) High-efficiency induction of soybean hairy-roots and propagation of the soybean cyst nematode. *Planta*, **210**: 195–204.

Damiani F, Mariotii D, Pezzoti M and Arcioni S (1985) Variation among plants regenerated from tissue culture of *Lotus corniculatus*. *Z. Pflanzenzuchtg.*, **94**: 332–339.

Damiani F, Nenz E, Paolocci F and Arcioni S (1993) Introduction of hygromycin resistance in *Lotus* ssp. through *Agrobacterium rhizogenes* transformation. *Transgenic Res.*, **2**: 330–335.

De Block M, Betterman J, Vandewiele M, Dockx J, Thoen C, Gossele V, Rao Movva N, Thompson C, van Montagu M and Leemans J (1987) Engineering herbicide resistance in plants with a detoxyfying enzyme. *EMBO J.*, **6**: 2513–2518.

de Cleene M and De Ley J (1976) The host range of crown gall. *Bot. Rev.*, **42**: 389–466.

de Clenne M and De Ley J (1981) The host range of infectious hairy-root. *Bot. Rev.*, **47**: 147–194.

Dellaporta S L, Wood J and Hicks J B (1983) A plant DNA minipreparation: version II. *Plant Mol. Biol. Rep.*, **1**: 19–21.

Dessaux Y, Elasri M, Glickmann E, Oger P, Petit A and Vaudequin-Dransart V (1995) The use of digoxigenin-labelled probes to detect DNA sequences specific for plant pathogenic bacteria. *Cell. Mol. Biol.*, **41**: 933–943.

Dessaux Y, Petit A and Tempe J (1992) Opines in *Agrobacterium* biology. In: *Molecular Signals in Plant-Microbe Communications* (Ed Verma D P S), CRC Press, Boca Raton, Florida, 109–136.

Ebunima H, Sugita K, Matsungaga E and Yamakado M (1997a) Selection of marker-free transgenic plants using the isopentenyl transferase gene as a selectable marker. *Proc. Natl. Acad. Sci, USA*, **94**: 2117–2121.

Ebunima H, Sugita K, Matsunaga E, Yamakado M and Komamine (1997b) Principle of MAT vectors. *Plant Biotechnol.*, **14**: 133–139.

Fraley R T, Rogers S G, Horsch R B, Eichholtz D A, Flick J S, Fink C L, Hoffmann N L and Sanders P R (1985) The SEV system – a new disarmed Ti-plasmid vector system for plant transformation. *Nature Biotech.*, **3**: 629–635.

Fraley R T, Rogers S G, Horsch R B, Sanders P R, Flick J S, Adams S P, Bittner M L, Brand L A, Fink C L, Fry J S, Gallupi G R and Goldberg S B (1983) Expression of bacterial genes in plant cells. *Proc. Natl. Acad. Sci. USA*, **80**: 4803–4807.

Frisch D A, Harris-Haller L W, Yokubaitis N T, Thomas T L, Hardin S H and Hall T C (1995) Complete sequence of the binary vector Bin19. *Plant Mol. Biol.*, **27**: 405–409.

Gamborg O L (1970) The effects of amino acids and ammonium on the growth of plant cells in suspension culture. *Plant Physiol.*, **45**: 372–375.

Gauthier P, Lumaret R and Bedecarrats A (1997) Chloroplast-DNA variation in the genus *Lotus* (*Fabaceae*) and further evidence regarding the maternal parentage of *Lotus corniculatus* L. *Theor. Appl. Genet.*, **95**: 629–636.

Gleave A P (1992) A versatile binary vector system with a T-DNA organisational structure conducive to efficient integration of cloned DNA into the plant genome. *Plant Mol. Biol.*, **20**: 1203–1207.

Gleave A P, Mitra D S, Mudge S R and Morris B A (1999) Selectable marker-free transgenic plants without sexual crossing: transient expression of *cre* recombinase and use of a conditional lethal dominant gene. *Plant Mol. Biol.*, **40**: 223–235.

Goddjin O J M, van der Dyun Shouten P M, Schilperoort R A and Hoge J H C (1993) A chimeric tryptophan decarboxylase gene as a novel selectable marker in plant cells. *Plant Mol. Biol.*, **22**: 907–912.

Goldsworthy A and Street H E (1965) The carbohydrate nutrition of tomato roots VIII: The mechanism of the inhibition by D-mannose of the respiration of excised roots. *Ann. Bot.*, **29**: 45–58.

Grant W F (1996) Seed pod shattering in the genus *Lotus* (*Fabaceae*): a synthesis of diverse evidence. *Can. J. Plant Sci.*, **76**: 447–456.

Grant W F and Marten G C (1985) Birdsfoot trefoil. In: *The Science of Grassland Agriculture* (Eds Heath M E, Barnes R F and Metcalfe D S), Iowa State University Press, Ames, 98–108.

Grant W F and Small E (1996) The origin of the *Lotus corniculatus* complex: a synthesis of diverse evidence. *Can. J. Bot.*, **74**: 975–989.

Guerineau F, Brooks L, Meadows J, Lucy A, Robinson C and Mullineaux P (1990) Sulfonamide resistance gene for plant transformation. *Plant Mol. Biol.*, **15**: 127–136.

Hamill J D, Prescott A and Martin C (1987) Assessment of the efficiency of co-transformation of the T-DNA of disarmed binary vector derived from *Agrobacterium tumefaciens. Plant Mol. Biol.*, **9**: 573–584.

Hamilton R H and Fall M Z (1971) The loss of tumor-initiating ability in *Agrobacterium tumefaciens* by incubation at high temperature. *Experimentia*, **27**: 229–230.

Handberg K and Stougaard J (1992) *Lotus japonicus*, an autogamous, diploid legume species for classical and molecular genetics. *Plant J.*, **2**: 487–496.

Hansen G and Wright M S (1999) Recent advances in the transformation of plants. *Trends Plant Sci.*, **4**: 226–231.

Hayford M B, Meldford J I, Hoffman N L, Rogers S G and Klee H J (1988) Development of a plant transformation selection system based on genes encoding gentamycin acetyltransferases. *Plant Physiol.*, **86**: 1216–1222.

Hellens R P, Edwards E A, Leyland N R, Bean S and Mullineaux P M (2000) pGreen: a versatile and flexible binary Ti vector for *Agrobacterium*-mediated plant transformation. *Plant Mol. Biol.*, **42**: 819–832.

Hernandez D and de la Fuenta M (1989) Mannose toxicity in Erlich ascities tumor cells. *Biochem. Cell Biol.*, **67**: 311–314.

Herrera-Estrella L, de Black M, Messens E, Hernalsteens J P, van Montagu M and Schell J (1983) Chimeric genes as dominant selectable markers in plant cells. *EMBO J.*, **2**: 987–995.

Herrera-Estrella L, Depicker A, van Montagu M and Schell J (1983) Expression of chimeric genes transferred into plant cell using a Ti-plasmid-derived vector. *Nature*, **303**: 209–213.

Hille J, Verheggen F, Roelvink P, Franssen H, van Kammen A and Zabe P (1986) Bleomycin resistance: A new dominant marker for plant cell transformation. *Plant Mol. Biol.*, **7**: 171–176.

Hoekema A, Hirsch P R, Hooykaas P J J and Schilperoort R A (1983) A binary plant vector based on the separation of *vir*- and T-region of the *Agrobacterium tumefaciens* Ti-plasmid. *Nature*, **303**: 179–180.

Hoffmann B, Trinh T H, Leung J, Kondorosi A and Kondorosi E (1997) A new *Medicago truncatula* line with superior *in vitro* regeneration, transformation, and symbiotic properties isolated through cell selection. *Mol. Plant-Microbe Interact.*, **10**: 307–315.

Hood E E, Gelvin S B, Melchers L S and Hoekema A (1993) New *Agrobacterium* helper plasmids for gene transfer to plants. *Transgenic Res.*, **2**: 208–218.

Hood E E, Helmer G L, Fraley R T and Chilton M D (1986) The hypervirulence of *Agrobacterium tumefaciens* A281 is encoded in a region of pTiBo 542 outside of T-DNA. *J. Bacteriol.*, **168**: 1291–1301.

Irdani T, Bogani P, Mengoni A, Mastromei G and Buiatti M (1998) Construction of a new vector conferring methotrexate resistance in *Nicotiana tabacum* plants. *Plant Mol. Biol.*, **37**: 1079–1084.

Jiang Q and Gresshoff P M (1997) Classical and molecular genetics of the model legume *Lotus japonicus. Mol. Plant-Microbe Interact.*, **10**: 59–68.

Joersbo M, Ddonaldson I, Kreiberg J, Peterson G S, Brunstedt J and Okkels F T (1998) Analysis of mannose selection used for transformation of sugar beet. *Mol. Breed.*, **4**: 111–117.

Jones J D, Shlumukov L, Carland F, English J, Schofield S R, Bishop G J and Harrison K (1992) Effective vectors for transformation, expression of heterologous genes, and assaying transposon excision in transgenic plants. *Transgenic Res.*, **1**: 285–297.

Jones J D G, Svab Z, Harper E C, Hurwitz C D and Maliga P (1987) A dominant nuclear streptomycin resistant marker for plant cell transformation. *Mol. Gen. Genet.*, **210**: 86–91.

Kojima S, Banno H, Yoshioka Y, Oka A, Machida C and Machida Y (1999) A binary vector plasmid for gene expression in plant cells that is stably maintained in *Agrobacterium* cells. *DNA Res.*, **6**: 407–410.

Komari T, Hiei Y, Murai N and Kumashiro T (1996) Vectors carrying two separate T-DNAs for co-transformation of higher plants mediated by *Agrobacterium tumefaciens* and segregation of transformants free from selection markers. *Plant J.*, **10**: 165–174.

Kononov M E, Bassuner B and Gelvin S B (1997) Integration of T-DNA binary vector 'backbone' sequences into the tobacco genome: evidence for multiple complex patterns of integration. *Plant J.*, **11**: 945–957.

Larson K (1958) Cytotaxonomical studies in *Lotus* IV. Somes cases of polyploidy. *Tidskr*, **54**: 44–56.

Lyon B R, Llewellyn D J, Huppatz J L, Dennis E S and Peacock W J (1989) Expression of a bacterial gene in transgenic tobacco plants confers resistance to herbicide 2,4-dichlorophenoxyacetic acid. *Plant Mol. Biol.*, **13**: 533–540.

McCormac A C, Elliott M C and Chen D F (1997) pBECKS. A flexible series of binary vectors for Agrobacterium-mediated plant transformation. *Mol. Biotech.*, **8**: 199–213.

McCormac A C, Elliott M C and Chen D F (1999) pBECKS2000: a novel plasmid series for the facile creation of complex binary vectors, which incorporates 'clean-gene' facilities. *Mol. Gen. Genet.*, **261**: 226–235.

McGraw R and Beuselink R P (1983) Growth and seed yield characteristics of birdsfoot trefoil. *Agron. J.*, **75**: 443–446.

Miller H N (1975) Leaf, stem, crown, and root galls induced in *Chrysanthemum* by *Agrobacterium tumefaciens. Phytopathology*, **65**: 805–811.

Montoya A L, Chilton M D, Gordon M P, Sciaki D and Nester E W (1977) Octopine and nopaline metabolism in *Agrobacterium tumefaciens* and crown gall tumor cells: role of plasmid genes. *J. Bacteriol.*, **129**: 101–107.

Morel G and Wetmore R H (1951) Fern callus culture. *Amer. J. Bot.*, **38**: 141–143.

Mozo T and Hooykaas P J (1992) Design of a novel system for the construction of vectors for *Agrobacterium*-mediated plant transformation. *Mol. Gen. Genet.*, **236**: 1–7.

Murashige T and Skoog F (1962) A revised medium for rapid growth and bioassays with tobacco tissue cultures. *Physiol. Plant.*, **15**: 473–497.

Naughn G W, Smith J, Mazur B and Somerville C (1988) Transformation of *Arabidopsis* acetolactate synthase gene renders tobacco resistant to sulfonylurea herbicides. *Mol. Gen. Genet.*, **211**: 266–271.

Nenz E, Pupilli F, Paolocci F, Damiani F, Cenci C A and Arcioni S (1996) Plant regeneration and genetic transformation of *Lotus angustissimus. Plant Cell Tiss. Org. Cult.*, **45**: 145–152.

Nikolic R, Mitic N and Neskovic M (1997) Evaluation of agronomic traits in tissue culture-derived progeny of bird's foot trefoil. *Plant Cell Tiss. Org. Cult.*, **48**.

Oger P, Petit A and Dessaux Y (1996). A simple technique for direct transformation and regeneration of the diploid legume species *Lotus japonicus. Plant Sci.*, **116**: 159–168.

Olszewski N E, Martin F B and Ausubel F M (1988) Specialized binary vector for plant transformation: expression of the *Arabidopsis thaliana* AHAS gene in *Nicotiana tabacum. Nucleic Acid Res.*, **16**: 10765–10782.

Ooms G, Hooykaas P J J, Moolenaar G and Schilperoort R A (1981) Crown gall plant tumors of abnormal morphology, included by *Agrobacterium tumefaciens* carrying mutated octopine Ti plasmids: analysis of T-DNA function. *Gene*, **14**: 33–50.

Palanichelvam K, Oger P, Clough S J, Cha C, Bent A and Farrand S K (2000) A second T-region on the chrysopine-type soybean-supervirulent Ti plasmid pTiChry5, and construction of a fully disarmed derivative. *Mol. Plant Microbe. Interact.*, **13**: 67–72.

Pasquali G, Ouwerkerk P B and Memelink J (1994) Versatile transformation vectors to assay the promoter activity of DNA elements in plants. *Gene*, **149**: 373–374.

explants to TDZ concentrations of up to 0.01 mg/L was stimulatory for somatic embryogenesis, but higher concentrations showed a decline in response towards somatic embryogenesis (Sankhla *et al.*, 1996; Khurana, 1998, respectively).

In vitro differentiation of plantlets both via organogenesis and embryogenesis in *Albizzia lebbeck* thus makes it a useful system for fundamental studies on the control of the two basic modes of differentiation in plant tissue cultures (Gharyal *et al.*, 1992) and for use as a suitable system for genetic transformation studies. Plantlets of several species have been transferred to field conditions either through a gradual acclimatization period under a controlled environment or directly to field conditions. Before transfer, special treatments to harden plantlets and improve root growth are necessary. Successful field establishment of regenerated trees has been reported for a limited number of species (Gharyal and Maheshwari, 1983, 1990; Detrez *et al.*, 1994; Rout *et al.*, 1995; Badji *et al.*, 1993).

Acacia sp. is another important member of Leguminosae, produces gum arabic, and is the woody species often used for reforestation in semi-arid zones. *Acacia koa* can be propagated by callus cultures derived from the shoot tips (Skolmen and Mapes, 1976). Plant regeneration via somatic embryogenesis was achieved from callus culture derived from immature cotyledons of *Acacia catechu* on Woody Plant Medium, and when supplemented with proline, resulted in somatic embryogenesis. However, in a close relative, *Acacia senegal*, only three to four micropropagules could be obtained from a uninodal explant for micropropagation (Badji *et al.*, 1993). In another arid and semi-arid leguminous tree, *Prosopis*, having proven value for forage production, human consumption and firewood production, micropropagation is reported for *P. cineraria* and *P. juliflora* (Goyal and Arya, 1981, 1984; Jordan, 1987; Kackar *et al.*, 1991; Shekhawat *et al.*, 1993).

In the black locust (*Robinia pseudoacacia*), another nitrogen-fixing, multipurpose tree, successful shoot regeneration has been reported from seedling-derived shoot callus (Han and Keathley, 1989) and callus derived from shoot cultures of mature trees (Han *et al.*, 1990). *In vitro* micropropagation has been reported from axillary buds (Chalupa, 1983; Davis and Keathley, 1987; Barghchi, 1987) and from leaf disks (Davis and Keathley, 1985). Along with this, regeneration via somatic embryogenesis has been reported (Merkle and Wiecko, 1989; Arrillaga *et al.*, 1994). Shoot regeneration from mature black locust is also reported by Han *et al.* (1993). In case of the Indian Rose Wood, *Dalbergia*, which contains approximately 25 species, clonal propagation has been reported in *D. sissoo, D. paniculata* and *D. latifolia* (Mukhopadhyay and Ram, 1981; Ravishankar and Chandra, 1988; Sharma and Chandra, 1988; Kumar *et al.*, 1991; Raghavaswamy *et al.*, 1992; Lakshmi Sita and Raghavaswamy, 1993; Das *et al.*, 1995, 1997; Gupta *et al.*, 1996).

2.4. INDUCTION OF ANDROGENESIS

Subsequent to the discovery of haploids by Guha and Maheshwari (1966) production of haploids has been extended to many different plant families. However, the technique has been exploited to a lesser extent in members of the Leguminosae, and especially the tree species where raising pure lines is severely hampered due to the long periods of juvenility (Bonga, 1987). Production of haploid plantlets in anther cultures of *Albizzia lebbeck* has been reported by Gharyal *et al.* (1983a). Other leguminous trees investigated for androgenesis were the ornamental *Cassia fistula* and a popular evergreen, avenue tree of India, *Cassia siamea*. Anthers of *Cassia siamea* when cultured on a medium supplemented with coconut milk, 2,4-D and kinetin, differentiated into a callus mass. The callus cells were analysed and shown to have a haploid chromosome number (n = 14) (Gharyal *et al.*, 1983b). Similar work has also been undertaken in two other ornamental leguminous trees, *Poinciana regia* (*Delonix regia*) and *Casssia fistula* by Bajaj and Dhanju (1983). As is common with tree species, in most of the above investigations on androgenesis, the problem of browning of the anthers and the emerging pollen callus is seen. Use of polyvinyl pyrrolidon (PVP), PVPP or activated charcoal has been beneficial to overcome this necrotic response of the parental tissue (Gharyal, 1983).

2.5. PROTOPLAST CULTURE IN TREE SPECIES

Since 1960, when E. C. Cocking first demonstrated enzymatic isolation of protoplasts, studies on protoplasts have developed immensely and an interest in genetic modification through protoplast culture has paved its way. Biotechnological utility of protoplasts with a direct application to tree improvement include somatic hybridization, somaclonal and gametoclonal variation and genetic transformation using suitable vectors. Fundamental to these research frontiers are successful procedures for the isolation and culture of protoplasts of important species and the regeneration of trees from protoplast derived cell and callus cultures.

Protoplasts can be prepared from the cells of many different plant tissues including leaf mesophyll, meristems, microspores and mega-gametophytes. In case of tree species, the most favourable source is the leaves. Various conditions including maintaining the osmoticum of the incubating solution, as well as giving a pre-plasmolysis treatment to the explant are important for successful isolation of the protoplasts. The response of a tissue (in terms of protoplast yield and viability) to different permutations and combinations of enzyme system, osmoticum and pH for successful protoplast isolation, is tissue specific and genotype dependent (Maheshwari *et al.*, 1992).

Since the first report of isolation of protoplasts from a tree species in *Acer pseudoplatanus* (Rona and Grignon, 1972) there have been many reports of protoplast isolation from tree species. However, leguminous trees have by and large remained untouched. Saxena and Gill (1985) reported protoplast culture in the leguminous tree species *Pithecellobium dulce*. Successful plantlet formation was reported via callusing and organogenesis from protoplasts isolated from leaves. In *Albizzia*, conditions were standardized for protoplast isolation and culture from the cotyledon and hypocotyls and cellular divisions leading to formation of microscopic calli were observed. Nonetheless, their further regeneration into plantlets was not realized (Gharyal, 1983) and remains an endeavour for the future. Thus, although protoplast to plant systems have been reported for a large number of herbaceous species, woody plants are generally regarded as being recalcitrant and could be the reason for the low success rates with the leguminous trees.

Protoplasts isolated from same or even diverse species can be fused, thus allowing mixing of the nuclei as well as the cytoplasmic material. However, the heterokaryon formed is not necessarily stable and may be accompanied with elimination of one or more chromosomes, or even an entire set of chromosomes. Two principal methods of protoplast fusion are either chemical, e.g. using polyethylene glycol (PEG) or by electrical means using chambers in which the cells are exposed to small electrical currents. These methods also allow the uptake of DNA molecules into the protoplasts. The formation of pores and reversible membrane discontinuities allow for the introduction of nucleic acid molecules into the cytoplasm and nucleus. Direct DNA uptake into protoplasts has its own special significance. This method allows for the study of transient gene expression and to verify various parameters for DNA uptake.

The exploitation of protoplast technology has thus opened up the possibility of creating agronomically useful genetic novelties. Plant regeneration from protoplasts can be a source of spontaneous somaclonal variation for certain genotypes and the same is true for *in vitro* selection based on protoplasts. Placing particular emphasis on selection of monogenic and allelogenic traits such as disease resistance and herbicide tolerance can lead to extremely desirable phenotypes. Homozygous plants obtained by somatic hybridization (homofusion) of protoplasts of haploid genotypes will prove helpful in the study of inheritance of horticultural traits. Finally, protoplasts are an ideal system for the production of novel transgenic trees. Genetic transformation of protoplasts followed by plant regeneration is of course the ultimate goal of various tree breeders.

3. Genetic transformation attempts

Transformation can be defined as a permanent genetic change in a cell following incorporation of foreign DNA. Plant transformation contributes enormously to progress in plant molecular biology research. It allows the development of techniques for unraveling the organization of plant cells at the molecular level. Plant genetic transformation has become a common technique both for basic research as well as for introduction of novel traits into commercially important species. Transformation of plants depends on two essential requirements: firstly, the ability to stably introduce a desired gene into the plant genome, and secondly, the ability to regenerate a fertile plant from transformed cells.

Tree species should be a major target for commercial genetic engineering and molecular breeding in the 21st century. The ability to manipulate forest tree species at the cellular and molecular level holds tremendous promise for circumventing limitations inherent in tree improvement programs such as the relatively long generation cycles, space requirements for large segregating populations and the lack of genetically pure lines (see Khurana and Khurana, 1999).

Different tools are now available to transform plants and the most commonly used are *Agrobacterium*-mediated co-cultivation and direct gene transfer by particle acceleration. *Agrobacterium*-mediated gene transfer has further revolutionized plant biotechnology. It allows the introduction into crop plants of specific characteristics that are of interest to agriculture and industry (Angenon *et al.*, 1992). Unfortunately, amongst the hundreds of forest tree species tested, only half developed tumors following inoculation with *A. tumefaciens* (De Cleene and De Ley, 1976), thus indicating the restricted host-plant compatibility. On the other hand, direct gene transfer into intact cells by microprojectile bombardment (Klein *et al.*, 1987) can overcome limitations imposed by the natural host range of *Agrobacterium*.

During the past few years, particle bombardment has proven to be an efficient method for plant cell transformation, especially in cases of restricted host range. Although a number of transient studies employing this technique have been undertaken in tree species, there are very few reports of stable transformation (Mc Cown *et al.*, 1990; Fitch *et al.*, 1990; Wilde *et al.*, 1992; Cabrera-Ponce *et al.*, 1994). The reasons being that several other factors have slowed down the adaptation of transformation techniques to tree species; regeneration protocols are still being developed for a number of woody plants, and the biology of *A. tumefaciens* and woody plant interaction is still not well understood!

The *in vitro* regeneration protocols in *Robinia pseudoacacia* were extended by Davis and Keathley (1987) and black locust was transformed with *Agrobacterium tumefaciens* strain GV3101 (pMP90) that harbored a binary vector that included genes for β-glucuronidase (GUS) and hygromycin phosphotransferase. Successful transformation was confirmed by histochemical and fluorometric assays of GUS activity in plant tissues and southern blot analysis. The transgenic plants were morphologically normal (Igasaki *et al.*, 2000). Transgenic black locust was also regenerated using Ri plasmid-mediated transformation (Han *et al.*, 1993, 1999). However, since the *A. rhizogenes* strains used for transformation were not disarmed, abnormal phenotypes were produced. Recently non-chimeric transformants of *Acacia mangium* were recovered on G418 containing medium from the rejuvenated stem tissue/buds inoculated with *A. tumefaciens* strain LBA4404 harbouring binary vector pBI121. The putative transformants were rooted in the presence of G418 and integration of *nptII* gene into plant genome was confirmed by Southern blot analysis (Xie and Hong, 2002).

As mentioned earlier, a good response towards regeneration by *in vitro* methods of micropropagation in *Albizzia lebbeck* encouraged studies towards its genetic transformation. Transient expression for the reporter gene GUS by particle bombardment has been reported for *Albizzia lebbeck* (Khurana and Khurana, 2000). A reporter gene encodes an enzyme which is readily assayable *in vitro*, and can be localized histochemically *in vivo* to provide both qualitative and quantitative information. Examples of other reporter genes include *cat* (chloramphenicol acetyl transferase), *lac*Z (β-galactosidase), *nptII* (neomycin phosphotransferase) and more recently the *luc* (luciferase) gene. The plasmid pCAMBIA2301, a 11.6 kb plasmid containing *gus* as the reporter marker and *nptII* as a selectable marker gene, was used for particle bombardment. Histochemical staining of the explants revealed high transient GUS expression in explants bombarded with this plasmid (Khurana and Khurana, 2000). Attempts at regenerating the bombarded explants are under way.

4. Conclusions and future prospects

Finally, it is envisaged that optimizing procedures for high rates of regeneration of various explants of different plant species by *in vitro* micropropagation holds an important place in plant sciences. Research in this field, especially for tree species can have important consequences including the genetic manipulation of tree species. Tree biotechnology has largely been less dealt with in comparison to the crop plants. The major areas of research where biotechnology can be important to tree species include:

- Micropropagation of 'elite' tree species.
- Somatic embryogenesis for the production of artificial seeds.
- Induction of haploidy for homozygosity of traits.
- Selection of useful somaclonal variants.
- Protoplast technology for somatic fusions, as well as obtaining cytoplasmic male sterile lines and increasing hybrid vigour.
- Finally, genetic transformation of tree species for incorporation of useful traits.

The intensive research in the vast field of plant biotechnology has enabled the identification and isolation of a number of genes responsible for varied characters. Based on this information some of the characters that are being prospected for tree biotechnology are for herbicide, pest, bacterial, viral and fungal resistance, resistance/tolerance to various biotic and abiotic stresses, improved nutrient uptake and enhanced photosynthetic efficiency, for altered products like sugar and starches, storage proteins, flavours and fragrances, pharmaceuticals, fibres and fruits. Besides these, altering plant form and architecture like plant height, branching, leaves and root structure, and controlling traits like self incompatibility and male sterility, which would have a direct bearing on hybrid production, are also envisaged.

References

Anand M and Bir S S (1984) Organogenetic differentiation in tissue cultures of *Dalbergia lanceolaria. Curr. Sci.*, **53**: 1305–1307.

Angenon G and Montagu V M (1992) Transgenic plants: *Agrobacterium* mediated transformation and its application in plant molecular biology research and biotechnology. In: *Biotechnology and Crop Improvement in Asia* (Ed Moss JP), pp. 181–199.

Anonymous (1985) In: *Wealth of India*, CSIR, New Delhi, 5, 6–8.

Arndt F, Rusch R and Stilfried H V (1976) SN 49537, a new cotton defoliant. *Plant Physiol.*, **57**: 99.

Arrillaga I, Tobolski J J and Merkle S A (1994) Advances in somatic embryogenesis and plant production of black locust (*Robinia pseudoacacia* L.). *Plant Cell Rep.*, **13**: 171–175.

Arya H C (1984) Tissue culture – an aid to forestry with special reference to Rajasthan. In: *National Seminar, Application of Science and Technology for Afforestation*, Jaipur, India, 60–63.

Badji S, Mairone Y, Ndiaye I, Merlin G, Danthu P, Neville P and Colonna J P (1992) *In vitro* propagation of gum Arabic tree (*Acacia senegal* (L.) Wild). Developing a rapid method for producing plants. *Plant Cell Rep.*, **12**: 629–633.

Bajaj Y P S and Dhanju M S (1983) Pollen embryogenesis in three ornamental trees *Cassia fistula, Jacaranda acutifolia* and *Poinciana regia. J. Tree Sci.*, **2**: 16–19.

Barghchi M (1987) Mass clonal propagation *in vitro of Robinia pseudoacacia* (Black Locust) cv. Jaszkiseri. *Plant Sci.*, **53**: 435–445.

Baweja K, Khurana J P and Gharyal-Khurana P (1995) Influence of light on somatic embryogenesis in hypocotyls of *Albizzia lebbeck. Curr. Sci.*, **68**: 544–546.

Berger K, Schaffner W (1995). *In vitro* propagation of the leguminous tree *Swartzia madagascariensis. Plant Cell Tiss. Org. Cult.*, **40**: 289–291.

Biddington N L (1992) The influence of ethylene in plant tissue culture. *Plant Growth Reg.*, **11**: 173–187.

Bonga J M (1977) Applications of tissue culture in forestry. In: *Applied and Fundamental Aspects of Plant Cell, Tissue, and Organ Culture* (Eds Reinert J and Bajaj Y P S), Berlin-Heidelberg-New York, Springer, 93–108.

Cabrera-Ponce J L, Vegas-Garcia A and Herrera-Estrella L (1994) Herbicide resistant transgenic papaya produced by an efficient particle bombardment transformation method. *Plant Cell Rep.*, **15**: 1–7.

Chalupa V (1983) *In vitro* propagation of willows (*Salix* spp.), European mountain Ash (*Sorbus aucuparia*) and Black Locust (*Robinia pseudoacacia* L.). *Biol. Plant.*, **25**: 305–307.

Chalupa V (1992) Tissue culture propagation of black locust. In: *Black Locust: Biology, Culture and Utilization. Proc. Int. Conf. Black Locust* (Eds Hanover J W, Miller K and Plesko S), East Lansing, MI, USA, 115–125.
Das A B, Rout G R and Das P (1995) *In vitro* somatic embryogenesis from callus cultures of the timber yielding tree *Hardwickia binata* Roxb. *Plant Cell Rep.*, **15**: 147–149.
Das P, Roberts A V and Rout G R (1997) *In vitro* somatic embryogenesis in *Dalbergia sissoo* Roxb. – a multipurpose timber-yielding tree. *Plant Cell Rep.*, **16**: 578–582.
Datta S K, Datta K and Pramanik T (1983) *In vitro* clonal multiplication of mature trees of *Dalbergia sissoo* Roxb. *Plant Cell Tiss. Org. Cult.*, **2**: 15–20.
Davis J M and Keathley D E (1985) Regeneration of shoots from leaf disk explants of black locust, *Robinia pseudoacacia* L. In: *Proc 4th North Central Tree Improvement Conf.*, East Lansing, MI, USA, 29–34.
Davis J M and Keathley D E (1987) Differential responses to *in vitro* bud culture in mature *Robinia pseudoacacia* L (black locust). *Plant Cell Rep.*, **6**: 431–434.
De Cleene M and De Ley J (1976) The host range of crown gall. *Bot. Rev.*, **42**: 389–466.
Debergh P, Aitken-Christie J and Cohen D (1992) Reconsideration of the term vitrification as used in micropropagation. *Plant Cell Tiss. Org. Cult.*, **30**: 135–140.
Detrez C, Ndiaye S and Dreyfus B (1994) *In vitro* regeneration of the tropical multipurpose leguminous tree *Sesbania grandiflora* from cotyledon explants. *Plant Cell Rep.*, **14**: 87–93.
Dewan A, Nanda K and Gupta S C (1992). *In vitro* micropropagation of *Acacia nilotica* subsp. *Indica* Brenan via cotyledonary nodes. *Plant Cell Rep.*, **12**: 18–21.
Duhoux E and Davies D (1985) Caulogenese a partir des bourgeons cotyledonaires d'*Acacia albida* et influence du saccharose sur la rhyzogenese. *J. Plant Physiol.*, **121**: 175–180.
Evans J M and Batty N P (1994) Ethylene precursors and antagonists increase embryogenesis in *Hordeum vulgare* Viv. Anther culture. *Plant Cell Rep.*, **13**: 678–679.
Fasolo F, Zimmerman R H and Fordham H (1989) Adventitious shoot formation on excised leaves of *in vitro* grown shoots of apple cultivars. *Plant Cell Tiss. Org. Cult.*, **16**: 75–79.
Fitch M M M, Manshardt R M, Gonsalves D, Slightom J L and Sanford J C (1990) Stable transformation of papaya via microprojectile bombardment. *Plant Cell Rep.*, **9**: 189–194.
Fluhr R and Mattoo A K (1996) Ethylene biosynthesis and perception. *Crit. Rev. Plant Sci.*, **15**: 479–523.
Gamborg O L, Miller R A and Ojima K (1968) Nutrient requirements of suspension cultures of soybean root cells. *Expt. Cell Res.*, **50**: 151–158.
Garg L, Bhandari N, Rani V and Bhojwani S S (1995) Somatic embryogenesis and regeneration of triploid plants in endosperm cultures of *Acacia nilotica. Plant Cell Rep.*, **15**: 855–858.
Gharyal P K (1983) *In vitro* differentiation in some leguminous species. Ph.D. Thesis, University of Delhi, Delhi, India.
Gharyal P K and Maheshwari S C (1981) *In vitro* differentiation of somatic embryoids in a leguminous tree, *Albizzia lebbeck* L. *Naturwissenchaften*, **68**: 379–380.
Gharyal P K and Maheshwari S C (1983) *In vitro* differentiation of plantlets from tissue cultures of *Albizzia lebbeck* L. *Plant Cell Tiss. Org. Cult.*, **2**: 49–53.
Gharyal P K and Maheshwari S C (1990) Differentiation in explants from mature leguminous trees. *Plant Cell Rep.*, **8**: 550–553.
Gharyal P K, Baweja K and Maheshwari S C (1992) *Albizzia lebbeck*: a model system for studying differentiation. In: *Biotechnology for Forest Tree Improvement*, BIOTROP Special Pub. No. 49, Southeast Asian Regional Centre for Tropical Biology, Bogor, Indonesia, 141–148.
Gharyal P K, Rashid A and Maheshwari S C (1983a) Production of haploid plantlets in anther cultures of *Albizzia lebbeck* L. *Plant Cell Rep.*, **2**: 308–309.
Gharyal P K, Rashid A and Maheshwari S C (1983b) Androgenic response from cultured anthers of a leguminous tree, *Cassia siamea* L. *Protoplasma*, **118**: 91–93.
Green B, Tabone T and Felker P (1990) A comparison of amide and ureide nitrogen sources in tissue culture of tree legume *Prosopis alba* clone B_2 V_{50}. *Plant Cell Tiss. Org. Cult.*, **21**: 83–86.
Guha S and Maheshwari S C (1966) Cell division and differentiation of embryos in *Datura in vitro. Nature*, **212**: 97–98.
Gupta N, Jain S K and Srivastava P S (1996) *In vitro* micropropagation of a multipurpose leguminous tree – *Delonix regia. Phytomorphology*, **46**: 267–275.
Han K H and Keathley D E (1989) Regeneration of whole plants from seedling-derived callus of black locust (*Robinia pseudoacacia* L.). *Nitrogen Fixing Tree Res. Rep.*, **7**: 112–114.
Han K H, Keathley D E, Davis J M and Gordon M P (1991) Regeneration of transgenic woody legume (*Robinia pseudoacacia*) and morphological alterations induced by *Agrobacterium rhizogenes* mediated transformation. *Plant Sci.*, **88**: 149–157.

Han K H, Gordon M P and Keathley D E (1999) Genetic transformation of black locust (*Robinia pseudoacacia* L.). In: *Biotechnology in Agriculture and Forestry, Vol 44, Transgenic Trees* (Ed. Bajaj Y P S), Springer-Verlag, Berlin and Heidelberg, 273–282.

Han K H, Keathley D E and Milton P G (1993) Cambial tissue culture and subsequent shoot regeneration from mature black locust (*Robinia pseudoacacia* L.) *Plant Cell Rep.*, **12**: 185–188.

Igasaki T, Mohri T, Ichikawa H and Shinohara K (2000) *Agrobacterium tumefaciens*-mediated transformation of *Robinia pseudoacacia. Plant Cell Rep.*, **19**: 448–453.

Jaiwal P K and Gulati A (1991) *In vitro* high frequency plant regeneration of a tree legume *Tamarindus indica* L. *Plant Cell Rep.*, **10**: 569–573.

Jordan M (1987) In vitro culture of *Prosopis* species. In: *Cell and Tissue Culture in Forestry. Vol 3* (Eds Bonga J M and Durzan D J), Martinus Nijhoff Publications, Boston, MA, 370–384.

Kackar N L, Vyas S C, Singh M and Solanki K R (1992) *In vitro* regeneration of *Prosopis cineraria* (L.) Druce using root as explant. *Indian J. Exp. Biol.*, **30**: 429–430.

Kaur K, Verma B and Kant U (1998) Plants obtained from the Khair tree (*Acacia catechu* Willd.) using mature nodal segments. *Plant Cell Rep.*, **17**: 427–429.

Khurana J (1998) Prospects for genetic manipulation of tree species (Part I) and *in vitro* differentiation and transformation studies on the leguminous tree – *Albizzia Lebbeck* L. (Part II), M.Sc. Thesis, University of Delhi, Delhi, India.

Khurana J and Khurana P K (2000) Biolistic-mediated DNA delivery into hypocotyls of a leguminous tree – *Albizzia lebbeck* L.: influence of biological and physical parameters. *J. Plant Biochem. and Biotech.*, **9**: 31–34.

Khurana P and Khurana J (1999) Applications of genetic transformation to tree biotechnology. *Indian J. Exp. Biol.*, **37**: 627–638.

Klein T M, Wolf E D, Wu R and Sanford J C (1987) High velocity microprojectiles for delivery of nucleic acids into living cells. *Nature*, **327**: 70–73.

Kumar A (1993) Micropropagation of a mature leguminous tree – *Bauhinia purpurea. Plant Cell Tiss. Org. Cult.*, **31**: 257–259.

Kumar A, Tandon P and Sharma A (1991) Morphogenetic responses of cultured cells of cambial origin of mature tree *Dalbergia sissoo* Roxb. *Plant Cell Rep.*, **9**: 703–705.

Kumar S, Sarkar A K and Kunhikannan C (1998) Regeneration of plants from leaflet explants of tissue culture raised safed siris (*Albizzia procera*). *Plant Cell Tiss. Org. Cult.*, **54**: 137–143.

Lakshmi Sita G, Chattopadhyay S and Tejawathi D M (1986) Plant regeneration from shoot callus of rosewood. (*Dalbergia latifolia* Roxb.) *Plant Cell Rep.*, **5**: 266–268.

Lakshmi Sita G and Raghavaswamy B V (1993) Regeneration of plantlets from leaf disc cultures of rosewood: control of leaf abscission and shoot tip necrosis. *Plant Sci.*, **88**: 107–110.

Lee S K and Rao A N (1980) Tissue culture of certain tropical trees. In: *Plant Cell Culture: Results and Prospectives* (Eds Sala F, Parisi B, Cella R and Cifferri O), Elsevier/North-Holland, Amsterdam and New York, 305–311.

Maheshwari S C, Gill R, Maheshwari N and Gharyal P K (1996) Isolation and regeneration of protoplasts from higher plants. In: *Differentiation of Protoplasts and of Transformed Plant Cells* (Eds Reinert J and Binding H), Springer-Verlag, Berlin, 3–36.

Martins-Loucao M A (1990) Carob (*Ceratonia siliqua* L). In: *Biotechnology in Agriculture and Forestry. Vol 10, Legumes and Oilseed Crops I* (Ed. Bajaj Y P S), Springer-Verlag, Berlin and Heidelberg, 658–667.

Mathur J and Mukuntha Kumar S (1992). Micropropagation of *Bauhinia variegata* and *Parkinsonia aculeate* from nodal explants of mature trees. *Plant Cell Tiss. Org. Cult.*, **28**: 119–121.

McCown D, McCabe D E, Russell D R, Robison D J, Barton K A and Raffa K F (1990) Stable transformation of *Populus* and incorporation of pest resistance by electric discharge particle acceleration. *Plant Cell Rep.*, **9**: 590–594.

Merkle S A, and Wiecko A T (1989) Regeneration of *Robinia pseudoacacia* via somatic embryogenesis. *Can. J. For. Res.*, **19**: 285–288.

Meyer M M and Kerns H R (1986) TDZ and *in vitro* shoot proliferation of *Celtis occidentalis* L. In: *Proc VI Intl. Congr Plant Tissue and Cell Culture*, Minneapolis, Abstract, 149.

Miller C O (1961) Kinetin related compounds in plant growth. *Annu. Rev. Plant Physiol.*, **12**: 395–408.

Mukhopadhyay A and Mohan Ram H Y (1986) Regeneration of plantlets from excised roots of *Dalbergia sissoo. Indian J. Exp. Biol.*, **19**: 1113–1115.

Murashige T and Skoog F (1962) A revised medium for rapid growth and bioassays with tobacco tissue cultures. *Physiol. Plant.*, **15**: 473–497.

Nandwani D and Ramawat K G (1992) High frequency plantlet regeneration from seedling explants of *Prosopis tamarugo. Plant Cell Tiss. Org. Cult.*, **29**: 56–60.

Newman M C, Preece J E, Van Sambeek J W and Gaffney G R (1993) Somatic embryogenesis and callus production from cotyledon explants of eastern black walnut (*Juglans nigra* L.). *Plant Cell Tiss. Org. Cult.*, **32**: 9–18.

Patri S, Bhatnagar S P and Bhojwani S S (1988) Preliminary investigations on micropropagation of a leguminous timber tree *Pterocarpus santalinus. Phytomorphology*, **38**: 41–45.

Phillips G C and Collins G B (1979) *In vitro* tissue culture of selected legumes and plant regeneration from callus cultures of red clover. *Crop Sci.*, **19**: 59–64.

Phillips G C and Collins G B (1980) Somatic embryogenesis from cell suspension cultures of red clover. *Crop Sci.*, **20**: 323–326.

Phukan M K, and Mitra G C (1982) *In vitro* regeneration of *Albizzia odoratissima* Benth, a shade tree for tea plantation of North East India. *Two and a Bud*, **30**: 54–58.

Purnhauser L, Medgyesy M, Czeko P J and Marton L (1987) Stimulation of shoot regeneration in *Triticum aestivum* and *Nicotiana plumbaginifolia* via tissue culture using ethylene inhibitor $AgNO_3$. *Plant Cell Rep.*, **6**: 1–4.

Raghavaswamy B V, Kimabindu K and Lakshmi Sita G (1992) *In vitro* micropropagation of elite rose wood (*Dalbergia latifolia* Roxb.). *Plant Cell Rep.*, **11**: 126–129.

Rahman S M, Hossain M, Biswas B K, Joarder O I and Islam (1993) Micropropagation of *Caesalpinia pulcherima* through nodal bud culture of mature tree. *Plant Cell Tiss. Org. Cult.*, **32**: 363–365.

Ranga Rao G V and Prasad M N V (1991) Plantlet regeneration from the hypocotyl callus of *Acacia auriculiformis* – multipurpose tree legume. *J. Plant Physiol.*, **173**: 625–627.

Rao M M and Lakshmi Sita G (1994) Direct somatic embryogenesis from immature embryos of rosewood (*Dalbergia latifolia* Roxb). *Plant Cell Rep.*, **15**: 355–359.

Ravishankar Rai V and Chandra K S (1989) *In vitro* regeneration of plantlets from shoot callus of mature trees of *Dalbegia latifolia. Plant Cell Tiss. Org. Cult.*, **13**: 77–83.

Rout G R, Samantary S and Das P (1995) Somatic embryogenesis and plant regeneration from callus cultures of *Acacia catechu* – a multipurpose leguminous tree. *Plant Cell Tiss. Org. Cult.*, **42**: 283–285.

Roy S K and Datta S K (1985) Clonal propogation of a legume tree *Albizzia procera* through tissue culture. *Bangladesh J. Bot.*, **14**: 127–131.

Sankhla D, Davis T D and Sankhla N (1996) *In vitro* regeneration of silk tree (*Albizzia julibrissin*) from excised roots. *Plant Cell Tiss. Org. Cult.*, **44**: 83–86.

Sankhla D, Sankhla N and Davis T D (1994) Thidiazuron-induced *in vitro* shoot formation form roots of intact seedlings regeneration of *Albizzia julibrissin. Plant Growth Reg.*, **14**: 267–272.

Sankhla D, Sankhla N and Davis T D (1995) Promotion of *in vitro* shoot formation from excised roots of silk tree (*Albizzia julibrissin*) by an oxime ether derivative and other ethylene inhibitors. *Plant Cell Rep.*, **15**: 115–118.

Saxena P K and Gill B (1987) Plant regeneration from mesophyll protoplasts of tree legume – *Pithecellobium dulce. Plant Sci.*, **53**: 257–262.

Schenk R U and Hildebrandt A C (1972) Medium and techniques for induction and growth of monocotyledonous and dicotyledonous plant cell cultures. *Can. J. Bot.*, **50**: 199–204.

Sharma S C and Chandra N C (1988) Organogenesis and plantlet formation *in vitro* in *Dalbergia sissoo* Roxb. *J. Plant Physiol.*, **132**: 145–147.

Shekhawat N S, Rathore T S, Singh R P, Deora N S and Rao S R (1993) Factors affecting *in vitro* clonal propagation of *Prosopis cineraria. Plant Growth Reg.*, **12**: 273–280.

Sinha R K and Mallick R (1993) Regeneration and multiplication of shoots in *Albizzia falcataria. Plant Cell Tiss. Org. Cult.*, **32**: 259–261.

Skolmen R G and Mapes M O (1976) *Acacia koa* – Gray plantlets from somatic callus tissue. *J. Hered.*, **67**: 114–115.

Thorpe T A, Harry I S and Kumar P P (1991) Application of micropropagation to forestry. In: *Micropropagation* (Eds Debergh P C and Zimmermann R H), Kluwer Academic Publishers, Dordrecht, The Netherlands, 311–336.

Tomar U K and Gupta S C (1988a) Somatic embryogenesis and organogenesis in callus cultures of a tree legume – *Albizzia richardiana. Plant Cell Rep.*, **7**: 70–73.

Tomar U K and Gupta S C (1988b) *In vitro* plant regeneration of leguminous trees (*Albizzia spp.*). *Plant Cell Rep.*, **7**: 385–388.

Upreti J and Dhar U (1996) Micropropagation of *Bauhinia vahii* Wight and Arnott – a leguminous liana. *Plant Cell Rep.*, **40**: 256–264.

Van Nieuwkerk J P, Zimmermann R H and Fordham I (1986) Thiadiazuron stimulation of apple shoot proliferation *in vitro. Hort. Sci.*, **21**: 516–518.

Villarreal M L and Rojas G (1995) *In vitro* propagation of *Mimosa tenuiflora* (Willd.) Poiret, a Mexican medicinal tree. *Plant Cell Rep.*, **16**: 80–82.
Wilde H D, Meagher R B and Merkle S A (1992) Expression of foreign genes in transgenic yellow poplar plants. *Plant Physiol.*, **98**: 114–120.
Xie D Y and Hong Y (2002) *Agrobacterium*-mediated genetic transformation of *Acacia mangium. Plant Cell Rep.*, **20**: 917–922.

REGENERATION AND GENETIC TRANSFORMATION OF TREE LEGUMES WITH SPECIAL REFERENCE TO *LEUCAENA* SPECIES

SMITA RASTOGI AND U. N. DWIVEDI*
Department of Biochemistry, Lucknow University, Lucknow – 226 007, India
**e-mail: upendradwivedi@hotmail.com*

Abstract

In the present times, when tissue culture techniques are reckoned as prerequisites for studies on genetic transformation, it is encouraging to note that a few important leguminous tree species such as *Dalbergia*, *Robinia*, *Sesbania*, *Bauhinia*, *Cassia*, *Cercis*, *Parkinsonia*, *Tamarindus*, *Acacia*, *Albizzia*, *Paraserianthus*, *Prosopis* and *Leucaena* can be mass propagated *in vitro*. Despite their utility as timber, pulpwood, firewood, forage, avenue trees and soil fertility enrichment, further improvement in their quality attributes is desirable through genetic transformation. However, since tree legumes are recalcitrant to genetic transformation, quality improvement programme for these tree legumes through tissue culture and genetic transformation is still in its preliminary stage of development. The present review gives a comprehensive account of the studies conducted so far on the regeneration and transformation of various tree legumes. Though *in vitro* regeneration for many tree legumes has been achieved, the genetic transformation has been achieved only in *Robinia pseudoacacia*, *Acacia mangium* and *Leucaena leucocephala*. A detailed protocol for plant regeneration and transformation of *Leucaena leucocephala* is presented here.

1. Introduction

Legumes are economically and ecologically important crops and have been the center of multifaceted scientific interest. Legumes constitute a major source of proteins for human food consumption (grain legumes), utilized as animal feed (forage or pasture legumes), have inherent capacity to supply their nitrogen demand from symbiotic association with *Rhizobium* and, finally, used for biomass production (forest tree legumes). Leguminous trees can be treated as crops for biomass production, economically important sources of timber, pulpwood or firewood and are suitable for agroforestry. They also find application in waste-land reclamation and as avenue trees (Singh and Singh, 1971). With the

P. K. Jaiwal and R. P. Singh (eds.), Applied Genetics of Leguminosae Biotechnology, 301–326.

expanding biomass requirements due to population explosion, the demand for quality and quantity of firewood and wood based products is continuously increasing, leading to gradual shrinking of forest cover. In this direction, an increasing effort is being devoted to the improvement and expansion of cultivation of tree legumes. Traditional breeding methods in tree are limited because of their large size and long life cycles. Moreover, these methods are slow and time consuming. Nowadays, micropropagation is being increasingly recognized as a promising tool with much potential in improving and multiplying economically important trees for their application in the field of agroforestry. Clonal forestry is, therefore, gaining increasing recognition as a quicker alternative in tree legume improvement (Mroginski and Kartha, 1984; Mascarenhas and Muralidharan, 1989; Kumar *et al.*, 1998; Lakshmi Sita and Raghavaswamy, 1998). It offers a rapid means of afforestation, multiplying woody biomass, generation of disease free stock and stress tolerant variants, selection of variants and mutants and for conserving elite and rare germplasm. The powerful technique of genetic engineering permits useful traits to be introduced into forest trees within an economically viable time frame. Thus, development of genetic transformation protocols offers great potential for quality improvement in these trees. Efforts are being made to develop efficient protocols for genetic transformation of tree legumes. The targets for tree legume engineering should include resistance to pests, herbicides, alteration of plant structure for ornamentals and manipulation of lignin content and composition. A reduction in lignin content and composition offers enhanced forage digestibility and increases the suitability of wood for pulp and paper industry.

Genetic engineering of forest trees to conform to desired traits has shifted the emphasis of forest tree improvement away from the traditional breeding programmes during the past decade. Unfortunately, quality improvement through biotechnological approach in the case of legumes has so far been restricted to herbaceous seed and forage crops, while the woody species which comprise some well-known ornamentals, medicinal and timber crops of the subtropics and tropics have generally been neglected. This is probably attributable either to lack of efficient gene transfer mechanism due to their recalcitrant nature or lack of *in vitro* culture system for plant regeneration.

Quite a lot of advances have been made in tissue culture methodologies related to leguminous tree species during the past three decades. However, the genetic transformation has not been achieved in tree legumes except for *Robinia pseudoacacia*, *Acacia mangium* and *Leucaena leucocephala*. The objective of this review is to provide a comprehensive account of regeneration and transformation of these leguminous trees with special reference to *Leucaena leucocephala*, a multipurpose nitrogen fixing tree legume.

2. Regeneration and transformation of tree legumes

A detailed account of studies conducted on plant regeneration and transformation of various tree legumes is described in the following sections.

2.1. PAPILIONACEAE

2.1.1. Dalbergia (Rosewood)

The genus *Dalbergia* contains about 25 species that are widely distributed throughout the tropics and subtropics and is commonly known as rosewood. It is an economically valued timber yielding forest tree and occupies a significant position in world market. These trees are slow growing and native to India and have good export value. Other applications include its utility as manure, fodder and tannin. The natural growing stock of this genus is dwindling due to indiscriminate cutting. Since the timber of rosewood is in great demand, application of tissue culture techniques to improve the quality and quantity of the tree on a global scale is greatly needed.

Dalbergia sissoo. Extensive research efforts have been done for *in vitro* plant regeneration of *Dalbergia sissoo* through organogenesis, employing various explants such as root explants from seedlings with or without apical meristem (Mukhopadhyay and Mohan Ram, 1981), hypocotyl (Bhargava *et al*., 1983), nodal explants of mature tree (Datta *et al*., 1983), cultured axillary buds of mature tree (Dawra *et al*., 1984), cultured cells of cambial origin of a mature tree (Kumar *et al*., 1991) and cotyledonary node explants (Pradhan *et al*., 1998a). Besides organogenesis, somatic embryogenesis potentially offers an alternative and efficient system for plant regeneration (Mukhopadhyay and Mohan Ram, 1981; Bhargava *et al*., 1983; Das *et al*., 1997). The induction of somatic embryogenesis was found to be strongly dependent on the proportion of auxin and cytokinin concentrations as well as on the age of zygotic embryo.

Dalbergia latifolia. *In vitro* multiplication of *Dalbergia latifolia* has been achieved from a variety of explants and tissues such as rooted axillary shoots (Mascarenhas *et al*., 1982), seedling explants (Rai and Chandra, 1988a; Nataraja and Sudhadevi, 1984), and shoot callus (Lakshmi Sita *et al*., 1986). Since plantlet regeneration with seedling explants is not desirable from a tree improvement point of view, researchers have also used nodal segments of mature trees, as old as 60–80 years, as explant source. Recently, an efficient plant regeneration system from cell suspension derived callus has been described (Pradhan *et al*., 1998b). Suspension cultures offer several distinct advantages over stationary cultures. When grown in liquid medium, the cells are evenly exposed to nutrients, vitamins and growth regulators. This allows a more precise manipulation of medium components and, consequently, a better control of growth and development.

Direct somatic embryogenesis from immature embryos (Rao and Lakshmi Sita, 1996), internodal segments (Rao, 1986; Lakshmi Sita and Raghavaswamy, 1992), leaf (Lakshmi Sita and Raghavaswamy, 1993) and shoot (Rai and Chandra, 1988b) of rosewood has also been demonstrated (Table 1).

Dalbergia lanceolaria. Plantlet regeneration through seedling explants has been described for *D. lanceolaria* (Anand and Bir, 1984).

2.1.2. Robinia

Robinia pseudoacacia (Black locust). *Robinia pseudoacacia* is a papilionaceous legume native to the South Western United States but has a worldwide naturalized range. It is

Table 1. Summary of regeneration studies reported for tree legumes belonging to Papilionaceae

Tree	Explant source	Explant(s)	Method	Result	References
Dalbergia latifolia	Mature tree (5 years)	Root, Nodal segment	Somatic embryogenesis	Plantlet, Soil transfer	Mascarenhas *et al.*, 1982
	Seedling	Various explants	Organogenesis	Plantlet	Nataraja and Sudhadevi, 1985
	Mature tree (5–6 years)	Internodal segment	Somatic embryogenesis	Plantlet	Rao, 1986
	Mature tree	Stem	Somatic embryogenesis	Plantlet, Soil transfer	Lakshmi Sita *et al.*, 1986
	Elite tree (50 years)	Shoot tip, Shoot segment	Somatic embryogenesis	Plantlet, Soil transfer	Rai and Chandra, 1988a
	Seedling	Hypocotyl, Shoot tip	Organogenesis	Plantlet, Soil transfer	Rai and Chandra, 1988b
	Elite tree (60–80 years)	Internodal segment, Nodal segment, Axillary bud	Somatic embryogenesis, Organogenesis	Plantlet, Field planting	Lakshmi Sita and Raghava-swamy, 1992
	Elite tree (60–80 years)	Leaf	Somatic embryogenesis	Plantlet, Soil transfer	Lakshmi Sita and Raghavaswamy, 1993
	Mature tree	Pod, Immature zygotic embryo	Somatic embryogenesis	Plantlet	Rao and Lakshmi Sita, 1996
	Seedling	Hypocotyl	Cell suspension	Plantlet, Soil transfer	Pradhan *et al.*, 1998b
Dalbergia sissoo	Seedling	Root	Somatic embryogenesis, Organogenesis	Plantlet, Field planting	Mukho-padhyay and Mohan Ram, 1981
	Mature tree (30 years)	Nodal segment	Organogenesis	Plantlet	Datta *et al.*, 1983
	Seedling	Hypocotyl	Somatic embryogenesis, Organogenesis	Callus	Bhargava *et al.*, 1983
	Mature tree (30–50 years)	Axillary bud	Organogenesis	Plantlet, Soil transfer	Dawra *et al.*, 1984

Table 1. (Continued)

Tree	Explant source	Explant(s)	Method	Result	References
Dalbergia sissoo	Elite tree	Cambial tissue	Cell suspension	Plantlet	Kumar *et al.*, 1991
	Mature tree	Pods, Immature zygotic embryo	Somatic embryogenesis	Plantlet	Das *et al.*, 1997
	Seedling	Cotyledonary node	Organogenesis	Plantlet, Soil transfer	Pradhan *et al.*, 1998a
Dalbergia lanceolaria	Seedling	Root, Hypocotyl, Cotyledon, Stem, Leaf	Somatic embryogenesis, Organogenesis	Plantlet	Anand and Bir, 1984
Robinia pseudo-acacia	Young plant (1–3 years)	Nodal segment, Shoot tip	Organogenesis	Plantlet	Chalupa, 1983
	Mature tree	Nodal segment	Organogenesis	Plantlet	Chalupa, 1983
	Seedling	Leaf	Organogenesis	Plantlet	Davis and Keathley, 1985
	Mature tree	Shoot tip	Organogenesis	Plantlet	Chalupa, 1987
	Mature tree	Axillary bud	Organogenesis	Plantlet	Barghchi, 1987
	Mature tree (20–30 years)	Axillary bud	Organogenesis	Plantlet, Soil transfer	Davis and Keathley, 1987
	Mature tree	Immature seed	Somatic embryogenesis	Plantlet, Soil transfer	Arrillaga *et al.*, 1994
Sesbania bispinosa	Seedling	Cotyledon, Leaf	Somatic embryogenesis	Plantlet	Sinha and Mallick, 1991
Sesbania speciosa	Seedling	Various explants	Organogenesis	Callus	Narmatha *et al.*, 1997
Sesbania grandiflora	Seedling	Hypocotyl, Cotyledon	Somatic embryogenesis, Organogenesis	Plantlet	Khattar and MohanRam, 1983
	Seedling	Hypocotyl, Cotyledon	Somatic embryogenesis, Organogenesis	Plantlet, Field planting	Shankar and MohanRam, 1990
Sesbania rostrata	Seedling	Hypocotyl, Cotyledon, Immature embryo	Somatic embryogenesis, Organogenesis	Plantlet	Vlachova *et al.*, 1987

a fast growing broad-leaved nitrogen-fixing tree. It finds use for timber, firewood, honey, fodder, soil conservation and erosion control (Keresztesi, 1983).

Several reports exist for regeneration of this species *in vitro* (Brown and Sommer, 1982; Chalupa, 1983, 1987; Barghchi, 1987; Davis and Keathley, 1985, 1987, 1988; Han and Keathley, 1989; Han *et al.*, 1993; Arrillaga *et al.*, 1994). A method for propagating shoots of black locust (Chalupa, 1983) has been reported which used shoot nodal segments as the primary explant. Another report (Brown and Sommer, 1982) indicated that whole black locust plantlets were regenerated from seedling derived callus tissue. Shoots have also been regenerated from leaf disks using the primary leaves of seedlings as explant source (Davis and Keathley, 1985). Later Davis and Keathley (1987) performed bud culture utilizing buds excised from the stems of dormant mature (20–30 year old) black locust trees. Barghchi (1987) has developed method for micropropagation of *R. pseudoacacia* using axillary bud explants obtained from clones propagated through stem cuttings and root cuttings and has also seen the effect of naphthalene acetic acid (NAA), 7-aza-indole (AZI), gibberellic acid (GA_3) and abscisic acid (ABA) on shoot growth and proliferation. Arrillaga *et al.* (1994) reported embryogenic culture initiation using seeds of different developmental stages. They obtained best results with seeds collected 2–3 weeks post anthesis and cultured for 3 weeks on modified Finer and Nagasawa medium containing 2,4-dichlorophenoxy acetic acid (2,4-D) and benzyladenine (BA). The plantlets recovered were acclimatized and successfully grown in greenhouse. This was a very efficient protocol for embryogenic culture initiation from immature seeds with regard to efficiency of embryogenesis, a major improvement over the earlier reports (Merkle and Wiecko, 1989; Merkle, 1992). Recently stable genetic transformation of *Robinia pseudoacacia* has been reported using *A. rhizogene* (Han *et al.*, 1993) and *A. tumefaciens* (Igasaki *et al.*, 2000) (Table 1).

2.1.3. Sesbania

Sesbania, a member of the Papilionaceae, constitutes a group of approximately 70 species, prevalent in the warmer areas. All *Sesbania* species have soil improving properties due to their unique symbiotic relationship with a specific strain of *Rhizobium* and are useful as ground cover and erosion prevention plants. Certain species are used for shadowing of plantations, production of fiber, paper pulp, fodder, food or as ornamentals and medicinal plants (Allen and Allen, 1981). *Sesbania* species are nodulated by nitrogen fixing bacteria and hence can grow rapidly on nitrogen-poor soil. It also improves their usefulness as green manure and intercrop with rice and wheat.

Sesbania rostrata. S. rostrata is an annual, fast growing species, which grows naturally in West Africa, Angola, Mozambique and Madagascar. A protocol for organogenesis from different explants of *S. rostrata* such as cotyledons, hypocotyls, immature embryos has been reported (Vlachova *et al.*, 1987). They also initiated experiments to test the susceptibility of *S. rostrata* to infection by *Agrobacterium tumefaciens* and *A. rhizogenes*, a step towards developing a protocol for genetic transformation of this tree. In their study, greenhouse grown plants as well as aseptically grown plants and explants were infected with a variety of bacterial strains. *Agrobacterium tumefaciens* GV3101 (pTiC58) (nopaline type) and B6S3 (octopine type) as well as *A. rhizogenes* 15834 induced the formation of tumours on wounded stems of all infected *S. rostrata* plants after 2–4 weeks, but unfortunately, no plants could be regenerated from these explants.

Sesbania grandiflora. Plantlet regeneration from tissue culture of *Sesbania grandiflora* has been reported (Khattar and MohanRam, 1983; Shankar and MohanRam, 1990). Differentiation of shoot buds from callus derived from cultured hypocotyl or cotyledonary explants of *Sesbania grandiflora* was achieved on B_5 medium.

Sesbania speciosa and *Sesbania bispinosa*. Callus induction and organogenesis from seedling explants have also been reported for *Sesbania speciosa* Tanbert ex. Eagl. and *Sesbania bispinosa* (Jacq.) (Narmatha *et al.*, 1997; Sinha and Mallick, 1991) (Table 1).

2.2. CAESALPINIACEAE

2.2.1. Bauhinia *(kachnar)*

Bauhinia is a fast growing ornamental plant and adapted to a wide range of soil types and climatic conditions.

Bauhinia purpurea. *B. purpurea*, a leguminous forest tree, is a good source of timber, fiber, fuel and tanning agent. It also serves as medicinal plant and avenue tree. It grows in sub-Himalayan tracts of India and South East Asia. It is seed propagated but due to low viability of seeds, *in vitro* propagation of *B. purpurea* is desirable (Anand and Bir, 1983; Kumar, 1992). Callus culture has been initiated by inoculating stem segments of 15–18 year old tree in liquid MS medium supplemented with 2,4-D (Kumar, 1992). These calli were subjected to caulogenesis and rhizogenesis by culturing on MS medium fortified with NAA and/or kinetin. Successful transfer of plantlets to soil was accomplished and the micropropagated plants were normal and identical.

Bauhinia variegata. Leaves and flower buds of *Bauhinia variegata* are eaten as vegetables and its bark is used for dyeing and tanning. Mathur and Mukunthakumar (1992) developed *in vitro* protocol for *B. variegata* using nodal explants from mature (6–8 year old) trees using MS medium fortified with BA. Regenerated shoots were subsequently subjected to rooting on MS medium containing IBA and the plantlets were successfully transferred to soil. Dasgupta and Bhattacharya (1995) developed a protocol for callus induction and *in vitro* regeneration of *B. variegata*.

Bauhinia vahlii. *B. vahlii*, an indigenous multipurpose species in Kumaun Himalaya is most suitable for plant propagation in mined, industrial waste and marginal lands as it is useful in increasing soil fertility. It is an important forest species for planting in degraded forest lands. *In vitro* regeneration system for *B. vahlii* Wight and Arnott has been described using nodal segment explants from mature trees (Upreti and Dhar, 1996; Dhar and Upreti, 1999) (Table 2).

2.2.2. Cassia

Micropropagation of two species of *Cassia*, i.e. *C. fistula* and *C. siamea*, has been described by Gharyal and Maheshwari (1990).

Cassia fistula. *C. fistula* is grown as an ornamental tree. Pulp of the fruits is used as purgative and laxative. Callus cultures of *C. fistula* were raised from stem and petiole explants of young sprouting shoots on B_5 basal medium supplemented with either

Table 2. Summary of regeneration studies reported for tree legumes belonging to Caesalpiniaceae

Tree	Explant source	Explant(s)	Method	Result	References
Bauhinia purpurea	Seedling	Nodal segment	Organogenesis	Plantlet	Anand and Bir, 1983
	Mature tree (15–18 year)	Stem segments	Somatic embryogenesis	Plantlet, Soil transfer	Kumar, 1992
Bauhinia variegata	Mature tree (6–8 year)	Nodal segment	Organogenesis	Plantlet, Soil transfer	Mathur and Mukunthakumar, 1992
Bauhinia vahlii	Mature tree	Nodal segment	Organogenesis	Plantlet	Upreti and Dhar, 1996
	Mature tree	Nodal segment	Organogenesis	Plantlet	Dhar and Upreti, 1999
Cassia fistula	Mature tree	Anthers	Embryogenesis, Organogenesis	Callus	Bajaj, 1983
	Mature tree	Stem, Petiole	Organogenesis	Plantlet, Soil transfer	Gharyal and Maheshwari, 1990
Cassia siamea	Mature tree	Anthers	Embryogenesis, Organogenesis	Plantlet	Gharyal and Maheshwari, 1983
	Mature tree	Stem, Petiole	Organogenesis	Plantlet, Soil transfer	Gharyal and Maheshwari, 1990
Cercis canadensis	Mature tree	Zygotic embryo	Somatic embryogenesis	Plantlet	Trigiano *et al.*, 1988
Parkinsonia aculeata	Mature tree (6–8 year)	Nodal segment	Organogenesis	Plantlet, Soil transfer	Mathur and Mukunthakumar, 1992
Tamarindus indica	Mature tree	Nodal segment	Organogenesis	Plantlet, Soil transfer	Mascarenhas *et al.*, 1981
	Seedling	Various explants	Organogenesis	Plantlet	Kulkarni *et al.*, 1981
	Mature tree	Anthers	Embryogenesis, Organogenesis	Plantlet	De and Rao, 1983
	Seedling	Shoot tip	Organogenesis	Plantlet, Soil transfer	Kopp and Nataraja, 1990
	Seedling	Cotyledon	Organogenesis	Plantlet, Soil transfer	Jaiwal and Gulati, 1991
	Mature tree	Shoot tip, Nodal explant	Organogenesis	Plantlet	Jaiwal and Gulati, 1992

Table 2. (Continued)

Tree	Explant source	Explant(s)	Method	Result	References
Tamarindus indica	Seedling	Hypocotyl	Organogenesis	Plantlet	Sonia *et al.*, 1998
—	Seedling	Cotyledonary node	Organogenesis	Plantlet, Soil transfer	Sonia *et al.*, 2000

NAA + BA or IAA + BA (Gharyal and Maheshwari, 1990). Plantlets were successfully rooted and transferred to soil. Bajaj (1983) has reported androgenesis in *C. fistula*.

Cassia siamea. C.siamea is an ornamental tree and its flowers are used as vegetable. Gharyal and Maheshwari (1990) reported organogenesis and plantlet formation from explants obtained from adult trees of *C. siamea* employing same medium composition as used for *C. fistula*. Gharyal and Maheshwari (1983a) have reported androgenesis for *C. siamea* (Table 2).

2.2.3. Cercis

Cercis canadensis. C. canadensis (redbud), a 5–10 m tall tree, is indigenous to the Eastern United States and is used extensively as an ornamental species. Commercial stock is usually propagated from seeds. Vegetative propagation of superior cultivars by rooted cuttings, budding or grafting is difficult and unreliable.

Trigiano *et al.* (1988) have reported somatic embryo development directly from 96–110 days post anthesis zygotic embryos explanted onto modified Schenk and Hildebrandt medium containing 2,4-D and ammonium ions. Zygotic embryos explanted 117 days after anthesis produced only callus and roots. A propagation scheme employing somatic embryogenesis can offer a more efficient means of vegetative propagation and a greater uniformity of plants than seed propagation. Micropropagation of *C. canadensis* has also been reported (Yusnita *et al.*, 1990) (Table 2).

2.2.4. Parkinsonia

Parkinsonia, like *Bauhinia*, is a member of the Caesalpiniaceae and is a fast growing ornamental tree. It is adapted to a wide range of climatic conditions and soil types.

Parkinsonia aculeata. P. aculeata is a native of Panama. *In vitro* regeneration system for this species has been described by Mathur and Mukunthakumar (1992). Nodal explants of mature trees (6–8 years old) were used for axillary shoot proliferation employing MS medium supplemented with BA. Efficient rooting was achieved in MS medium containing IBA and the regenerants were successfully transferred to soil (Table 2).

2.2.5. Tamarindus (imli)

Tamarindus indica. T. indica is a multipurpose leguminous tree. It finds wide application for timber, firewood, food, medicine, confectionery, textile and paper industries. It is also grown as a carminative and laxative. Unripe fruits are a rich source of tartaric acid, which is used in various foods, chemical and pharmaceutical industries.

In vitro multiplication of *T. indica* has been reported through seedling explants (Kulkarni *et al.*, 1981), mature nodes (Mascarenhas *et al.*, 1981), anthers (De and Rao, 1983), shoot tip of nodal explants (Kopp and Nataraja, 1990; Jaiwal and Gulati, 1991, 1992), hypocotyl cultures (Sonia *et al.*, 1998) and cotyledonary nodes (Sonia *et al.*, 2000) (Table 2).

2.3. MIMOSACEAE

2.3.1. Acacia

The genus *Acacia* constitutes about 1200 species and is primarily distributed between Africa and Australia. In comparison to such a large number of species available, only few species have been selected for *in vitro* regeneration studies. Thus, for example, *A. koa* (Skolmen and Mapes, 1976), *A. mangium* (Crawford and Hartway, 1986; Bon *et al.*, 1998), *A. stenophylla* (Crawford and Hartway, 1986), *A. senegal* (Agarwal and Prasad, 1980; Dave *et al.*, 1980; Hustache *et al.*, 1986), *A. nilotica* (Mathur and Chandra, 1983; Dewan *et al.*, 1992), *A. auriculiformis* (Mittal *et al.*, 1989; Das *et al.*, 1993) and *A. catechu* (Kaur *et al.*, 1998) have been regenerated *in vitro.*

Acacia senegal. A. senegal is of considerable importance in the struggle against desertification and is a rich source of polysaccharide, gumarabic. Tissue culture of *A. senegal* has been successfully demonstrated (Agarwal and Prasad, 1980; Dave *et al.*, 1980; Hustache *et al.*, 1986). Hustache *et al.* (1986) suggested the use of Knop and Ball medium with 2,4-D or IAA for proliferation of cambial and phloem cells which when transferred to Schenk and Hildebrandt gave rise to several cell lines. These cell lines when allowed to grow on solid medium gave rise to friable callus similar to initial colonies.

Acacia auriculiformis. A. auriculiformis is a fast growing fuel wood tree and is also a good source of paper pulp. Tissue culture of *A. auriculiformis* employing axillary node of seedling explant was demonstrated by Mittal *et al.* (1989). Later Das *et al.* (1993) demonstrated the production of clones of *A. auriculiformis* through cotyledonary callus tissue raised on MS medium supplemented with 2,4-D, kinetin and BA.

Acacia koa. A. koa, a large endemic forest tree of Hawaii, is valued for its timber. Clones of such a superior tree are needed for reforestation. A clone of *A. koa* was grown from shoot tip callus tissue which differentiated into shoot primordia on a basal medium supplemented with coconut water followed by a medium with BA (Skolmen and Mapes, 1976).

Acacia nilotica and *Acacia catechu. A. nilotica* is a rich source of fodder, gum and finds use in confectionery and tanning. *In vitro* propagation of *A. nilotica* employing nodal segments of mature tree has been established (Mathur and Chandra, 1983). Dewan *et al.* (1992) have demonstrated plantlet regeneration employing cotyledonary node explant from seedling. Regeneration of *A. catechu* (Khair tree) employing mature nodal explants has been described only recently (Kaur *et al.*, 1998).

Acacia mangium and *Acacia stenophylla*. *A. mangium*, tropical nitrogen fixing tree legume, has capability to restore the fertility of soil. Studies have been conducted to determine the influence of macronutrient solutions and growth regulators on micropropagation of *A. mangium* (Bon *et al*., 1998). Crawford and Hartway (1986) reported the successful micropropagation of *A. mangium* and *A. stenophylla*. Recently the procedures for *in vitro* regeneration of *Acacia mangium* through somatic embryogenesis (Xie and Hong, 2001a), organogenesis (Xie and Hong, 2001b) and *Agrobacterium*-mediated genetic transformation have been developed (Xie and Hong, 2002) (Table 3).

Table 3. Summary of regeneration studies reported for tree legumes belonging to Mimosaceae

Tree	Explant source	Explant(s)	Method	Result	References
Acacia auriculiformis	Seedling	Axillary bud	Organogenesis	Plantlet	Mittal *et al*., 1989
	Seedling	Cotyledon	Somatic embryogenesis	Plantlet	Das *et al*., 1993
Acacia nilotica	Mature tree	Nodal segment, Internodal segment	Organogenesis, Embryogenesis	Plantlet	Mathur and Chandra, 1983
	Seedling	Cotyledonary node	Organogenesis	Plantlet, Soil transfer	Dewan *et al*., 1992
Acacia senegal	Seedling	Various explants	Organogenesis	Plantlet	Dave *et al*., 1980
	Mature tree	Cambial and phloem cells	Organogenesis	Callus	Hustache *et al*., 1986
Acacia koa	Mature tree	Shoot tip	Somatic embryogenesis	Plantlet	Skolmen and Mapes, 1976
Acacia mangium	Seedling (10–12 d)	Nodal segment	Organogenesis	Plantlet	Bon *et al*., 1998
Acacia catechu	Mature tree	Nodal segment	Organogenesis	Plantlet	Kaur *et al*., 1998
Albizzia amara	Seedling	Hypocotyl	Somatic embryogenesis, Organogenesis	Plantlet	Tomar and Gupta, 1986
	Seedling (10–12 d)	Hypocotyl	Organogenesis	Plantlet, Soil transfer	Tomar and Gupta, 1988a
Albizzia lebbeck	Seedling (10–25 d)	Hypocotyl, Root, Cotyledon, Leaflet	Somatic embryogenesis, Organogenesis	Plantlet	Gharyal and Maheshwari, 1981
	Mature tree	Anther	Androgenesis	Plantlet haploid	Gharyal *et al*., 1983
	Seedling (10–15 d)	Hypocotyl, Root, Cotyledon, Leaflet	Somatic embryogenesis, Organogenesis	Plantlet, Soil transfer	Gharyal and Maheshwari, 1983b

Table 3. (Continued)

Tree	Explant source	Explant(s)	Method	Result	References
Albizzia lebbeck	Seedling (10 d)	Hypocotyl, Root, Cotyledon	Somatic embryogenesis, Organogenesis	Plantlet	Upadhyay and Chandra, 1983
	Mature tree	Anther	Organogenesis	Callus	De and Rao, 1983
	Seedling (10–25 d)	Hypocotyl, Leaf, Internodal segment	Somatic embryogenesis, Organogenesis	Plantlet, Soil transfer	Rao and De, 1987
	Mature tree	Stem, Petiole	Organogenesis	Plantlet, Soil transfer	Gharyal and Maheshwari, 1990
Albizzia procera	Seedling	Various explants	Somatic embryogenesis	Plantlet	Anand and Bir, 1983
	Seedling	Various explants	Somatic embryogenesis, Organogenesis	Plantlet	Anand and Bir, 1983
	Seedling	Various explants	Somatic embryogenesis, Organogenesis	Plantlet	Gharyal and Maheshwari, 1990
	Seedling (10 d)	Hypocotyl, Nodal segment	Organogenesis	Plantlet, Soil transfer	Ahlawat, 1997
Albizzia lucida	Seedling	Hypocotyl	Organogenesis	Plantlet	Tomar and Gupta, 1988a
Albizzia richardiana	Seedling	Various explants	Somatic embryogenesis, Organogenesis	Plantlet	Tomar and Gupta,1986
	Seedling (12 d)	Hypocotyl	Somatic embryogenesis, Organogenesis	Plantlet	Tomar and Gupta, 1988c
Paraserianthus falcataria	Seedling	Nodal segment	Organogenesis	Plantlet	Bon *et al.*, 1998
Prosopis cineraria	Seedling (7–10 d)	Hypocotyl	Organogenesis	Plantlet, Soil transfer	Goyal and Arya, 1981
	Mature tree	Lateral bud	Organogenesis	Plantlet	Goyal and Arya, 1984a
	Mature tree	Lateral bud	Organogenesis	Plantlet, Soil transfer	Goyal and Arya, 1984b
	—	Gall	Organogenesis	Callus	Kant and Ramani, 1985
	Elite tree (4–5 year)	Nodal segment	Organogenesis	Plantlet, Soil transfer	Kackar *et al.*, 1991

Table 3. (Continued)

Tree	Explant source	Explant(s)	Method	Result	References
Prosopis cineraria	Plus tree	Stem segment	Organogenesis	Plantlet Soil transfer	Shekhawat *et al.*, 1993
Prosopis juliflora	Mature tree	Nodal segment	Organogenesis	Plantlet	Wainright and England, 1987
	Mature tree	Nodal segment	Organogenesis	Plantlet	Yi *et al.*, 1989
	Seedling	Apical bud, Inflorescence, Hypocotyl, Cotyledonary node	Somatic embryogenesis, Organogenesis	Callus	Nandwani and Ramawat, 1991
	Mature tree (10 year)	Nodal segment	Somatic embryogenesis, Organogenesis	Plantlet	Nandwani and Ramawat, 1991
Prosopis tamarugo	Mature tree	Shoot tip	Organogenesis	Plantlet	Jordan *et al.*, 1985
	Seedling	Hypocotyl, Cotyledon, Nodal segment	Somatic embryogenesis, Organogenesis	Plantlet	Nandwani and Ramawat, 1992
Leucaena leucocephala	Seedling	Hypocotyl	Organogenesis	Plantlet	Glovack and Greatbach, 1980
	Seedling	Hypocotyl	Organogenesis	Plantlet	Peasley and Collins, 1980
	Seedling	Hypocotyl	Organogenesis	Plantlet	Nagamani and Venkateswaran, 1982
	Seedling	Hypocotyl	Organogenesis	Plantlet	Nagamani and Venkateswaran, 1983
	Seedling	Hypocotyl	Organogenesis	Plantlet	Ravishankar *et al.*, 1983
	Mature tree	Nodal segment	Organogenesis	Plantlet, Soil transfer	Dhawan and Bhojwani, 1984
	Mature tree	Nodal segment	Organogenesis	Plantlet	Kulkarni *et al.*, 1984
	Mature tree (2–3 m)	Lateral bud	Organogenesis	Plantlet, Field planting	Goyal *et al.*, 1985
	Seedling (14 d)	Cotyledonary node, Nodal segment	Organogenesis	Plantlet, Field planting	Dhawan and Bhojwani, 1985
	Mature tree	Nodal segment	Organogenesis	Plantlet, Soil transfer	Dhawan and Bhojwani, 1985

nodulation of micropropagated plants by *Rhizobium*. The objective of this study was to enhance transplantation efficiency of micropropagated plants. Two different methods, viz. Gibson's tube method and aseptic flask method, were satisfactory because of high survival rates (90%). Nodule formation, however, could not be observed up to 3 weeks after inoculation, but before the end of 3 weeks, 80% of the plants produced nodules. This delay in nodulation of the micropropagated plants may be due to poor photosynthetic efficiency of these plants (Dhawan and Bhojwani, 1987) or because of the fact that at the time of *Rhizobium* application, the micropropagated plants of *Leucaena* lacked lateral roots which are the sites of nodule formation. Another study was conducted by Puthur *et al.* (1998) who transferred nodulated *in vitro* raised *Leucaena* to garden soil by laying inoculum of either *Glomus fasiculatum* or *G. macrocarpum*, vesicular arbuscular mycorrhizal fungi, around their roots in order to increase the resistance of *L. leucocephala* to the transplantation shock. Mycorrhization of micropropagated *L. leucocephala* was also performed by Naqvi *et al.* (1998).

Recently, production of somaclonal variants of *L. leucocephala* with high CO_2 assimilating potential have been described by Pardha Saradhi and Alia (1995). Hypocotyl explants of 15 day old seedlings, when inoculated on modified B_5 medium supplemented with 10^{-7} M 2,4-D, led to calli formation. The presence of NAA induced callus and also promoted root induction while IAA was least effective. The variation in callus inducing efficiency with different auxins was due to differential affinities of various auxins to available auxin binding sites in cells of hypocotyls and/or the presence of different hormone binding sites for different auxins. Caulogenesis was achieved from this hypocotyl callus only in the presence of BA. A combination of BA with NAA (10^{-7} or 10^{-6} M) in B_5 medium, led to maximum shoot induction. Good root induction was achieved in the presence of 10^{-6} M IBA. The plantlets thus obtained developed healthy root nodules in soilrite inoculated with *Rhizobium* in the presence of any form of nitrogen other than atmospheric nitrogen and were successfully transferred to soil. Rapid cell division occurring during callus proliferation caused a number of genetic aberrations. The variants differed in terms of leaf size, green pigmentation, growth rate, extent of branching, net photosynthesis capacity (measured as net CO_2 fixing potential in μmol m^{-2} s^{-1}), intracellular CO_2 concentration (ppm), stomatal conductance (mol m^{-2} s^{-1}) and rate of transpiration (mol m^{-2} s^{-1}). Few variants showed high CO_2 assimilation capacity. These somaclonal variants can be of tremendous importance for afforestation/reforestation programmes and hence in developing forests which can act as strong CO_2 sinks and ultimately can be employed as a safe biological method for curtailing global warming.

3. Genetic transformation of *Leucaena leucocephala*

To our knowledge, no one else except us, has worked on genetic transformation of *L. leucocephala*. In our laboratory, we have developed an *Agrobacterium tumefaciens*-mediated DNA transformation protocol. Our transformation protocol is of immense significance for improving its quality when it is used as fodder for cattle or as wood for paper manufacture. The forage from *L. leucocephala* is highly palatable and nutritious. Its protein content amounts to 224–344 g Kg^{-1} dry matter basis. However, its optimal use as livestock fodder is limited due to its high lignin content and presence of a toxin, mimosine. Lignin also presents an obstacle for optimal utilization of plant biomass in pulp and paper industry, where its removal from the cellulose fibers requires vigorous energy and chemical intensive processes (Dwivedi *et al.*, 1994). Mimosine, a toxic amino acid leads to poor growth, dermatitis, swollen hooves, lameness, inhibition of protein synthesis, goiter, alopecia, teratogenecity, reduced feed intake and ultimately death in both ruminants and non-ruminants. Thus, improvement in the quality of *L. leucocephala* with reduced lignin and mimosine contents is desirable for its efficient utilization for paper manufacturing as well as for producing nutritious fodder.

Here we describe our *Agrobacterium tumefaciens*-mediated transformation protocol for *L. leucocephala* in detail. This transformation and regeneration protocol would provide the basis for raising transgenic plants of *L. leucocephala* with improved quality attributes such as exhibiting down-regulation of lignin and mimosine biosynthesis.

3.1. TRANSFORMATION PROTOCOL

Nodal explants from field grown plants of *L. leucocephala* cv. K-8 and K-29, containing axillary buds were taken and surface sterilized using detergent (labolene), absolute alcohol (5 seconds), 1% NaOCl (15 min) and 0.5% (w/v) Hg_2Cl_2 (5 min). These explants were decapitated, proximal meristematic ends removed, wounded with scalpel at various regions and used as a source of explant for transformation and regeneration studies.

For seedling derived explants, seeds of K-8 and K-29 were soaked in boiling water for 5 min and then treated in the same way as described for field explants and allowed to germinate aseptically for 2 days in sterile petriplate with damp filter paper at $29 \pm 0.2\,^{\circ}C$ in dark. For transformation, cotyledonary nodes with amputated radicles were used as explants.

An *Agrobacterium tumefaciens* binary vector p35SGUSINT (derived from pBIN19) containing β-glucuronidas (GUS) gene *uidA* as a reporter gene with an intron, under the control of CaMV35S promoter and polyadenylation signal and *nptII* gene under the control of nos promoter and polyadenylation signal, has been used to transform the explants. The binary vector was mobilized into *Agrobacterium tumefaciens* MP90 C58 strain by Freeze Thaw method (Sambrook *et al.*, 1992).

Shoot regeneration medium comprised of full strength Murashige and Skoog (1962) medium (MS media) supplemented with 3% (w/v) sucrose, 100 mg/L myoinositol, 100 mg/L glutamine, 5 mg/L benzyl adenine (BA) and 1 mg/L naphthalene acetic acid (NAA), solidified with 0.8% (w/v) agar at pH 5.7. The regenerating potentials of explants from nodal region containing axillary buds from mature plants and seedling raised cotyledonary node explants were assessed prior to the application of genetic transformation. It was found that nodal explants from field grown plants gave significantly higher regeneration frequencies than that of the seedling raised cotyledonary node explants. The use of BA (5 mg/L) and NAA (1 mg/L) in the shoot regeneration medium prompted quite rapid and prolific multiple shoot induction. Among the cultivars tested, greater number of shoots per explant were achieved with K-29. About 6–7 shoots per explant were achieved in case of K-29, whereas 3–5 shoots per explant in case of K-8, within 3–4 weeks of culture. Supplementation of glutamine at a concentration of 100 mg/L in the medium prevented precocious leaf fall, which was otherwise a problem earlier. Adenine and phloroglucinol have also been reported to prevent premature leaf fall (Dhawan and Bhojwani, 1985). After 3–4 weeks, regenerated shoots, 4–6 cm high, were excised from the explants and sub-cultured directly onto the shoot regeneration medium. These shoots were then transferred to root induction medium. The half-strength MS medium containing 100 mg/L myoinositol, 2% (w/v) sucrose and fortified with 3 mg/L IBA and kinetin (0.05 mg/L) solidified with 0.8% (w/v) agar at pH 5.7 was found to be very conducive for rooting (80–90%). For both the varieties (K-8 and K-29), root initiation was observed within 15–21 days of culture. No root induction was observed with other auxins such as IAA or 2,4-D. Rooted plantlets were allowed to acclimatize to soil conditions for 2 weeks inside the culture room, in a pot covered with polythene sheet having pinholes for passage of air. After hardening, the plantlets were transferred to soil in greenhouse, where they successfully adapted to natural environmental conditions.

For genetic transformation of *L. leucocephala*, *Agrobacterium tumefaciens* MP90 C58 strain was grown at 28°C overnight in liquid Luria Bertani (LB) medium (40 mL) containing 100 μg/mL kanamycin. Bacterial cells were pelleted at 4000 rpm for 5 min and resuspended in liquid MS medium (150 mL). Wounded nodal or seedling derived explants were co-cultivated with diluted culture of *Agrobacterium* for 5 min in a sterile petriplate. Explants were blotted dry on sterile filter papers and cultured horizontally on plant regeneration medium for 2 days, so as to allow the phenolics to leach out. After 2 days, the explants were transferred on regeneration and propagation medium containing 250 mg/L cefotaxime to eliminate the *Agrobacteria*. After 7 days, cefotaxime was omitted from the medium and subsequently the explants were placed on selection medium which was basically the regeneration and propagation medium as described above, supplemented with 25–50 mg/L kanamycin depending on the cultivars. A genotypic variation in the resistance towards kanamycin was observed in the two cultivars of *L. leucocephala*. Thus the cultivar K-29 variety could resist higher dosage of kanamycin (50 mg/L) as compared to K-8 (25 mg/L). Shoot elongation and multiplication was allowed under a photoperiod of 16 hours (at a light intensity of about 3000 lux) and 55% relative humidity at 29 ± 0.2°C. Within 3–4 weeks of co-cultivation, 5–7 transformed shoots were obtained in case of field derived explants. On the other hand, in case of seedling raised cotyledonary nodal explants, only 2–3 shoots per explant were obtained after 4 weeks of culture. Some callusing was found to occur at the base of shoots. Three weeks after culturing in the presence of kanamycin, the green shoots along with yellowish white shoots were transferred to fresh selection medium, few shoots became chlorotic, suggesting the elimination of untransformed tissues while green and chimeric shoots were sub-cultured on fresh medium containing kanamycin. The green surviving shoots were thus subjected to 2–3 passages of selection by repeated excision of branches and their exposure to selection medium. On the basis of selection of shoots on kanamycin containing medium, it was observed that the transformation efficiency was of the order of 20% in case of cv. K-29 while it was about 11% in case of cv. K-8. Selection of kanamycin resistant plants was further confirmed by survival of leaflets on medium containing kanamycin. Rooting was induced in kanamycin resistant green shoots in root inducing medium, as described above, without kanamycin. The rooted plantlets were successfully transferred to greenhouse after hardening. The greenhouse grown transgenic plants exhibited a phenotype indistinguishable from normal untransformed *Leucaena* plants maintained under same growth conditions, though some exhibited early flowering and fruiting as compared to normal control plants. Integration and expression of the *nptII* and *uidA* genes into plant genome was established by their growth in the presence of otherwise lethal dose of kanamycin and by histochemical and fluorogenic GUS positive assay of leaf and stem sections of these putative transformants, respectively. Kanamycin resistant and *gus* positive transformants were analyzed by Southern hybridization using specific probe for *uidA* gene which revealed a strong hybridization band of 2.8 kb in case of transformants.

4. Conclusions and future prospects

Plantlet regeneration has been worked out for tree legumes but commercially viable technologies with successful transfer of plantlets to soil and their establishment in the field has been developed only in few instances. This step is, however, significant to determine the applicability of the technique for large-scale micropropagation. Commercialization of *in vitro* multiplication techniques of tree legumes is gaining interest, especially for the pulp and paper industry. For the process to be commercially viable, economic feasibility and clonal fidelity are important. A great potential exists for the improvement of tree species with respect to pulp yield, timber quality, disease resistance, stress tolerance. Unfortunately, tree improvement programmes for these tree legumes are still in their infancy. It is important to develop suitable protocols to improve the quality of these legumes. Therefore, in future, research efforts should be focussed on improving quality and productivity of tree legumes through genetic engineering tools, especially in a country like India with expanding population. The method of plant transformation, as described here for *Leucaena*, can be extended to other tree legumes.

Acknowledgements

Financial support from Department of Biotechnology, New Delhi (for *Leucaena leucocephala* work) is gratefully acknowledged. Also financial support of CSIR, New Delhi in the form of SRF to SR is acknowledged.

References

Agarwal S C and Prasad B N (1980) In: *Proc. 3rd All India Bot. Conf.*, Dec. 28–30.

Ahlawat S P (1997) Regeneration of tree legume – *Albizzia procera* (Roxb.) Benth. through tissue culture. *Ind. For.*, **123**: 1058–1066.

Allen O N and Allen E K (1981) *The Leguminosae*. A source book of characteristics, uses and nodulation, Wisconsin Univ., Wisconsin Press, 812.

Anand M and Bir S S (1983) *Dalbergia lanceolaria*. In: *Proceedings of Natl. Symp. on Advances in Biotechnol. Plant Sci.*, Jodhpur, 123.

Anand M and Bir S S (1984) Organogenetic differentiation in tissue cultures of *Dalbergia lanceolaria. Curr. Sci.*, **53**: 1305–1307.

Arrillaga I, Tobolski J J and Merkle S A (1994) Advances in somatic embryogenesis and plant production of black locust (*Robinia pseudoacacia* L.). *Plant Cell Rep.*, **13**: 171–175.

Bajaj Y P S (1983) Pollen embryogenesis in three ornamental trees – *Cassia fistula*, *Jacaranda acutifolia* and *Poinciana regia*. *J. Tree Sci.*, **2**: 16.

Barghchi M (1987) Mass clonal propagation *in vitro* of *Robinia pseudoacacia* L. (Black locust) cv. Jaszkisert. *Plant Sci.*, **53**: 183–189.

Bhargava S, Upadhaya S, Garg K and Chandra N (1983) In: *Plant Cell Culture in Crop Improvement* (Eds Sen S K and Giles K L), Plenum Press, New York, 431.

Bon M C, Bonal D, Goh D K and Monteuuis O (1998) Influence of different macronutrient solutions and growth regulators on micropropagation of juvenile *Acacia mangium* and *Paraserianthus falcataria* explants. *Plant Cell Tiss. Org. Cult.*, **53**: 171–177.

Brewbaker J L (1982) Systematics, self incompability, breeding systems and genetic improvement of *Leucaena* species. In: *Leucaena Research in the Asian Pacific Region*, Proceedings of Workshop in Singapore, Int. Dev. Res. Center, Ottawa Publishers, 17–22.

Brewbaker J L and Hutton E M (1979) *Leucaena* – versatile tropical legume. In: *New Agricultural Crops* (Ed. Ritchie G A), Am. Assoc. for Advancement of Science, Westview Press, Boulder CO, 207–259.

Brown C L and Sommer H E (1982) In: *Tissue Culture in Forestry* (Eds Bonga J M and Durzan D J), New York, Martinus Nijhoff, 109–149.

Chalupa V (1983) *In vitro* propagation of Willows (*Salix* spp.), European Mountain Ash (*Sorbus aucuparia* L.) and Black Locust (*Robinia pseudoacacia* L.). *Biol. Plant.*, **25**: 305–307.

Chalupa V (1987) Effect of benzylaminopurine and thidiazuron on *in vitro* shoot proliferation of *Tilia cordata* MILL., *Sorbus aucuparia* L. and *Robinia pseudoacacia* L. *Biol. Plant.*, **29**: 425–429.

Crawford D F and Hartway U J (1986) Micropropagation of *Acacia mangium* and *Acacia stenophylla.* In: *Australian Acacias in Developing Countries* (Ed. Turnbull J W), ACIAR, Proc. No. 16, Proc. Int. Workshop at the Forestry Training Center, Gympie, Queensland, Australia, 64–65.

Crizaldo E N (1980) Tissue culture of fast growing trees. *Silva. Trop.*, **5**: 123–137.

Das P K, Chakravarti V and Maity S (1993) Plantlet formation in tissue culture from cotyledon of *Acacia auriculiformis* A. Cunn. Ex. Benth. *Indian J. For.*, **16**: 189–192.

Das P, Samantaray S, Roberts A V and Rout G R (1997) *In vitro* somatic embryogenesis of *Dalbergia sissoo* Roxb. a multipurpose timber yielding tree. *Plant Cell Rep.*, **16**: 578–582.

Dasgupta S and Bhattacharya S (1995) Callus induction and plant regeneration of *Bauhinia variegata. Environ. Ecol.*, **13**: 646–647.

Datta K and Datta S K (1985) Auxin + KNO_3 induced regeneration of leguminous tree – *Leucaena leucocephala* through tissue culture. *Curr. Sci.*, **54**: 248–250.

Datta S K (1987) In: *Agroforestry for Rural Needs*, Vol. 1 (Eds Khosla P K and Khurana D K), *Ind. Soc. Tree Scientists*, 234.

Datta S K, Datta K and Pramonik T (1983) *In vitro* clonal multiplication of mature trees of *Dalbergia sissoo. Plant Cell Tiss. Org. Cult.*, **2**: 15–20.

Dave V S, Goyal Y, Vaishnawa G R, Surana N M and Arya H C (1980) Leguminous trees. In: *Proc. 3rd All India Bot. Conf.*, Lucknow, India, 14.

Davis J M and Keathley D E (1985) *Robinia pseudoacacia.* In: *Proc. 4th North Central Tree Imp. Conf.*, East Lansing, MI, 29–34.

Davis J M and Keathley D E (1987) Differential responses to *in vitro* bud culture in mature *Robinia pseudoacacia* L. (Black locust). *Plant Cell Rep.*, **6**: 431–434.

Davis J M and Keathley D E (1988) In: *Nitrogen Fixing Tree Res. Rep.*, **6**: 65–67.

Dawra S, Sharma D R and Chowdhury J B (1984) Clonal propagation of *Dalbergia sissoo* Roxb. through tissue culture. *Curr. Sci.*, **53**: 807–809.

De D N and Rao P V L (1983) In: *Plant Cell Culture in Crop Improvement* (Eds Sen S K and Giles K L), Plenum Press, New York, 469.

Dewan A, Nanda K and Gupta S C (1992) *In vitro* micropropagation of *Acacia nilotica* subsp. indica Brenan via cotyledonary nodes. *Plant Cell Rep.*, **12**: 18–21.

Dhar U and Upreti J (1999) *In vitro* regeneration of mature leguminous liana (*Bauhinia vahlii* Wight and Arnott). *Plant Cell Rep.*, **18**: 664–669.

Dhawan V (1993) Micropropagation of nitrogen fixing trees. In: *Micropropagation of Woody Plants* (Ed. Ahuja M R), Kluwer Academic Publishers, Dordrecht, The Netherlands, 303–315.

Dhawan V and Bhojwani S S (1984) Reduction in cost of tissue culture of *Leucaena leucocephala* (Lam) deWit. by replacing AR grade sucrose by sugar cubes. *Curr. Sci.*, **53**: 1159–1161.

Dhawan V and Bhojwani S S (1985) *In vitro* vegetative propagation of *Leucaena leucocephala* (Lam) deWit. *Plant Cell Rep.*, **4**: 315–318.

Dhawan V and Bhojwani S S (1986) *Leucaena reucocephala*. In: *Proc. Cong. on Plant Tissue Cell Culture*, Minnesota, 113.

Dhawan V and Bhojwani S S (1987a) Hardening *in vitro* and morphological changes in the leaves during acclimatization of micropropagated plants of *Leucaena leucocephala* (Lam) deWit. *Plant Sci.*, **53**: 65–72.

Dhawan V and Bhojwani S S (1987b) *In vitro* nodulation of seedlings and micropropagated plants of the tropical tree legume *Leucaena leucocephala* by *Rhizobium*. *Proc. Indian Natl. Sci. Acad.*, **53B**: 351–357.

Dianeles M, Romero A and Cruz A M (1998) Preliminary study on indicators for the *in vitro* propagation of *Leucaena leucocephala* cv. Peru. *Cuban J. Agric. Sci.*, **32**: 393–397.

Dwivedi U N, Campbell W H, Yu J, Datla R S S, Bugos R C, Chiang V L and Podila G K (1994) Modification of lignin biosynthesis in transgenic *Nicotiana* through expression of an antisense O-methyltransferase gene from *Populus*. *Plant Mol. Biol.*, **26**: 61–71.

Gharyal P K and Maheshwari S C (1981) *In vitro* differentiation of somatic embryoids in a leguminous tree *Albizzia lebbeck* L. *Naturossenschaften*, **67**: 379.

Gharyal P K and Maheshwari S C (1983a) Androgenic response from cultured anthers of a leguminous tree. *Protoplasma*, **118**: 91–93.

Gharyal P K and Maheshwari S C (1983b) *In vitro* differentiation of plantlets from tissue culture of *Albizzia lebbeck*. *Plant Cell Tiss. Org. Cult.*, **2**: 49–53.

Gharyal P K and Maheshwari S C (1990) Differentiation in explants from mature leguminous trees. *Plant Cell Rep.*, **8**: 550–553.

Gharyal P K, Rashid A and Maheshwari S C (1983) Production of haploid plants in anther cultures of *Albizzia lebbeck* L. *Plant Cell Rep.*, **2**: 308–309.

Glovack L and Greatbach W (1980) Successful tissue culture of *Leucaena*. *Leucaena Res. Rep.*, **3**: 81.

Goyal Y and Arya H C (1981) Differentiation in cultures of *Prosopis cineraria* Linn. *Curr. Sci.*, **50**: 468–469.

Goyal Y and Arya H C (1984a) Effects of sugars, nitrogen, amino acids and vitamins on shoot differentiation from single bud *in vitro* culture of *Prosopis cineraria* L. *Indian J. Exp. Biol.*, **22**: 592–595.

Goyal Y and Arya H C (1984b) Tissue culture of desert trees. I. Clonal multiplication of *Prosopis cineraria* by bud culture. *J. Plant Physiol.*, **115**: 183–189.

Goyal Y, Bingham R L and Felker P (1985) Propagation of the tropical tree, *Leucaena leucocephala* K-67 by *in vitro* bud culture. *Plant Cell Tiss. Org. Cult.*, **4**: 3–10.

Han H K and Keathley D E (1989) Regeneration of whole plants from seedling derived callus of black locust. *Nitrogen Fixing Tree Res. Rep.*, **7**: 129–131.

Han H K, Keathley D E and Gordan M P (1989) Cambial tissue culture and subsequent shoot regeneration from mature black locust (*Robinia Pseudoacacia* L.). *Plant Cell Rep.*, **12**: 185–188.

Han K H, Keathley D E, Davis J M and Gordan M P (1993) Regeneration of transgenic woody legume (*Robinia pseudoacacia* L. black locust) and morphological alterations induced by *Agrobacterium rhizogenes*-mediated transformation. *Plant Sci.*, **88**: 149–157.

Hustache G, Barnoud F and Joseleau J P (1986) Callus formation and induction of a cell suspension culture from *Acacia senegal*. *Plant Cell Rep.*, **5**: 365–367.

Igasaki T, Mohri T, Ichikawa H and Shinohara K (2000) *Agrobacterium tumefaciens* mediated transformation of *Robinia pseudoacacia*. *Plant Cell Rep.*, **19**: 448–453.

Jaiwal P K and Gulati A (1991) *In vitro* high frequency plant regeneration of a tree legume *Tamarindus indica* L. *Plant Cell Rep.*, **10**: 569–573.

Jaiwal P K and Gulati A (1992) Micropropagation of *Tamarindus indica* L. from shoot tip and nodal explants. *Natl. Acad. Sci. Lett.*, **15**: 63–67.

Jordan M (1988) *In vitro* culture of *Prosopis* species. In: *Cell and Tissue Culture in Forestry* (Eds Bonga J M and Durzan D J), Martinus Nijhoff Pubs., Boston, MA, 370–384.

Jordan M, Cortes I and Goreus A (1987) Potentialities of cell and callus tissue culture to regenerate two mesquite species (*Prosopis tamarugo*) and (*P. chilensis*). *Gartenbauwissenschaft*, **52**: 166–169.

Jordan M, Pedraza J and Goreux A (1985) *In vitro* propagation studies of three *Prosopis* species (*P. alba*, *P. chilensis* and *P. tamarugo*) through shoot tip culture. *Gartenbauwissenschaft*, **50**: 265–267.

Kackar N L, Solanki K R, Singh M and Vyas S C (1991) Micropropagation of *Prosopis cineraria*. *Indian J. Expt. Biol.*, **29**: 65–67.

Kant U and Ramani V (1985) Cytomorphological studies of gall callus cells of *Prosopis cineraria* (Linn.) druce. *J. Indian Bot. Soc.* (suppl.), **64**: 22–23.

Kaur K, Verma B and Kant U (1998) Plants obtained from the Khair tree (*Acacia catechu* Willd.) using mature nodal segments. *Plant Cell Rep.*, **17**: 427–429.